AF895609

Properties and Testing of Fiber-Reinforced Polymers

Properties and Testing of Fiber-Reinforced Polymers

Yasushi Miyano and Masayuki Nakada

WILEY VCH

Authors

Yasushi Miyano
Materials System Research Laboratory
Kanazawa Institute of Technology
3-1 Yatsukaho Hakusan, Ishikawa 924-0838,
Japan

Masayuki Nakada
Materials System Research Laboratory
Kanazawa Institute of Technology
3-1 Yatsukaho Hakusan, Ishikawa 924-0838,
Japan

Cover Design: Wiley
Cover Image: © Peter Burdon/unsplash

Typesetting Lumina Datamatics

Library of Congress Card No.: applied for

British Library Cataloguing-in-Publication Data A catalogue record for this book is available from the British Library.

Bibliographic information published by the Deutsche Nationalbibliothek
The Deutsche Nationalbibliothek lists this publication in the Deutsche Nationalbibliografie; detailed bibliographic data are available on the Internet at <http://dnb.d-nb.de>.

Print ISBN: 9783527355150
ePDF ISBN: 9783527852178
ePub ISBN: 9783527852161
oBook ISBN: 9783527852185

© 2026 WILEY-VCH GmbH,
Boschstraße 12, 69469 Weinheim, Germany

All rights reserved, including those of translation into other languages, text and data mining and training of artificial technologies or similar technologies. No part of this book may be reproduced in any form – by photoprinting, microfilm, or any other means – nor transmitted or translated into a machine language without written permission from the publishers. Registered names, trademarks, etc. used in this book, even when not specifically marked as such, are not to be considered unprotected by law.

The manufacturer's authorized representative according to the EU General Product Safety Regulation is WILEY-VCH GmbH, Boschstr. 12, 69469 Weinheim, Germany, e-mail: Product_Safety@wiley.com.

While the publisher and the authors have used their best efforts in preparing this work, including a review of the content of the work, neither the publisher nor the authors make any representations or warranties with respect to the accuracy or completeness of the contents of this work and specifically disclaim all warranties, including without limitation any implied warranties of merchantability or fitness for a particular purpose. No warranty may be created or extended by sales representatives, written sales materials or promotional statements for this work. The fact that an organization, website, or product is referred to in this work as a citation and/or potential source of further information does not mean that the publisher and authors endorse the information or services the organization, website, or product may provide or recommendations it may make. This work is sold with the understanding that the publisher is not engaged in rendering professional services. The advice and strategies contained herein may not be suitable for your situation. You should consult with a specialist where appropriate. Further, readers should be aware that websites listed in this work may have changed or disappeared between when this work was written and when it is read. Neither the publisher nor authors shall be liable for any loss of profit or any other commercial damages, including but not limited to special, incidental, consequential, or other damages.

Printed and bound by CPI Group (UK) Ltd,
Croydon, CR0 4YY

C9783527355150_180226

Contents

Preface

The carbon-fiber-reinforced polymers (CFRPs) discussed in this book comprise carbon fibers and a resin matrix. Carbon fibers are nearly elastic until they fracture, exhibiting extremely high fracture strength, but with wide scatters of strength. Furthermore, these mechanical properties change little with time or temperature. By contrast, the matrix resin, which is orders of magnitude softer than carbon fibers and which can keep much greater ultimate strain, exhibits viscoelastic behavior, with stress–strain relations that change with time and temperature. The ultimate strain and fracture stress also change with time and temperature. Whereas CFRP comprises carbon fibers and matrix resins with extremely different mechanical properties, this book is intended to describe scientific elucidation of the roles that the mechanical properties of carbon fibers and matrix resins play in CFRP strength under static loads as well as creep and fatigue loads. This book also elucidates the actual strength achieved under various conditions, including carbon fiber and resin types, load configurations, load directions, temperatures, and aqueous environments, thereby addressing the fundamental durability of CFRP for structural applications.

This book comprises three parts. Part I, titled "Accelerated Testing Methodology," details the modified time–temperature superposition principle, which is necessary for evaluation and long-term prediction of the viscoelastic properties of matrix resins, which are of paramount importance in evaluating the durability of CFRPs and their structures. The results are used to evaluate the long-term durability of several CFRPs and their structures. Part II, titled "Advanced Accelerated Testing Methodology," uses Christensen's viscoelastic crack kinetic theory to formulate the static, fatigue, and creep strengths of CFRPs and their structures, including their scatter, and uses the results to evaluate the long-term durability of various CFRPs and their structures. Part III, titled "Integrated Accelerated Testing Methodology," further develops Christensen's viscoelastic crack kinetic theory to produce a formula integrating the static, creep, and fatigue strengths of CFRPs. The results are used to clarify the role that the mechanical properties of carbon fibers and matrix resins play in determining the strength of CFRPs.

The first author, Miyano, who studied viscoelasticity as a Ph.D. student at Keio University, worked on the development of FRP structures at Hitachi, Ltd for eight years, and has worked consistently on evaluation of CFRP durability and CFRP structures since becoming a professor at Kanazawa Institute of Technology in 1977.

In 1986, the second author, Nakada, began working on the same topic under Miyano as a graduate student at Kanazawa Institute of Technology. The two have been collaborating ever since. During this time, when Miyano was a visiting professor at the Department of Aeronautics and Astronautics at Stanford University in 1995, he began collaboration with Professors Stephen Tsai and Richard Christensen of the same department. Between 1998 and 2015, the "Composites Durability Workshop" was held 20 times around the world as a forum for discussion. The term "Accelerated Testing Methodology" was named by Professor Tsai. Since 2001, we have received awards from the Office of Naval Research (ONR). During 2005–2017, we presented our research results at the ONR Solid Mechanics Program, held annually near Washington, D.C.

We published "Durability of Fiber-Reinforced Polymers" in 2018 through Wiley-VCH. Please note that its contents are included as part of this book.

Finally, we would like to express our gratitude to Professor Stephen Tsai and Professor Richard Christensen of Stanford University and to Dr. Yapa Rajapakse of ONR for their trusted advice on CFRP durability. All data included herein are the results of research conducted during the past 50 years by students and researchers of Materials System Research Laboratory at Kanazawa Institute of Technology. Once again, we would like to extend our deepest gratitude to them.

March, 2026

Yasushi Miyano
Masayuki Nakada

Part 1

Accelerated Testing Methodology

Introduction

Fiber-reinforced polymers (FRPs), which combine a fiber as a reinforcing material and a polymer as the matrix, are used as structural materials in many industrial fields. Carbon-fiber-reinforced polymers (CFRPs), which offer excellent properties, such as high strength, high rigidity, and lightweight, have been used as structural materials since the 1980s for applications such as main structural members of aircraft, ships, automobiles, and other devices that must maintain high reliability during long operating periods. Research on the durability evaluation methods for FRP, especially for CFRP, has been pursued actively. The accelerated testing methodology (ATM) should be established for long-term prediction of the useful life of CFRP and its structures, as shown in Figure 1. For ATM, data collection is conducted using accelerated testing. From the obtained data, design for durability becomes possible, that is, highly reliable CFRP structures can be developed.

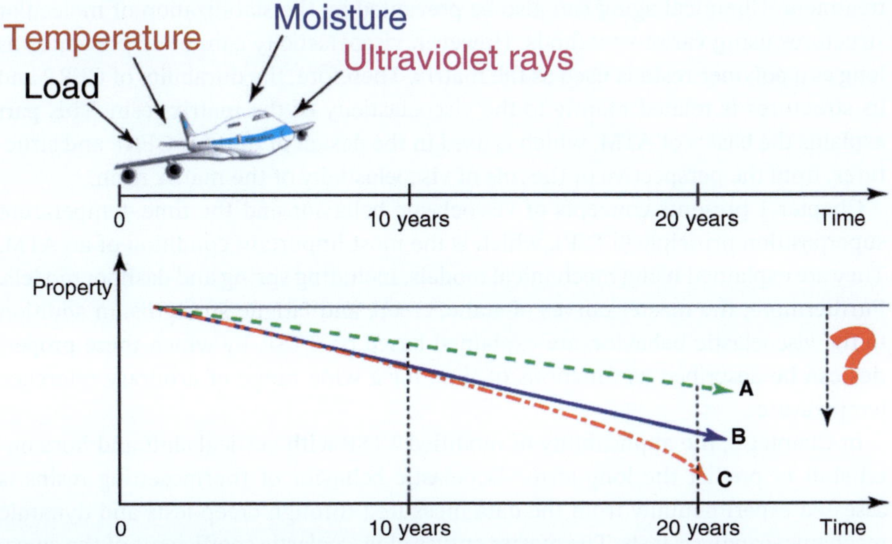

Figure 1 Necessity for accelerated testing.

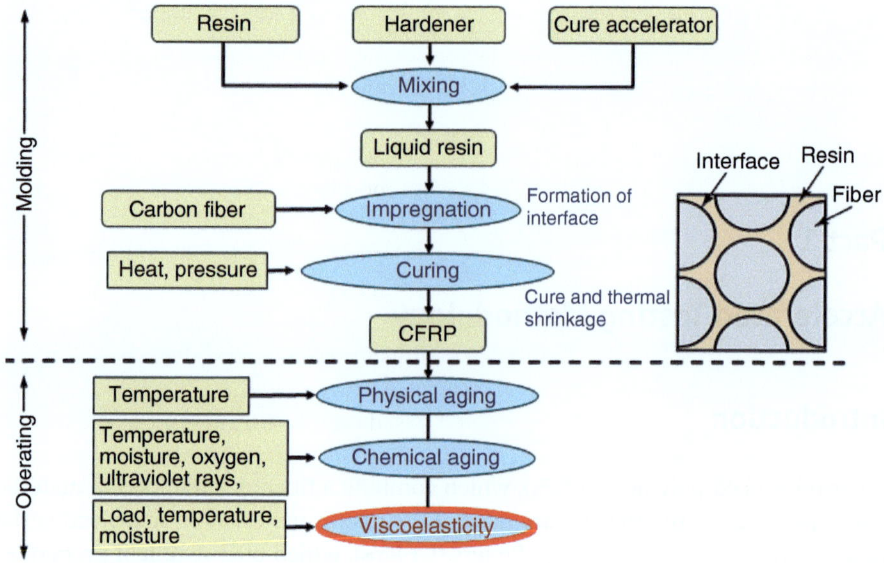

Figure 2 Role of matrix resin on CFRP.

Figure 2 shows the role of the resin matrix during molding and operation of CFRPs. The mechanical behavior of the resin matrix changes drastically from liquid to solid by curing during molding process, whereas carbon fibers remains stable during molding process. The behaviors of physical aging, chemical aging, and viscoelasticity, all of which influence CFRPs' durability, are generated in the resin matrix and the interface during operating processes. Carbon fibers are perfectly stable during operation, as shown in Figure 2. The most important characteristic is the viscoelasticity of the matrix resin because physical aging is preventable by pre-aging treatment. Chemical aging can also be prevented by the stabilization of molecular structures using various methods. However, viscoelasticity cannot be prevented as long as a polymer resin is used as the matrix. Therefore, the durability of CFRP and its structures is related mainly to the viscoelasticity of the matrix resin. This part explains the basics of ATM, which is used in the design of durable CFRP and structures, from the perspective of the role of viscoelasticity of the matrix resin.

Chapter 1 presents concepts of viscoelastic behavior and the time–temperature superposition principle (TTSP), which is the most important condition of an ATM. They are explained using mechanical models, including spring and dashpot models. Furthermore, the master curves of static, creep, and fatigue strengths, in addition to the viscoelastic behavior, are explained based on TTSP, by which these properties can be described as functions of time for a wide range of arbitrary reference temperatures.

In Chapter 2, the applicability of modified TTSP with vertical shift and horizontal shift to predict the long-term viscoelastic behavior of thermosetting resins is assessed experimentally from the data measured through creep tests and dynamic mechanical analysis tests. The master curves of viscoelastic coefficients of the epoxy resin as an actual polymer resin are obtained. Furthermore, the accurate relaxation modulus of matrix resin in the wide range of time at a reference temperature and the accurate time-temperature shift factor are the most important items for reliable CFRP life prediction. The process of determining these properties using tests of two kinds is also proposed.

In constructing the master curve for the viscoelastic properties of the resin, the TTSP is used in Chapters 4 and 6, while the modified TTSP is used in Chapters 3 and 5.

In Chapter 3, the roles of the mixture for ascertaining the mechanical and thermal properties of carbon fiber and matrix resin from a unidirectional CFRP are assessed and presented for an actual CFRP.

In Chapter 4, the static and fatigue strengths for various load directions of unidirectional CFRP are measured at various strain rates and temperatures. The master curves of these strengths are constructed based on TTSP for the matrix resin. It is clarified experimentally that the same TTSP for the matrix resin holds for these strengths.

Two applications of ATM are introduced in Chapters 5 and 6 as follows. In Chapter 5, the master curves of static and fatigue flexural strengths of various woven FRP laminates are obtained under water absorption conditions. The influence of water absorption on the flexural strengths of FRP laminates combined with fibers and matrix resins of various kinds are discussed. In Chapter 6, life prediction of CFRP bolted joints is performed based on ATM.

1

Viscoelasticity

1.1 Introduction

Thermosetting resins used as the matrix of fiber-reinforced polymers (FRPs) show viscoelastic behavior: it is nondestructive time- and temperature-dependent mechanical behavior. Furthermore, the time–temperature superposition principle (TTSP) holds for the viscoelasticity of thermosetting resins. A master curve showing the viscoelasticity of a resin over a wide range of reduced time at a reference temperature can be constructed based on the TTSP. In addition, the long-term nondestructive mechanical behavior of the resin can be predicted from this master curve.

It can be readily inferred that FRPs also show viscoelastic behavior as a result of the influence of the viscoelasticity of the matrix resin, and that the same TTSP as for the matrix resin also holds for FRP viscoelasticity. Furthermore, it can be inferred that the same TTSP as that for matrix resin holds for the static, creep, and fatigue strengths of FRP, and that master curves of these strengths can be constructed to predict the long-term degradation of these strengths.

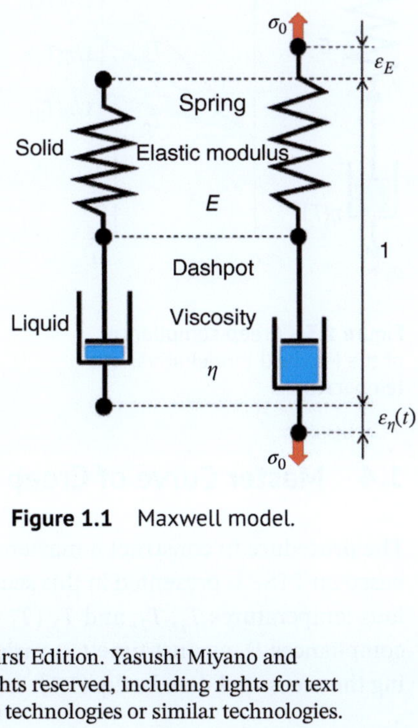

Figure 1.1 Maxwell model.

1.2 Concept of Viscoelastic Behavior

The concept of linear viscoelastic behavior is explained using a Maxwell model with a spring and dashpot, as shown in Figure 1.1. The spring is a solid having an elastic

Properties and Testing of Fiber-Reinforced Polymers, First Edition. Yasushi Miyano and Masayuki Nakada. © 2026 Wiley-VCH GmbH. All rights reserved, including rights for text and data mining and training of artificial intelligence technologies or similar technologies.
Published 2026 by Wiley-VCH GmbH.

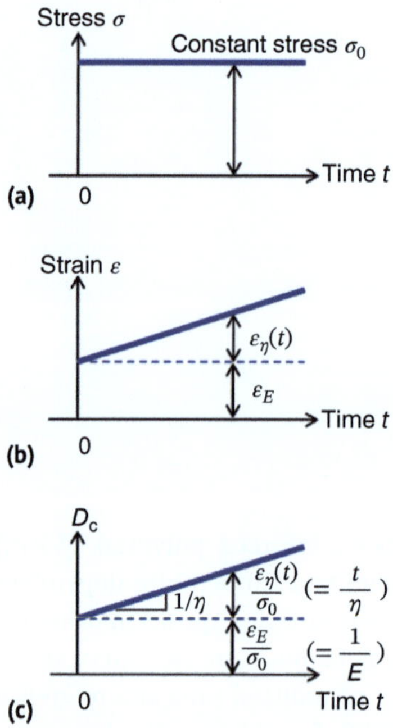

Figure 1.2 Creep compliance of the Maxwell model.

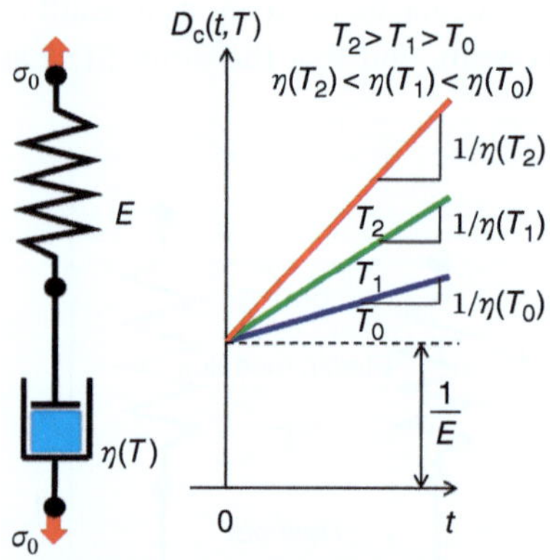

Figure 1.3 Creep compliance of the Maxwell model at various temperatures.

modulus E. The dashpot is a liquid having viscosity η. When this Maxwell model is loaded using a constant stress σ_0, as shown in Figure 1.2a, the total strain $\varepsilon(t) = \varepsilon_E + \varepsilon_\eta(t)$ is generated, as shown in Figure 1.2b. Creep compliance $D_c(t)$ shown in Figure 1.2c is definable by the following equation:

$$D_c(t) = \frac{\varepsilon(t)}{\sigma_0} = \frac{\varepsilon_E}{\sigma_0} + \frac{\varepsilon_\eta}{\sigma_0} = \frac{1}{E} + \frac{t}{\eta} \tag{1.1}$$

1.3 Concept of TTSP

Although the elastic modulus E of the spring in the Maxwell model does not change with temperature, the viscosity η of dashpot decreases drastically with increasing temperature, as shown in Figure 1.3. Figure 1.4 presents the creep compliance of the Maxwell model with various temperatures against time. Each creep compliance keeps a constant value of $1/E$ in the short time range, and each maintains a constant slope of 1 over a long time range. These creep compliances at various temperatures are superimposed on each other by a horizontal shift, as shown in Figure 1.4. The amount of horizontal shift is defined as the time–temperature shift factor $a_{T_0}(T)$, which is shown by the following equation and Figure 1.5:

$$a_{T_0}(T_i) = \frac{t_i}{t_0}, \qquad \log a_{T_0}(T_i) = \log t_i - \log t_0, \ (i = 1, 2) \tag{1.2}$$

The time–temperature shift factor can be regarded as the rate of acceleration by increasing temperature.

1.4 Master Curve of Creep Compliance of Matrix Resin

The procedure to construct a master curve of creep compliance of the matrix resin based on TTSP is presented in this section. First, the creep tests are conducted at various temperatures T_1, T_2, and T_3 ($T_1 < T_2 < T_3$), as shown in Figure 1.6. The creep compliances D_c against time t at various temperatures are determined by substituting the measured data in Eq. (1.1), as shown on the left side of Figure 1.7.

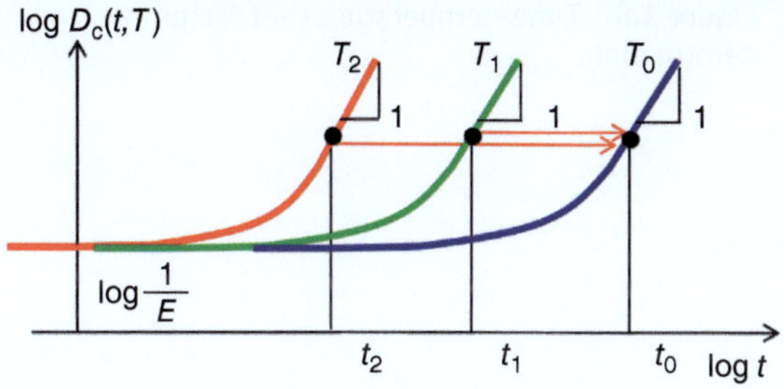

Figure 1.4 Superposition of creep compliances at various temperatures by shifting.

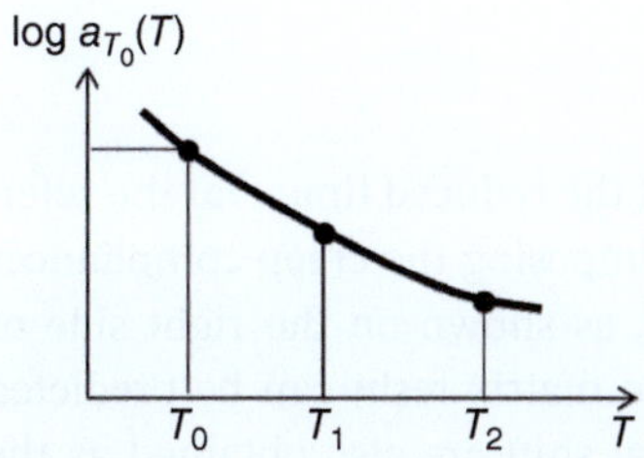

Figure 1.5 Time–temperature shift factor.

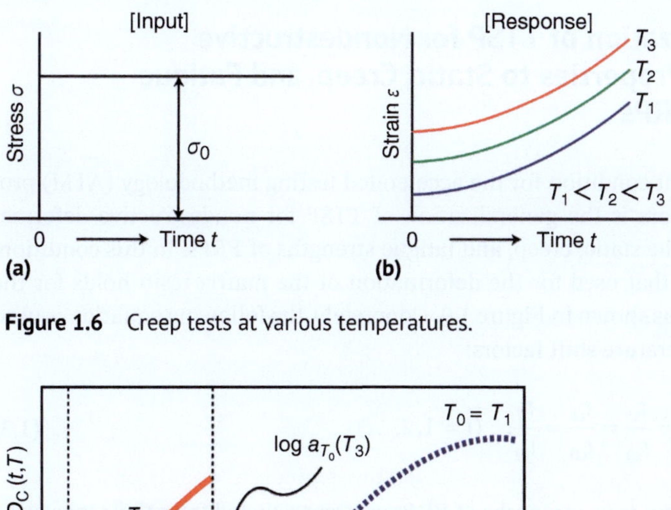

Figure 1.6 Creep tests at various temperatures.

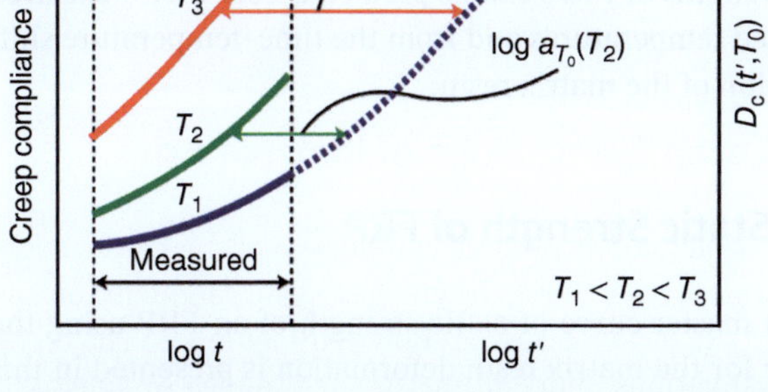

Figure 1.7 Creep compliance at various temperatures and the master curve.

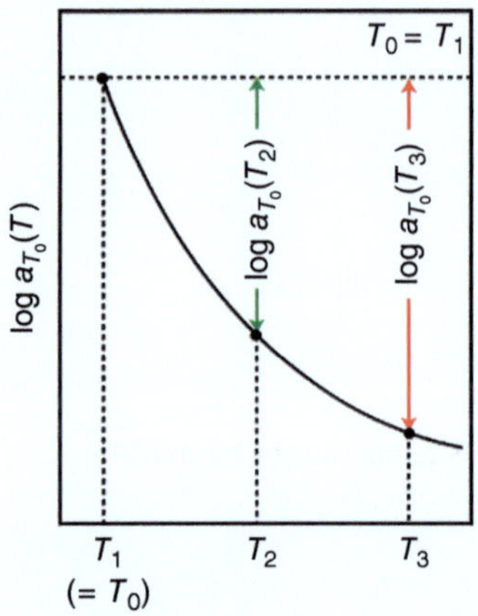

Figure 1.8 Time–temperature shift factor for deformation.

The master curve of creep compliance D_c against the reduced time t' at the reference temperature T_0 (= T_1) is obtainable by superimposing the creep compliances D_c at various temperatures by horizontal shifting, as shown on the right side of Figure 1.7. The long-term creep compliance of the matrix resin can be predicted from this master curve. The amounts of horizontal shift are also obtained as the time–temperature shift factor, as shown in Figure 1.8.

1.5 Generalization of TTSP for Nondestructive Deformation Properties to Static, Creep, and Fatigue Strengths of FRPs

The most important condition for the accelerated testing methodology (ATM) proposed by the authors is the generalization of TTSP for nondestructive deformation properties to the static, creep, and fatigue strengths of FRPs. In this condition, the same TTSP as that used for the deformation of the matrix resin holds for the strengths of FRPs, as shown in Figure 1.9. Concretely, the following equation is used for the time–temperature shift factors:

$$a_{T_0}(T_i) = \frac{t_i}{t_0} = \frac{t_{si}}{t_{s0}} = \frac{t_{ci}}{t_{c0}} = \frac{t_{fi}}{t_{f0}}, \quad (i = 1, 2, \ldots) \tag{1.3}$$

Therefore, the long-term strengths of FRPs can be predicted from their measured short-term strengths at elevated temperatures and from the time–temperature shift factor $a_{T_0}(T)$ for the deformation of the matrix resin.

1.6 Master Curve of Static Strength of FRP

The procedure to construct a master curve of static strength of an FRP using the time–temperature shift factor for the matrix resin deformation is presented in this section. First, the static tests are conducted at various temperatures T and strain

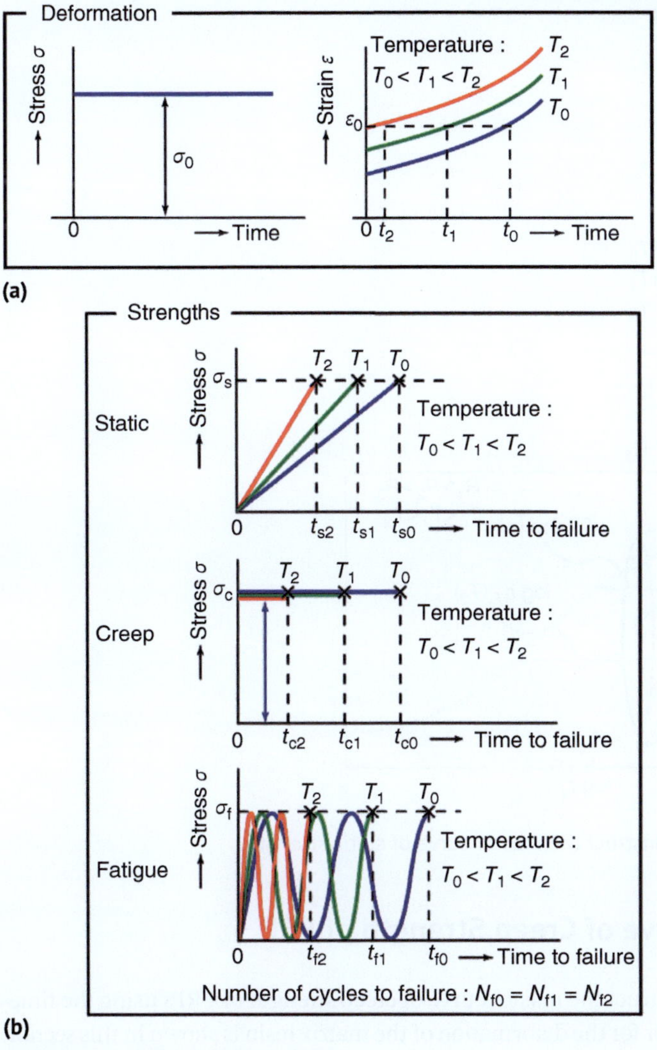

Figure 1.9 Generalization of TTSP for nondestructive deformation properties to static, creep, and fatigue strengths of FRP.

rates R, as shown in Figure 1.10a,b. The static strengths σ_s against the failure time t_s at various temperatures are determined from the measured data, as shown on the left side of Figure 1.10c, where the failure time t_s is defined as the period from the beginning of loading to the failure load.

The master curve of static strength σ_s against the reduced failure time t'_s at reference temperature T_0 is obtainable by horizontally shifting of the static strength σ_s at various temperatures, as shown on the right side of Figure 1.10c. The long-term static strength of the FRP can be predicted from this master curve. The amount of horizontal shift is the time–temperature shift factor for the matrix resin deformation.

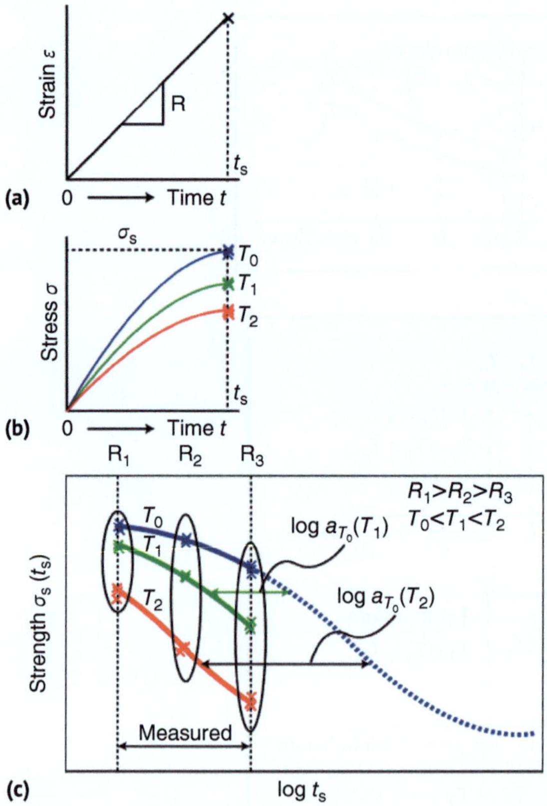

Figure 1.10 How to construct the master curve of static strength.

1.7 Master Curve of Creep Strength of FRP

The procedure to construct the master curve of creep strength of FRPs using the time–temperature shift factor for the deformation of the matrix resin is shown in this section. First, the creep tests are conducted at various temperatures T and constant stresses σ_c, as shown in Figure 1.11a. The creep strength σ_c against failure time t_c at various temperatures is determined from measured data, as shown on the left side of Figure 1.11b.

The master curve of creep strength σ_c against reduced failure time t_c' at the reference temperature T_0 is obtainable by shifting the creep strength σ_c horizontally at various temperatures, as shown on the right side of Figure 1.11b. The long-term creep strength of the FRP can be predicted from this master curve. The amount of horizontal shift is the time–temperature shift factor for the matrix resin deformation.

1.8 Master Curve of Fatigue Strength of FRP

The procedure used to construct the master curve of fatigue strength of FRPs using the time–temperature shift factor for the matrix resin deformation is shown in this section. First, the fatigue tests are conducted at various temperatures T and various

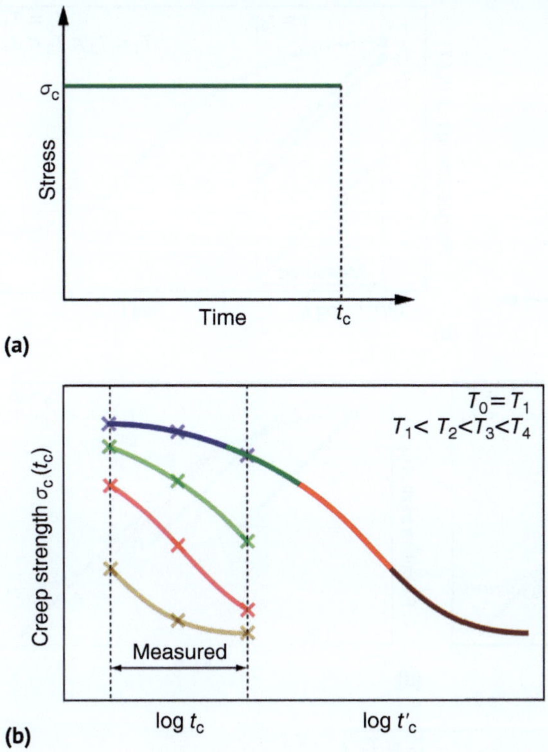

Figure 1.11 How to construct the master curve of creep strength.

maximum stresses σ_f under a constant frequency f_0, as shown in Figure 1.12a. The fatigue strengths σ_f against the failure time t_f at various temperatures under a constant frequency f_0 are determined from the measured data, as shown on the left side of Figure 1.12b. The master curve of static strength in this figure can be regarded as the fatigue strength at the number of cycles to failure $N_f = 1/2$.

The fatigue strength σ_f against the reduced failure time t'_f at the reference temperature T_0 under various corresponding frequencies f_i is obtainable by shifting the fatigue strength σ_f horizontally at various temperatures, as shown on the right side of Figure 1.12c. The corresponding frequency f_i for T_i is obtained as shown in the following equation:

$$\frac{f_0}{f_i} = a_{T_0}(T_i), \quad (i = 1, 2, \ldots) \tag{1.4}$$

The long-term fatigue strength σ_f against the reduced failure time t'_f at the reference temperature T_0 for an arbitrary frequency f can be predicted from this figure. The amount of horizontal shift is the time–temperature shift factor for matrix resin deformation.

The master curves of fatigue strength σ_f at various numbers of cycles to failure N_{fi} are obtained by connecting the fatigue strength σ_f at various frequencies f_i at

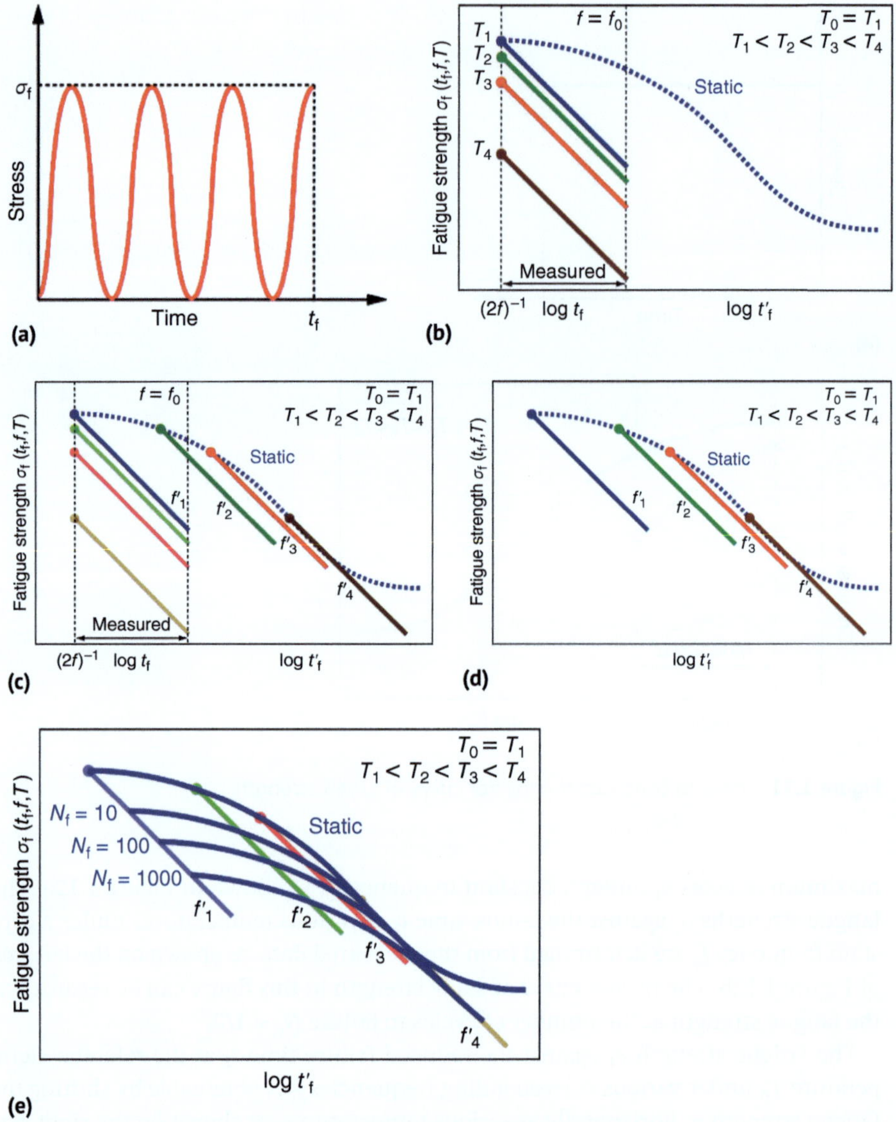

Figure 1.12 How to construct the master curve of fatigue strength.

the same number of cycles to failure N_{fi}, as shown in Figure 1.12d,e. The long-term fatigue strength σ_{f} against the reduced failure time t'_{f} at reference temperature T_0 for an arbitrary number of cycles to failure N_{f} can be predicted from Figure 1.12e.

1.9 Conclusion

The concepts of viscoelasticity and TTSP were explained using the Maxwell model. The generalization of TTSP for nondestructive deformation properties of the matrix resin to static, creep, and fatigue strengths of FRPs was introduced as the most important condition for the ATM. Procedures to obtain the master curves of static, creep, and fatigue strengths of FRPs were explained. Readers who want more details related to viscoelasticity may refer to [1].

Reference

1 Christensen, R.M. (2003) *Theory of Viscoelasticity*, 2nd edn, Dover Publications, Inc., Mineola, New York.

2

Master Curves of Viscoelastic Coefficients of Matrix Resin

2.1 Introduction

In the previous chapter, we proposed the accelerated testing methodology (ATM) to predict the long-term static, creep, and fatigue strengths of fiber-reinforced polymer (FRPs) based on the time–temperature superposition principle (TTSP) for the viscoelastic behavior of the matrix resin. Based on TTSP, the master curve of viscoelastic behavior versus reduced time at a reference temperature is constructed by shifting the measured data at elevated temperatures along the log time axis (horizontal axis). However, the measured data at elevated temperatures are not superimposed smoothly and solely by horizontal shifting. In this chapter, the applicability of modified TTSP with vertical shift and horizontal shift to predict the long-term viscoelastic behavior of thermosetting resins is assessed experimentally [1].

A long-term creep test is suitable for predicting long-term viscoelastic behavior accurately. Nevertheless, it is a time-consuming method. Even the short-term creep test will take several hours to obtain data for each temperature. In this chapter, a rapid and simplified determination method using dynamic mechanical analysis (DMA) is proposed to elucidate the long-term viscoelastic behavior of amorphous resin. The short-term viscoelastic behavior is measured using DMA under various frequencies and temperatures. The automatic shifting method is proposed to obtain a smooth master curve and accurate time–temperature shift factors. The validity of the simplified determination method is discussed by comparison of the creep compliance master curves obtained from DMA and creep tests [2].

Furthermore, the accurate relaxation modulus of matrix resin in the wide range of time at a reference temperature and the accurate time–temperature shift factor are the most important items for reliable carbon-fiber-reinforced polymers (CFRP) life prediction. The process of determining these properties using two kinds of tests is also proposed [3].

When assessing the durability of FRPs used for actual structures, both chemical and physical aging should be considered. Additionally, the effects of water absorption and ultraviolet radiation should be considered. These effects are explained in a later chapter.

Properties and Testing of Fiber-Reinforced Polymers, First Edition. Yasushi Miyano and Masayuki Nakada. © 2026 Wiley-VCH GmbH. All rights reserved, including rights for text and data mining and training of artificial intelligence technologies or similar technologies.
Published 2026 by Wiley-VCH GmbH.

2.2 Master Curve of Creep Compliance Based on Modified TTSP

Based on the TTSP mentioned in Chapter 1, the master curve of viscoelastic behavior versus reduced time at a reference temperature is constructed by shifting the measured data at elevated temperatures along the log time axis (horizontal axis). However, the measured data at elevated temperatures are not superimposed smoothly merely by horizontal shifting, as shown in Figure 2.1a. Figure 2.1b shows that a smooth master curve might be obtained by shifting the measured data vertically and horizontally. This vertical shifting is well known as the thermal correction based on the entropy elasticity at temperatures higher than the glass transition temperature T_g.

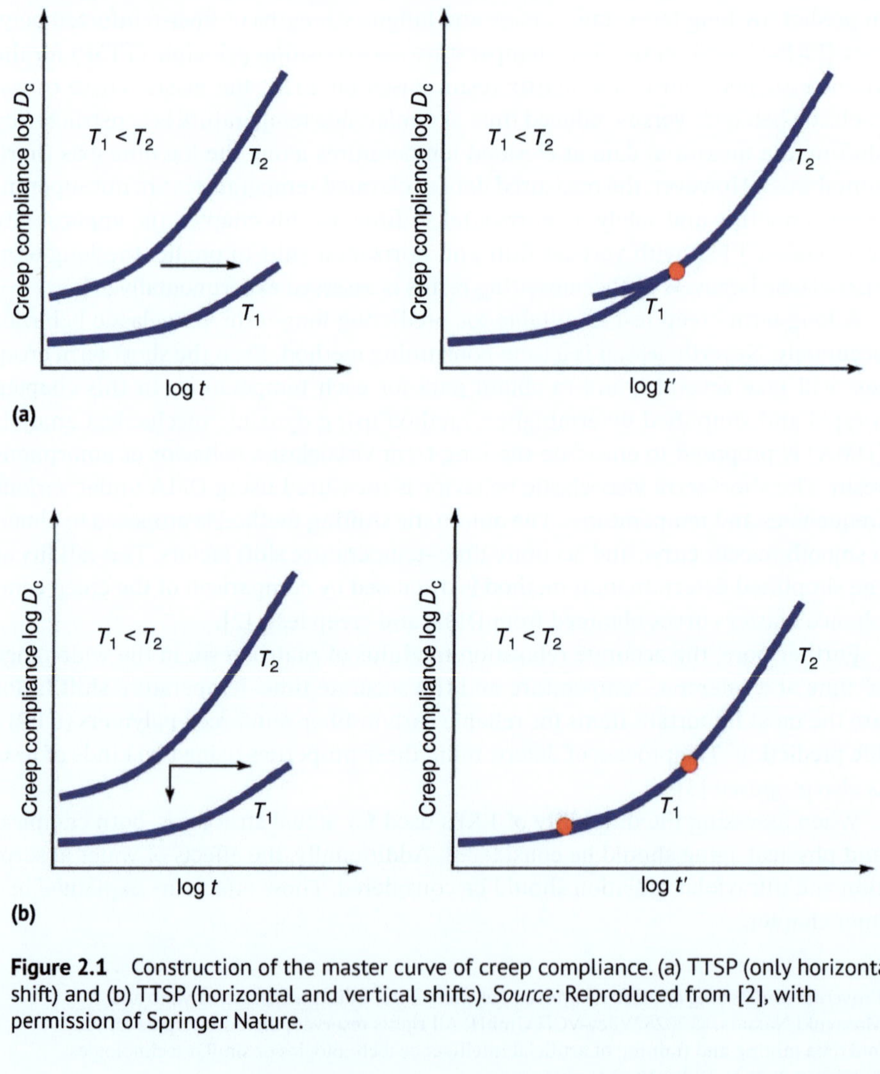

Figure 2.1 Construction of the master curve of creep compliance. (a) TTSP (only horizontal shift) and (b) TTSP (horizontal and vertical shifts). *Source:* Reproduced from [2], with permission of Springer Nature.

This section describes experimental confirmation of the applicability of modified TTSP with a vertical shift and horizontal shift for the prediction of long-term viscoelastic behavior of thermosetting resin at temperatures below and above T_g. We measured the long-term creep compliance D_c at elevated temperatures just as we did for short-term and medium-term D_c. The master curves of D_c constructed from measured data by short-term and medium-term creep tests are compared with those measured by long-term creep tests based on the modified TTSP with a vertical shift and a horizontal shift. The influence of the range of measuring times of D_c on the master curve and the shift amount is discussed.

2.2.1 Experimental Procedures

The material used for this study is a general-purpose epoxy resin. Details of its composition are shown in Table 2.1. The hardener and cure accelerator used are, respectively, MHAC-P and 2-ethyl-4-methylimidazole. The epoxy resin plates were molded by casting and were cured at 70 °C for 12 hours. The creep test specimens cut from epoxy resin plates were post-cured at 150 °C for 4 hours and at 190 °C for 2 hours. Then the cured specimens were cooled to room temperature at a rate of 0.5 °C/min.

In the glassy state, the free volume of matrix resin depends on its thermal history. The free volume decreases gradually. The thermodynamic equilibrium state is reached with the passage of time. This phenomenon is called physical aging. To minimize physical aging during creep tests, heat treatment of the cured specimens was conducted at 100 °C for 167 hours in a constant-temperature chamber.

Three-point bending creep tests for the epoxy resin were conducted using a creep testing machine in a constant-temperature chamber, as shown in Figure 2.2. The long-term, medium-term, and short-term creep tests were conducted at various temperatures, as shown in Table 2.2. The creep compliance D_c was calculated from deflection δ at the center of the specimen using the following equation:

$$D_c = \frac{4bh^3\delta}{P_0 L^3} \tag{2.1}$$

where P_0 is the dead weight, for which it was decided that the surface strain at the center of the specimen for each temperature should not exceed 1%. Also, L is the span, and b and h, respectively, represent the width and thickness of the specimen. The nominal span, width, and thickness of specimens are, respectively, 50, 25, and 3 mm.

Table 2.1 Composition and cure schedule of epoxy resin.

Composition		Weight ratio
Epoxy	Epikote 828 (Yuka Shell Epoxy)	100
Hardener	MHAC-P (Hitachi Chemical)	103.6
Cure accelerator	2-Ethyl-4-methylimidazole (Shikoku Chemicals)	1
Cure schedule		
70 °C × 12 h + 150 °C × 4 h + 190 °C × 2 h + (−0.5 °C/min to room temperature)		

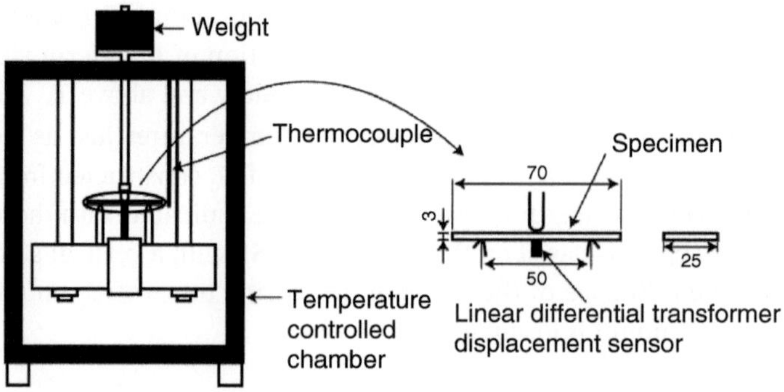

Figure 2.2 Specimen configuration and test apparatus.

Table 2.2 Creep test conditions.

Test type	Test period *t* (h)	Test temperature *T* (°C)
Long term	1000	50, 80, 100
Medium term	24	50, 60, 70, 80, 90, 100, 110
Short term	3	50, 60, 70, 80, 90, 100, 110

2.2.2 Reliable Long-term Creep Compliance of Matrix Resin

As shown in Figure 2.3a, the D_c versus time curves measured at temperatures T_2 and T_3 are superimposed on that at T_1 by shifting the D_c curves horizontally and vertically to form a single master curve at a reference temperature $T_0(= T_1)$ against reduced time t'. The horizontal and vertical shift amounts are called, respectively, the time–temperature shift factor $a_{T_0}(T)$ and the temperature shift factor $b_{T_0}(T)$, as shown in Figure 2.3b,c. The shift factors $a_{T_0}(T)$ and $b_{T_0}(T)$ are defined as

$$a_{T_0}(T) = \frac{t}{t'} \tag{2.2}$$

and

$$b_{T_0}(T) = \frac{D_c(t, T)}{D_c(t', T_0)} \tag{2.3}$$

The D_c versus time t at various temperatures T for an epoxy resin measured by long-term, medium-term, and short-term creep tests are shown on the left side of Figure 2.4. The values of D_c at various temperatures were shifted horizontally until they overlapped, as shown in Figure 2.1a. The D_c values at various temperatures are not superimposed, as shown in the center of Figure 2.4. However, the smooth master curves of D_c can be constructed by shifting the D_c's at various temperatures vertically and horizontally, as shown on the right side of Figure 2.4. These master curves of D_c constructed from the measured data by medium-term and short-term creep tests agree well with those

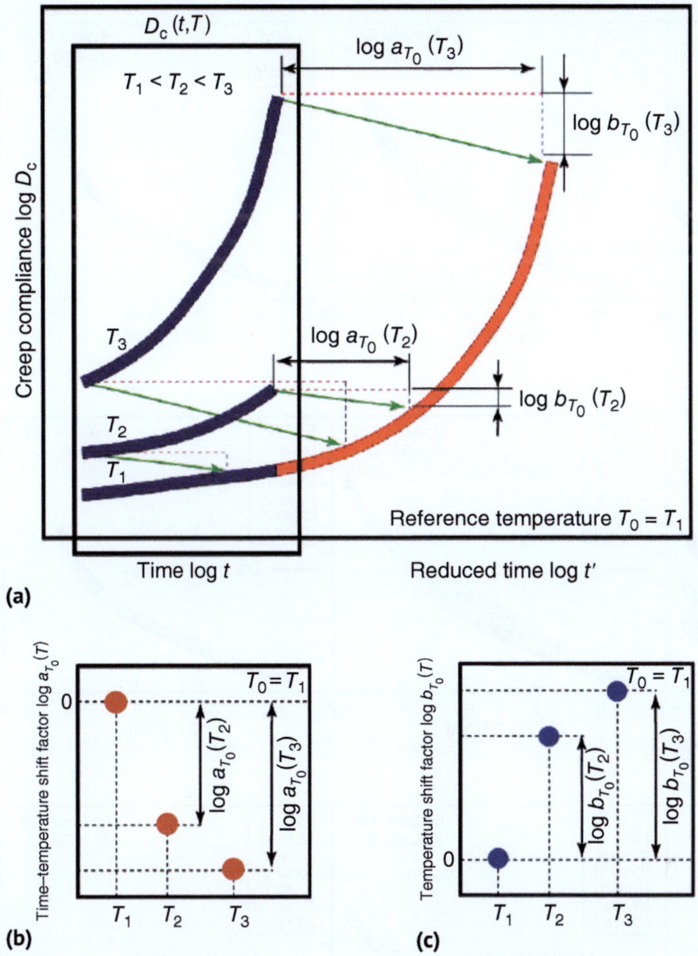

Figure 2.3 (a) Master curve of creep compliance, (b) time–temperature shift factor, and (c) temperature shift factor.

measured by long-term creep tests at $T = 50°C$, indicated, respectively, by the solid curves shown on the right sides of Figures 2.4b,c.

Figure 2.5a shows $a_{T_0}(T)$ obtained by constructing the master curves shown at the center of Figure 2.4. $a_{T_0}(T)$ obtained from long-term, medium-term, and short-term creep tests do not agree with each other because the smooth master curves are not obtained merely by horizontal shifting. Figure 2.5b shows $a_{T_0}(T)$ and $b_{T_0}(T)$ obtained from constructing the master curves by vertical and horizontal shifting shown on the right side of Figure 2.4. $a_{T_0}(T)$ and $b_{T_0}(T)$ obtained from long-term, medium-term, and short-term creep tests agree well with each other. Furthermore, the master curves of D_c constructed from the measured data by long-term, medium-term, and short-term creep tests agree well with each other, as shown in Figure 2.6.

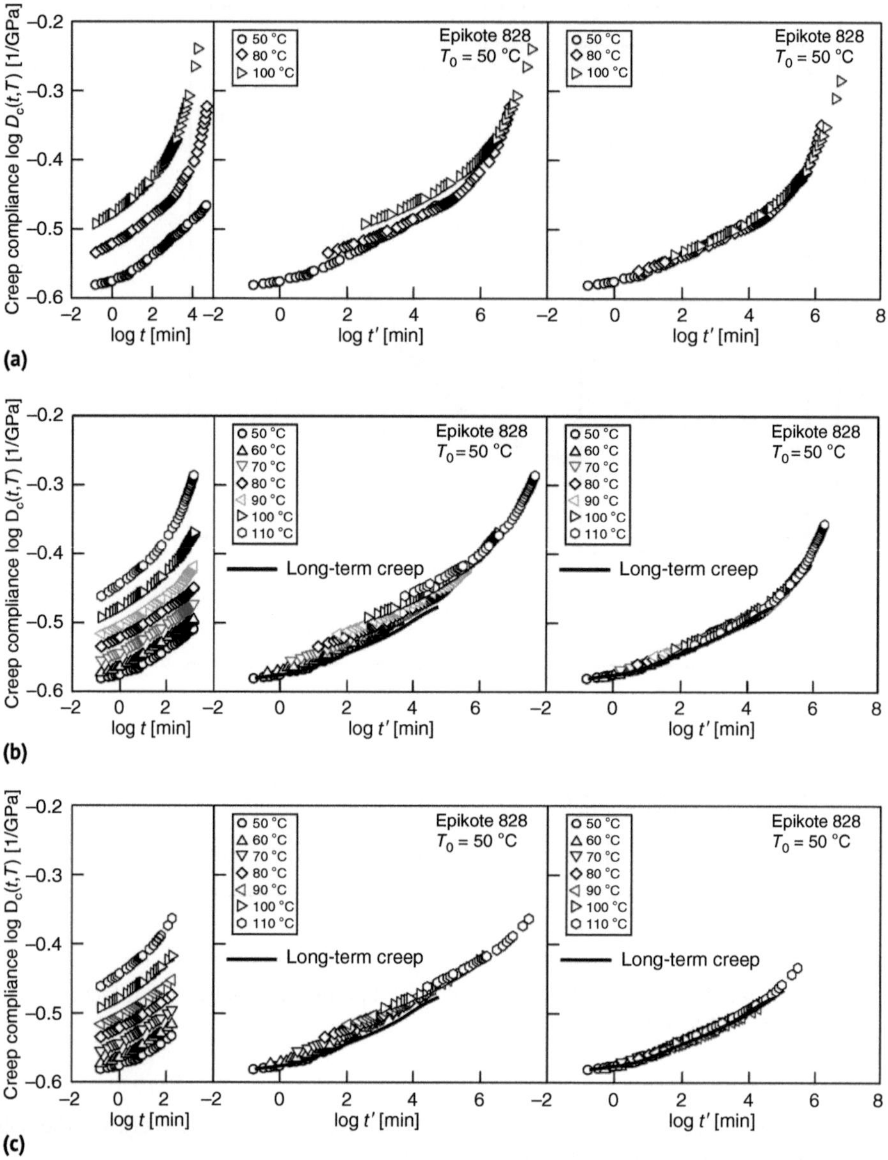

Figure 2.4 Master curves of creep compliance for epoxy resin. (a) Long-term creep, (b) medium-term creep, and (c) short-term creep.

These facts clarify that the long-term D_c can be evaluated accurately using short-term D_c at elevated temperatures by modified TTSP using vertical and horizontal shifts. Therefore, the long-term viscoelastic behavior at temperatures below T_g can be predicted accurately by measuring the short-term viscoelastic behavior at elevated temperatures based on the modified TTSP with a vertical shift and a horizontal shift.

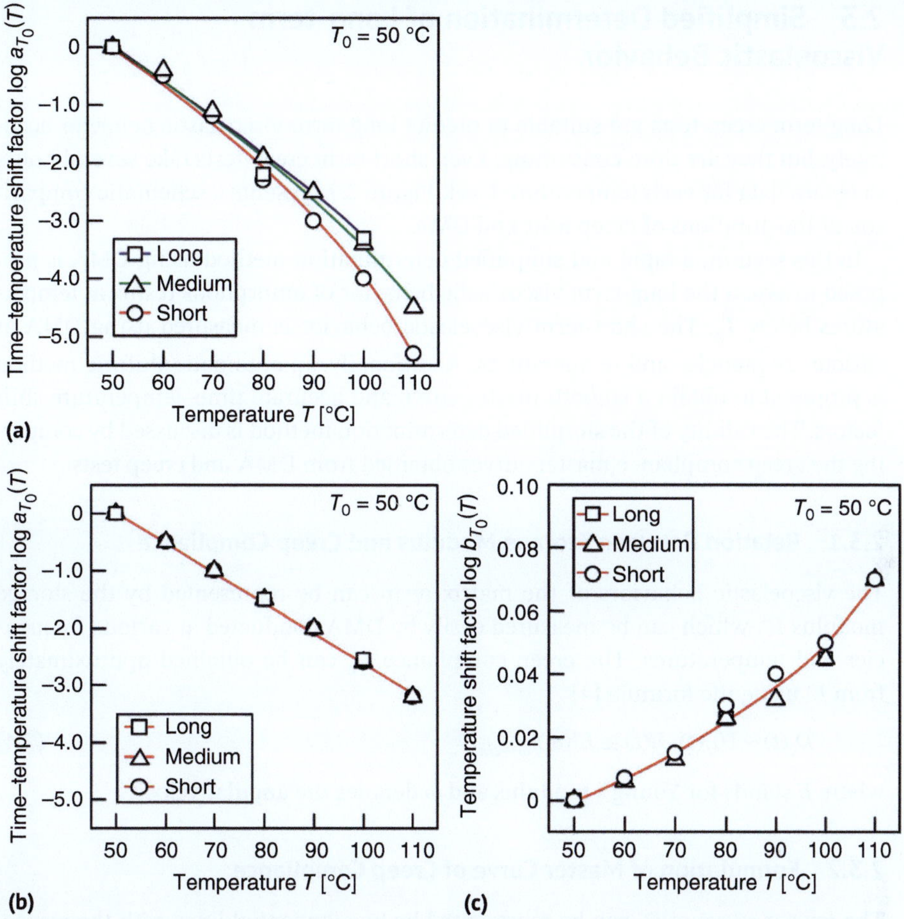

Figure 2.5 Comparison of shift factors obtained from long-, medium-, and short-term creep compliances for epoxy resin. (a) Shift factor $a_{T_0}(T)$ obtained by only horizontal shift, and (b) and (c) shift factors $a_{T_0}(T)$ and $b_{T_0}(T)$ obtained by horizontal and vertical shifts.

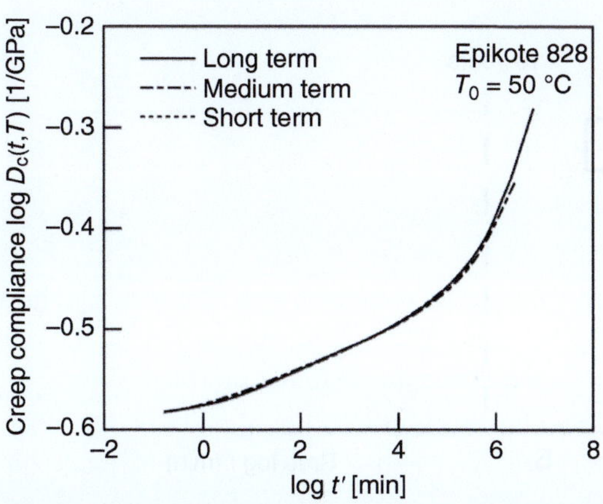

Figure 2.6 Comparison of master curves for creep compliance by TTSP with horizontal and vertical shifts.

2.3 Simplified Determination of Long-term Viscoelastic Behavior

Long-term creep tests are suitable to predict long-term viscoelastic behavior accurately, but they are time-consuming. Even short-term creep tests take several hours to obtain data for each temperature level. Figure 2.7 presents a schematic comparison of the durations of creep tests and DMA.

In this section, a rapid and simplified determination method using DMA is proposed to assess the long-term viscoelastic behavior of amorphous resins at temperatures below T_g. The short-term viscoelastic behavior is measured using DMA at various frequencies and temperatures. Additionally, an automatic shifting method is proposed to obtain a smooth master curve and accurate time–temperature shift factors. The validity of the simplified determination method is discussed by comparing the creep compliance master curves obtained from DMA and creep tests.

2.3.1 Relation Between Storage Modulus and Creep Compliance

The viscoelastic behaviors of the matrix resin can be represented by the storage modulus E', which can be measured easily by DMA conducted at various frequencies and temperatures. The creep compliance D_c can be obtained approximately from E' using the formula [4]

$$D_c(t) \sim 1/E(t),\ E(t) \cong E'(\omega)|_{\omega \to 2/\pi t} \tag{2.4}$$

where E stands for Young's modulus and ω denotes the angular velocity.

2.3.2 Formulation of Master Curve of Creep Compliance

The master curve of D_c can be represented by two tangential lines with the respective slopes of m_g and m_r, as shown in Figure 2.8a. The reduced time at the intersection of the tangential lines is designated as the reduced glass transition time t'_g at

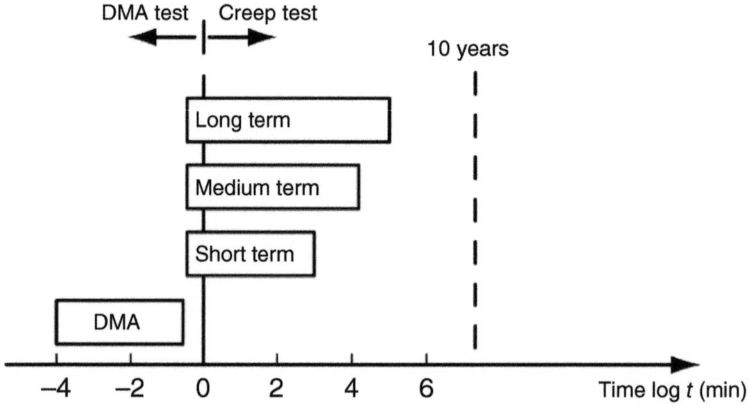

Figure 2.7 Comparison of test periods between the DMA test and the conventional creep test. *Source:* Reproduced from [2], with permission of Springer Nature.

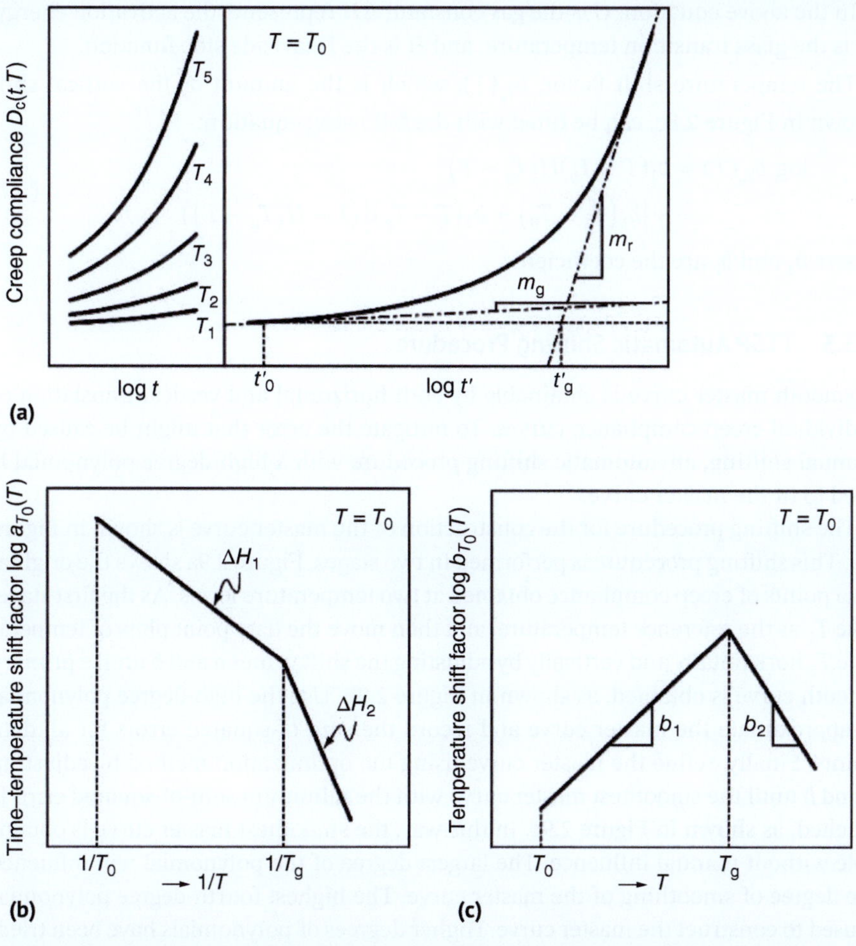

Figure 2.8 Concept of formulation: (a) master curve of creep compliance, (b) time–temperature shift factor, and (c) temperature shift factor. *Source:* Reproduced from [2], with permission of Springer Nature.

the reference temperature T_0. With these parameters, the master curve of D_c can be fitted with the following formula:

$$\log D_c = \log D_{c0}(t'_0, T_0) + \log\left[\left(\frac{t'}{t'_0}\right)^{m_g} + \left(\frac{t'}{t'_g}\right)^{m_r}\right] \tag{2.5}$$

Here, D_{c0} is the initial creep compliance at the initial reduced time t'_0 at reference temperature T_0.

The time–temperature shift factor $a_{T_0}(T)$, which is the amount of the horizontal shift shown in Figure 2.8b, can be fitted with the following equation:

$$\begin{aligned}\log a_{T_0}(T) = {}& \frac{\Delta H_1}{2.303G}\left(\frac{1}{T} - \frac{1}{T_0}\right)H\left(T_g - T\right) \\ & + \left[\frac{\Delta H_1}{2.303G}\left(\frac{1}{T_g} - \frac{1}{T_0}\right) + \frac{\Delta H_2}{2.303G}\left(\frac{1}{T} - \frac{1}{T_g}\right)\right]\left(1 - H\left(T_g - T\right)\right)\end{aligned} \tag{2.6}$$

In the above equation, G is the gas constant, ΔH represents the activation energy, T_g is the glass transition temperature, and H is the Heaviside step function.

The temperature shift factor, $b_{T_0}(T)$, which is the amount of the vertical shift shown in Figure 2.8c, can be fitted with the following equation:

$$\begin{aligned}\log b_{T_0}(T) &= b_1(T - T_0)H(T_g - T) \\ &\quad + [b_1(T_g - T_0) + b_2(T - T_g)](1 - H(T_g - T))\end{aligned} \tag{2.7}$$

where b_1 and b_2 are the coefficients.

2.3.3 TTSP Automatic Shifting Procedure

A smooth master curve is obtainable by both horizontal and vertical translation of individual creep compliance curves. To mitigate the error that might be caused by manual shifting, an automatic shifting procedure with a high-degree polynomial is used to fit the master curve.

The shifting procedure for the construction of the master curve is shown in Figure 2.9. This shifting procedure is performed in two stages. Figure 2.9a shows the original data points of creep compliance obtained at two temperature levels. As the first stage, take T_1 as the reference temperature, and then move the data point plots of temperature T_2 horizontally and vertically by adjusting the shift values a and b until a primary smooth curve is obtained, as shown in Figure 2.9b. Use the high-degree polynomial to approximate the master curve and record the sum-of-squared errors for all data points. Finally, refine the master curve using the optimization method by adjusting a and b until the smoothest master curve with the minimum sum-of-squared error is reached, as shown in Figure 2.9c. In this way, the smoothest master curve is obtainable without manual influence. The largest degree of the polynomial will influence the degree of smoothing of the master curve. The highest fourth-degree polynomial is used to construct the master curve. Higher degrees of polynomials have been tried, but they produce almost identical results to the fourth-degree polynomial. Therefore, the fourth-degree polynomial is taken as sufficient for the master curve construction.

This procedure is repeated for subsequent data points of higher temperature levels. Thus, the complete master curve can be constructed. Moreover, the time–temperature shift factors are obtainable by constructing the master curve. The same procedure is applicable to the data measured using DMA.

2.3.4 Experimental Procedures

The test specimens for DMA were cut from epoxy resin plates and were post-cured at 150 °C for 4 hours and at 190 °C for 2 hours. Then, the cured specimens were cooled at 0.5 °C/min. To minimize physical aging effects during the tests, heat treatment for the cured specimens was applied at 100 °C for 167 hours in a constant-temperature chamber before the test.

The specimen used for DMA had width, thickness, and length, respectively, of 6.4, 1.6, and 50 mm. The span was 38.6 mm. The temperature range was 25–140 °C. The strain amplitude of 0.06% by a sinusoidal wave with frequencies of 0.005–10 Hz was applied to the test specimen using dual cantilever bending.

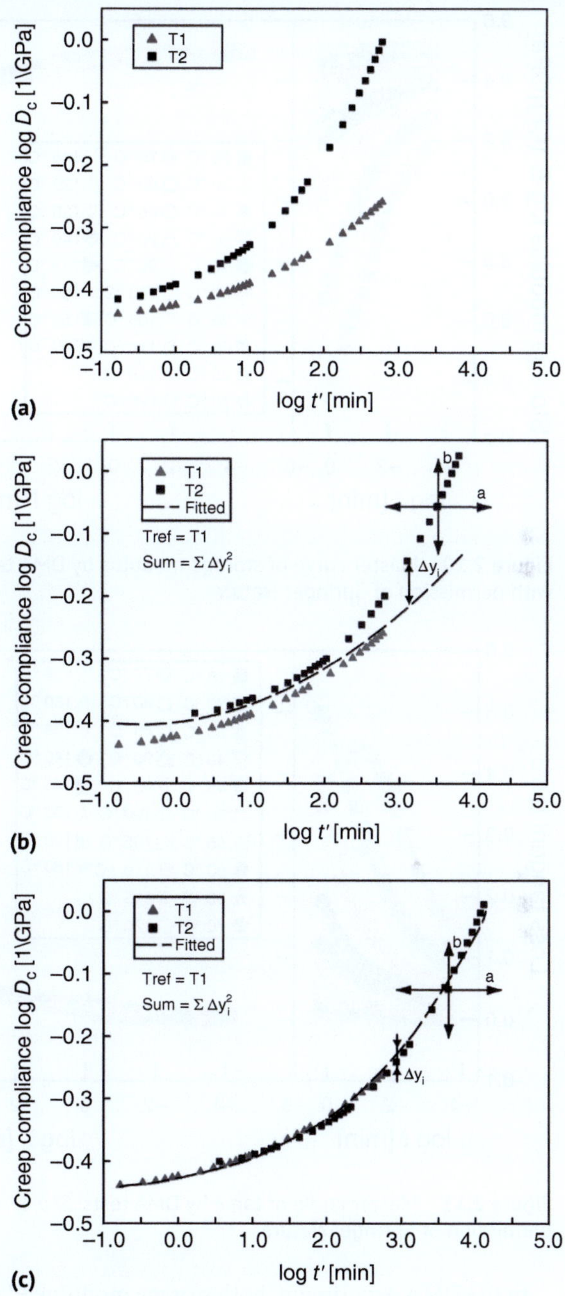

Figure 2.9 Automatic shifting procedure for the construction of the master curve of creep compliance. (a) Original experiment data points. (b) Primary shift horizontally and vertically. (c) Refinement of the shift by optimization. *Source:* Reproduced from [2], with permission of Springer Nature.

2.3.5 Master Curve of Storage Modulus by DMA

The left side of Figure 2.10 shows the storage modulus E' versus testing time $t(= 1/f)$ at various temperatures T. The master curve of E' versus reduced time t' was constructed by shifting E' at various constant temperatures along the log scale of t and log scale of E', as shown on the right side of Figure 2.10. E' at various temperatures can be superimposed smoothly. Therefore, modified TTSP is applicable to E'.

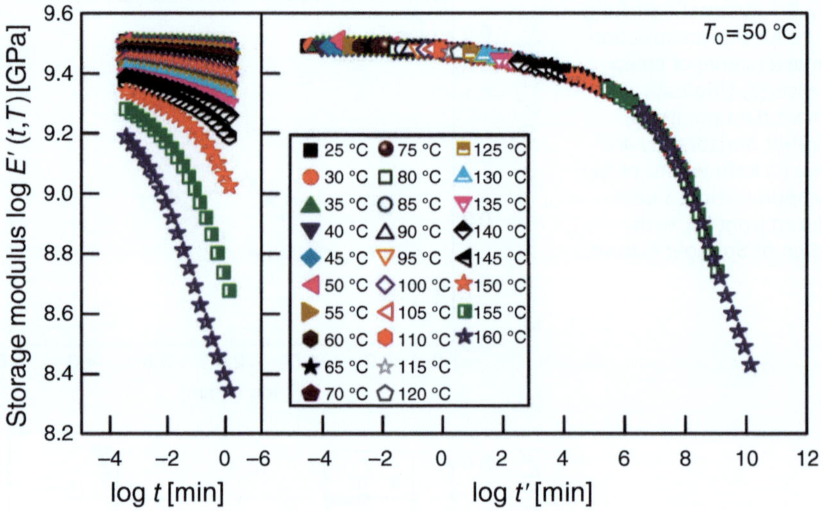

Figure 2.10 Master curve of storage modulus by DMA tests. *Source:* Reproduced from [2], with permission of Springer Nature.

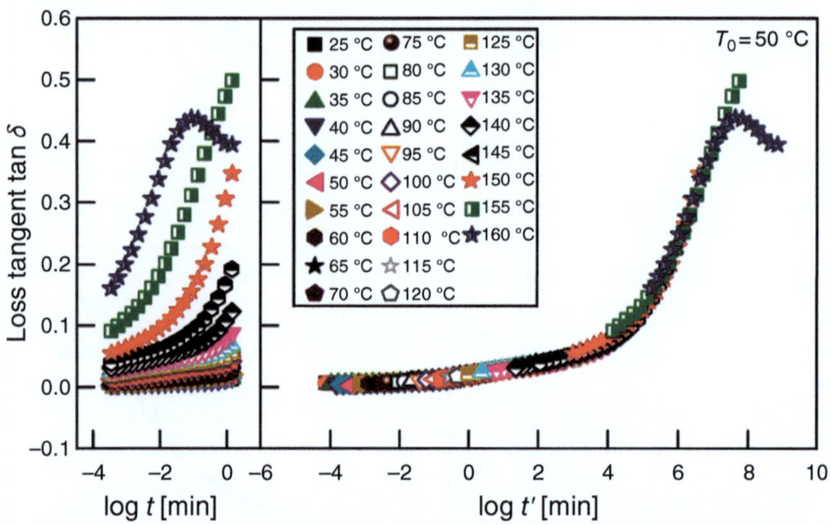

Figure 2.11 Master curve of tan δ by DMA tests. *Source:* Reproduced from [2], with permission of Springer Nature.

In the DMA experiment, both storage modulus E' and loss modulus E'' were measured. From E' and E'', we can calculate the tan δ (ratio of E'' to E') relative degree of damping of the material. Because the tan δ is the ratio of E'' to E', no vertical shift is needed for constructing the tan δ master curve. A smooth master curve of tan δ can be constructed easily merely by horizontal shifting, as shown in Figure 2.11.

Figure 2.12 shows the time–temperature shift factor a_{T_0} and the temperature shift factor b_{T_0} obtained by constructing the master curve of E' shown on the right side of Figure 2.10. The a_{T_0} obtained by constructing the master curve of tan δ is also shown by

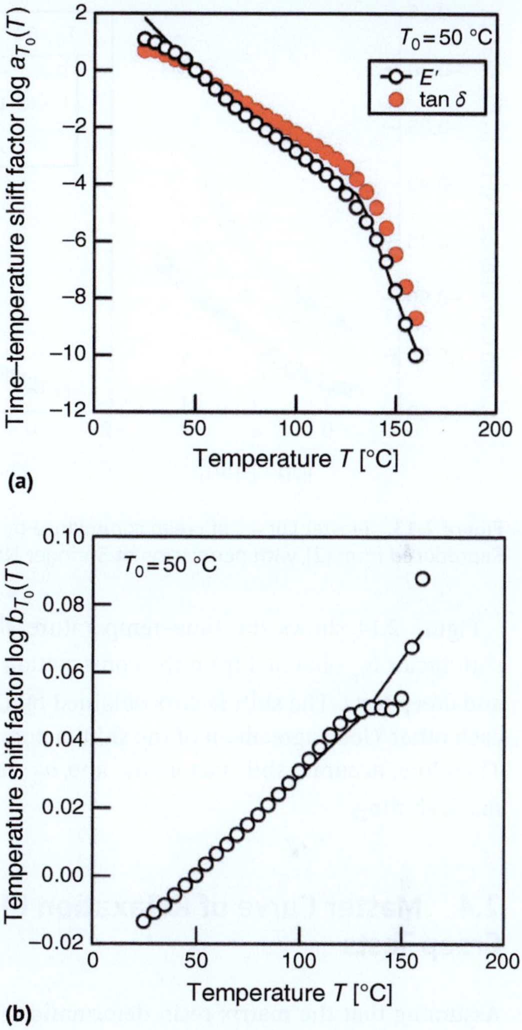

Figure 2.12 Time–temperature shift factor (a) and temperature shift factor (b) for the storage modulus shown by an open circle and loss tangent shown by a red solid circle. *Source:* Reproduced from [2], with permission of Springer Nature.

the solid symbols in Figure 2.12a. Good agreement of a_{T_0} obtained from E' and tan δ is apparent. Therefore, a_{T_0} can be evaluated accurately from E' with automatic shifting.

2.3.6 Comparison of Master Curves of Creep Compliance

The left side of Figure 2.13 shows the creep compliance D_c versus time t at various temperatures T measured using creep tests. A smooth master curve of D_c can be constructed by shifting D_c at various constant temperatures vertically and horizontally using automatic shifting, as shown on the right side of Figure 2.13. The master curve of D_c obtained by DMA is shown as a solid curve, where D_c obtained by DMA is calculated from E' using Eq. (2.4). The master curves of D_c from DMA and creep tests agree well with each other. Results show that, based on our simplified determination method, the same master curve is obtainable from DMA and creep tests.

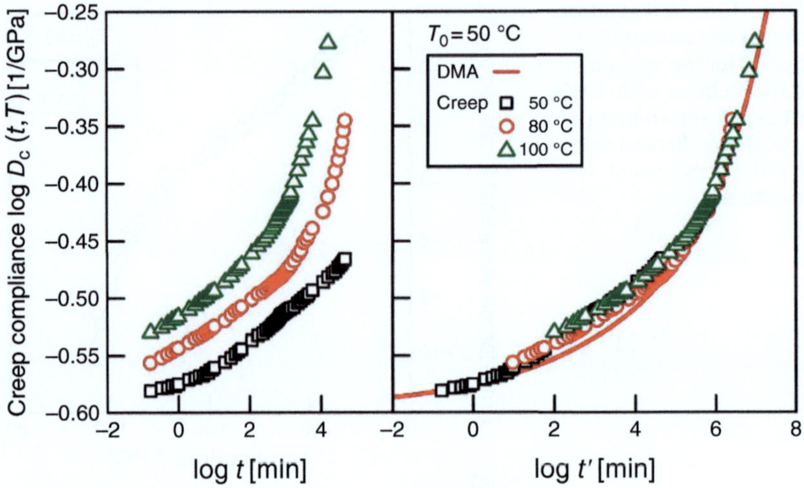

Figure 2.13 Master curves of creep compliance by DMA and creep tests. *Source:* Reproduced from [2], with permission of Springer Nature.

Figure 2.14 shows the time–temperature shift factor a_{T_0} and the temperature shift factor b_{T_0} obtained from the construction of the master curves of D_c by DMA and creep tests. The shift factors obtained by DMA and creep tests agree well with each other. Good agreement of the shift factors by DMA and creep tests is apparent. Therefore, accurate shift factors a_{T_0} and b_{T_0} can be evaluated by DMA with automatic shifting.

2.4 Master Curve of Relaxation Modulus by DMA and Creep Tests

Assuming that the matrix resin deformation in CFRP structure during loading is constrained perfectly by carbon fiber rigidity as presented in Figure 2.15, the time-dependent strength of CFRP structure is controlled by the relaxation modulus of the matrix resin.

2.4.1 Determination Procedure of Relaxation Modulus of Matrix Resin

The accurate relaxation modulus of matrix resin in the wide range of time at a reference temperature and the accurate time–temperature shift factor are the most important items for reliable CFRP life prediction. For this study, the process of determining these properties using tests of two kinds is proposed as follows.

The short-term relaxation modulus of the matrix resin E_r as a function of time t and temperature T is obtainable by substituting the storage modulus E' measured at various frequencies f and temperatures T shown on the left side of Figure 2.16 using DMA into the following equation [4].

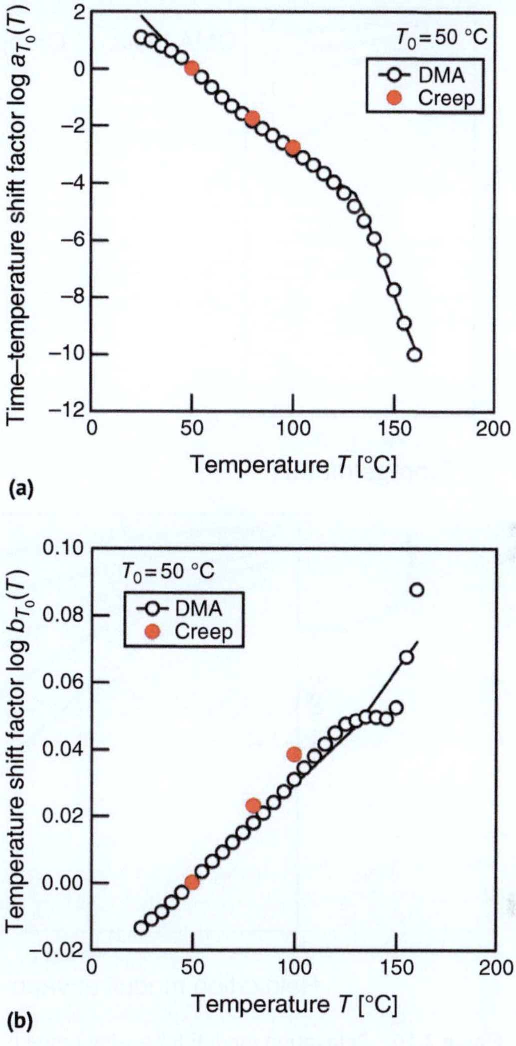

Figure 2.14 Time–temperature shift factor (a) and temperature shift factor (b) for creep compliance by DMA and creep tests. *Source:* Reproduced from [2], with permission of Springer Nature.

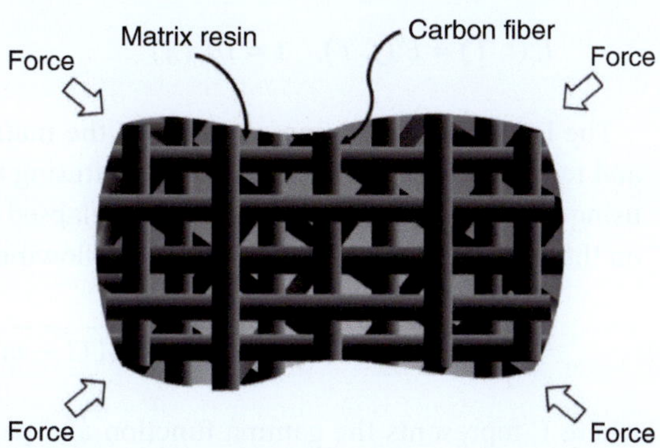

Figure 2.15 Mechanical model of CFRP.

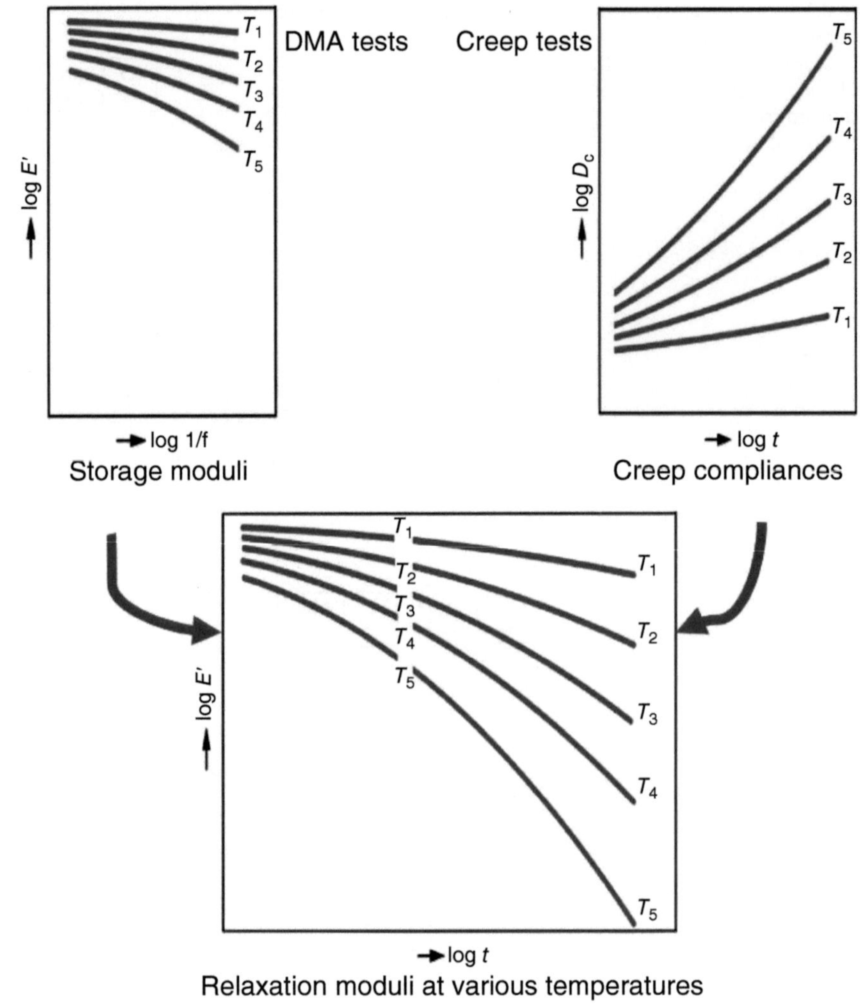

Figure 2.16 Relaxation moduli for a wide range of times.

$$E_r(t, T) = E'(f, T), \quad t = 1/(\pi^2 f) \tag{2.8}$$

The long-term relaxation modulus of the matrix resin E_r as a function of time t and temperature T is obtainable by substituting the creep compliance D_c measured using a creep testing machine at various elapsed times t and temperatures T shown on the right side of Figure 2.16 into the following equation as

$$E_r(t, T) = \frac{a}{D_c(t, T)}, \quad a = \frac{1}{\Gamma(1 + m)\Gamma(1 - m)} \tag{2.9}$$

where Γ represents the gamma function and parameter m represents the slope of the curve for log D_c versus log t. The relation of Eq. (2.9) is readily obtainable using linear viscoelastic theory.

Relaxation moduli E_r obtained at various temperatures for short-term using a DMA are combined with E_r obtained from long-term use of a creep testing machine at various temperatures. Then, E_r value is obtainable for the wide range of times at various temperatures depicted in Figure 2.16.

The time–temperature shift factor a_{T_0} for the modified TTSP can be determined by horizontal shifting of the loss tangents tan δ at various temperatures T as depicted in Figure 2.17. The temperature shift factor b_{T_0} can be determined by shifting the relaxation moduli E_r vertically at various temperatures T as portrayed in Figure 2.18. These shift factors are formulated as Eqs. (2.6) and (2.7).

The master curve of relaxation modulus E_r at a reference temperature T_0 can be formulated as Eq. (2.10) as portrayed on the left of Figure 2.18 [3].

$$\log E_r(t, T_0) = \log E_r(t_0, T_0) - \log\left[\left(\frac{t}{t_0}\right)^{m_1} + \left(\frac{t}{t_g}\right)^{m_2}\right] \tag{2.10}$$

Therein, t_0 and t_g represent reference and glass transition times, respectively. In addition, m_1 and m_2 are slopes in short and long time regions, respectively.

The viscoelasticity of matrix resin can be expressed by the 3 equations Eqs. (2.6), (2.7) and (2.10) and 10 parameters included in these equations.

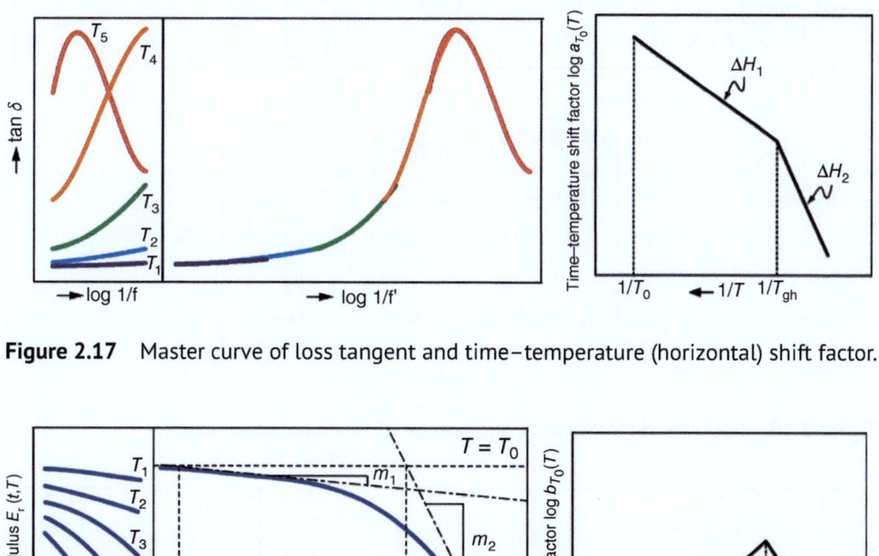

Figure 2.17 Master curve of loss tangent and time–temperature (horizontal) shift factor.

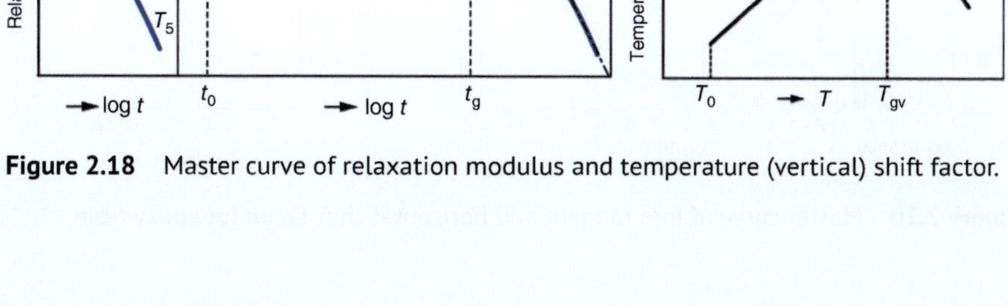

Figure 2.18 Master curve of relaxation modulus and temperature (vertical) shift factor.

2.4.2 Master Curve of Relaxation Modulus of Epoxy Resin

Based on the modified TTSP described above, the master curve of relaxation modulus and time–temperature (horizontal) and the temperature (vertical) shift factors are determined using the storage moduli versus frequency measured at various temperatures using DMA tests and using the creep compliances versus time measured at various temperatures using creep tests.

Figure 2.19 shows the relaxation moduli of epoxy resin for a wide range of time at various temperatures obtained using the procedure of Figure 2.16 from the results of DMA and creep tests. Figure 2.20 shows the master curve of loss tangent and the horizontal shift factor of epoxy resin measured from DMA tests. Figure 2.21 shows the master curve of relaxation modulus and the vertical shift factor of epoxy resin.

All parameters in Eqs. (2.6), (2.7), and (2.10) expressing the master curve of relaxation modulus of epoxy resin at the reference time 1 minute and the reference temperature 25 °C are shown in Table 2.3.

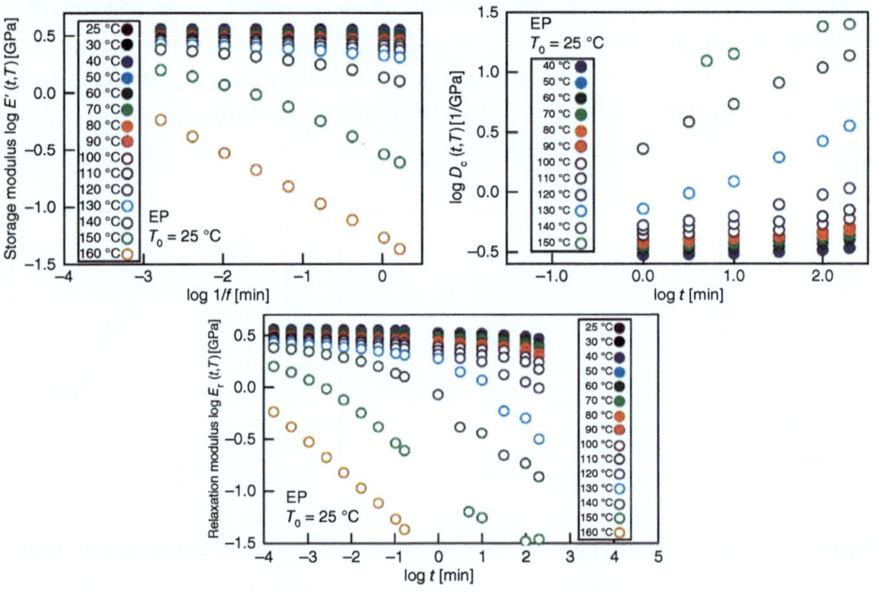

Figure 2.19 Relaxation moduli for a wide range of times for epoxy resin.

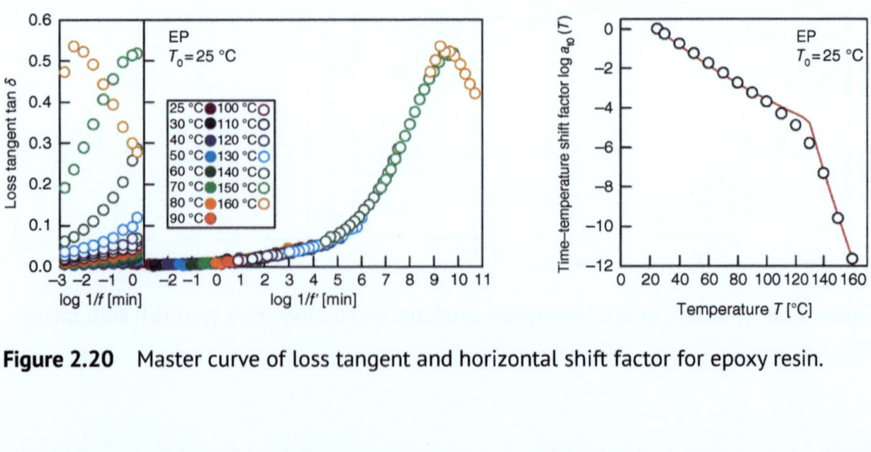

Figure 2.20 Master curve of loss tangent and horizontal shift factor for epoxy resin.

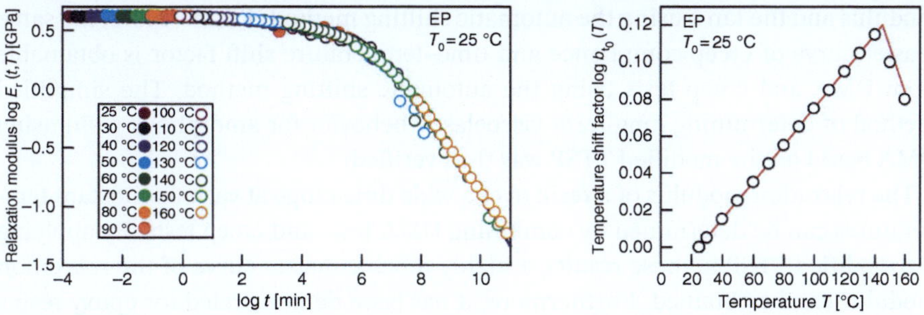

Figure 2.21 Master curve of relaxation modulus and vertical shift factor for epoxy resin.

Table 2.3 All parameters expressing the master curve of the relaxation modulus of epoxy resin.

$E_r(t_0,T_0)$ (GPa)	3.40
m_1	0.007
m_2	0.440
t_g (min)	4.0×10^6
ΔH_1 (kJ/mol)	102
ΔH_2 (kJ/mol)	770
T_{gh} (°C)	130
b_1	0.001
b_2	−0.002
T_{gv} (°C)	145

2.5 Conclusion

Long-term creep compliance D_c was measured at elevated temperatures, as were short- and medium-term D_c. The master curves of D_c can be constructed from measured data by vertical and horizontal shifting. Master curves of D_c constructed from data measured during short- and medium-term creep tests agree well with those measured during long-term creep tests. Furthermore, horizontal and vertical shift factors obtained from constructing the master curve are independent of the creep test time period. Therefore, the long-term viscoelastic behavior at a given temperature can be predicted accurately from measurement of the short-term viscoelastic behavior at elevated temperatures based on the modified TTSP with vertical shift and horizontal shift.

A simplified determination method was proposed for measuring the long-term viscoelastic behavior of amorphous resin using DMA, which was conducted under various frequencies and temperatures for epoxy resin. The creep tests were also conducted for various temperatures. Results show that the time–temperature shift factors are obtainable accurately and easily using DMA by measuring the storage modulus and the tan δ using the automatic shifting method. Furthermore, the same master curve of creep compliance and time–temperature shift factor is obtainable from DMA and creep tests using the automatic shifting method. The simplified method of determining long-term viscoelastic behavior for amorphous resin using DMA based on the modified TTSP was thus verified.

The relaxation modulus of a resin over a wide time range at various constant temperatures can be determined by combining DMA tests and creep tests. By applying the modified TTSP to these results, a highly reliable master curve of the relaxation modulus can be obtained. Furthermore, it has been demonstrated for epoxy resins that if this master curve can be formulated, the viscoelastic properties of the resin can be expressed by several parameters.

References

1 Nakada, M., Miyano, Y., Cai, H. and Kasamori, M. (2011) Prediction of long-term viscoelastic behavior of amorphous resin based on the time–temperature superposition principle. *Mechanics of Time-Dependent Materials*, **15**, 309–316.

2 Cai, H., Nakada, M. and Miyano, Y. (2013) Simplified determination of long-term viscoelastic behavior of amorphous resin. *Mechanics of Time-Dependent Materials*, **17**, 137–146.

3 Miyano, Y. and Nakada, M. (2024) Integrated accelerated testing methodology for CFRP durability. *Composites Part B*, **281**, 111527. https://doi.org/1016/j.compositesb.2024.111.527.

4 Christensen, R.M. (1982) *Theory of Viscoelasticity*, 2nd edn, Dover Publications, Inc., Mineola, New York, p.142.

3

Nondestructive Mechanical Properties of Fiber-reinforced Polymers

3.1 Introduction

The mechanical and thermal properties of fiber-reinforced polymers (FRPs) are determined by the properties of the matrix resin and fiber, and the configuration of fibers distributed within the resin matrix: the volume fraction and weave. The mechanical and thermal properties of FRPs necessarily show heterogeneous and nonisotropic behaviors. Many books describe the mechanical and thermal properties of FRPs [1]. This chapter introduces the rule of the mixture, showing the relation between these properties of FRPs and those of the fiber and the matrix resin. This role is applied concretely to unidirectional carbon-fiber-reinforced polymers (CFRPs) as the most basic configuration of such polymers.

3.2 Rule of Mixture

The approximate averaging method is one of several rules of the mixture for determining the mechanical and thermal properties of FRPs from the properties of the matrix resin and fibers [2]. In this method, the contiguity factor C is used as shown in Figure 3.1, where the unidirectional FRP consists of two models: Models I and II. These models are weighted at $(1 - C)$ and $C(0 < C < 1)$. The elastic coefficients for the longitudinal and transverse directions X and Y of the fibers are shown below:

$$E_{XX} = E_{fXX}V_f + E_m(1 - V_f) \tag{3.1}$$

$$E_{YY} = (1 - C)\frac{E_{fYY}E_m}{E_mV_f + E_{fYY}(1 - V_f)} + C\{E_{fYY}V_f + E_m(1 - V_f)\} \tag{3.2}$$

$$\nu_{XY} = (1 - C)\{\nu_{fXY}V_f + \nu_m(1 - V_f)\} + C\frac{\nu_{fXY}E_{fYY}V_f + \nu_mE_m(1 - V_f)}{E_{fYY}V_f + E_m(1 - V_f)} \tag{3.3}$$

Properties and Testing of Fiber-Reinforced Polymers, First Edition. Yasushi Miyano and Masayuki Nakada. © 2026 Wiley-VCH GmbH. All rights reserved, including rights for text and data mining and training of artificial intelligence technologies or similar technologies. Published 2026 by Wiley-VCH GmbH.

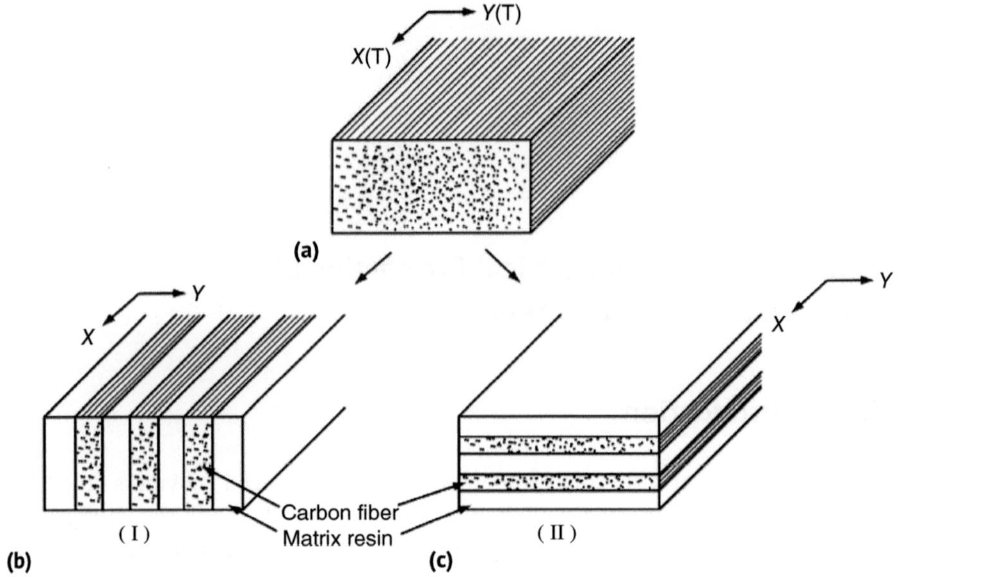

Figure 3.1 Two constitutive element models for determining the elastic moduli of a unidirectional FRP.

$$\nu_{YX} = \nu_{XY}(E_{YY}/E_{XX}) \tag{3.4}$$

$$G_{XY} = (1 - C)\frac{G_{fXY}G_m}{G_m V_f + G_{fXY}(1 - V_f)} + C\{G_{fXY}V_f + G_m(1 - V_f)\} \tag{3.5}$$

$$C = 0.4V_f - 0.025 \tag{3.6}$$

In the aforementioned equations, V_f is the volume fraction of fibers, and E, ν, G, and α denote the longitudinal elastic modulus, Poisson's ratio, the transverse elastic modulus, and coefficient of thermal expansion, respectively. Subscripts f and m, denote the fiber and the matrix resin, respectively.

Assuming that the mechanical and thermal coefficients of the Z-direction (the transverse direction to the fiber direction) are the same as those in the Y direction, the following equations are obtained:

$$E_{ZZ} = E_{YY} \tag{3.7}$$

$$\nu_{XZ} = \nu_{XY} \tag{3.8}$$

$$\nu_{YZ} = (1 - C)\{\nu_{fYZ}V_f + \nu_m(1 - V_f)\} + C\frac{\nu_{fYZ}E_{fYY}V_f + \nu_m E_m(1 - V_f)}{E_{fYY}V_f + E_m(1 - V_f)} \tag{3.9}$$

$$G_{XZ} = G_{XY} \tag{3.10}$$

$$G_{YZ} = G_{fYZ}V_f + G_m(1 - V_f) \tag{3.11}$$

$$\alpha_{XX} = \frac{\alpha_{fXX} V_f E_{fXX} + \alpha_m (1 - V_f) E_m}{V_f E_{fXX} + (1 - V_f) E_m} \tag{3.12}$$

$$\alpha_{YY} = \alpha_{fYY} V_f (1 + \nu_{fYZ}) + \alpha_m (1 - V_f)(1 + \nu_m) - \alpha_{XX} \nu_{XY} \tag{3.13}$$

$$\alpha_{ZZ} = \alpha_{YY} \tag{3.14}$$

3.3 Mechanical and Thermal Properties of Unidirectional CFRPs, Fibers, and Matrix Resins

Table 3.1 shows the mechanical and thermal properties of the carbon fiber Torayca® T800S (Toray Industries, Inc.) calculated using the properties of the unidirectional CFRP T800S/3900-2B (Toray Industries, Inc.) and the matrix resin 3900-2B measured at room temperature [3]. The mechanical and thermal properties of unidirectional CFRP are remarkably nonisotropic; those of carbon fiber are also remarkably nonisotropic.

3.4 Master Curves of Creep Compliance of Matrix Resin

The left side of Figure 3.2 shows the tan δ and the storage modulus at various temperatures for the transverse direction of the unidirectional CFRP T800S/3900-2B. The right side of this figure shows the master curves of tan δ and storage modulus at reference temperature $T_0 = 25\,°\text{C}$. Figure 3.3 presents the shift factors of two kinds for the master curves: the time–temperature shift factor $a_{T_0}(T)$ and the temperature shift factor $b_{T_0}(T)$.

Table 3.1 Mechanical and thermal properties of unidirectional CFRP, matrix resin, and carbon fiber at room temperature.

CFRP T800S/3900-2B

E_{XX}	152 [GPa]
E_{YY}	8 [GPa]
E_{ZZ}	E_{YY}
ν_{XY}	0.34
ν_{YZ}	0.45
ν_{XZ}	ν_{XY}
G_{XY}	4 [GPa]
G_{XZ}	G_{XY}
G_{YZ}	2.75 [GPa]
α_{XX}	0.36×10^{-6} [1/K]
α_{YY}	37.8×10^{-6} [1/K]
α_{ZZ}	α_{YY}
V_f	58[%]

Matrix resin 3900-2B

E_m	3.3 [GPa]
ν_m	0.35
G_m	1.22 [GPa]
α_m	57.6×10^{-6} [1/K]

Carbon fiber T800S

E_{fXX}	303 [GPa]
E_{fYY}	15.2 [GPa]
E_{fZZ}	E_{fYY}
ν_{fXY}	0.2
ν_{fYZ}	0.2
ν_{fXZ}	ν_{fXY}
G_{fXY}	9.65 [GPa]
G_{fXZ}	G_{fXY}
G_{fYZ}	6.32 [GPa]
α_{fXX}	0×10^{-6} [1/K]
α_{fYY}	8.3×10^{-6} [1/K]
α_{fZZ}	α_{fYY}

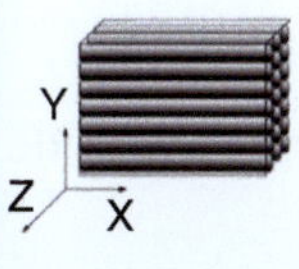

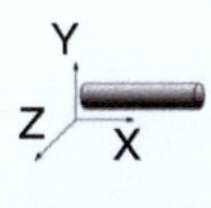

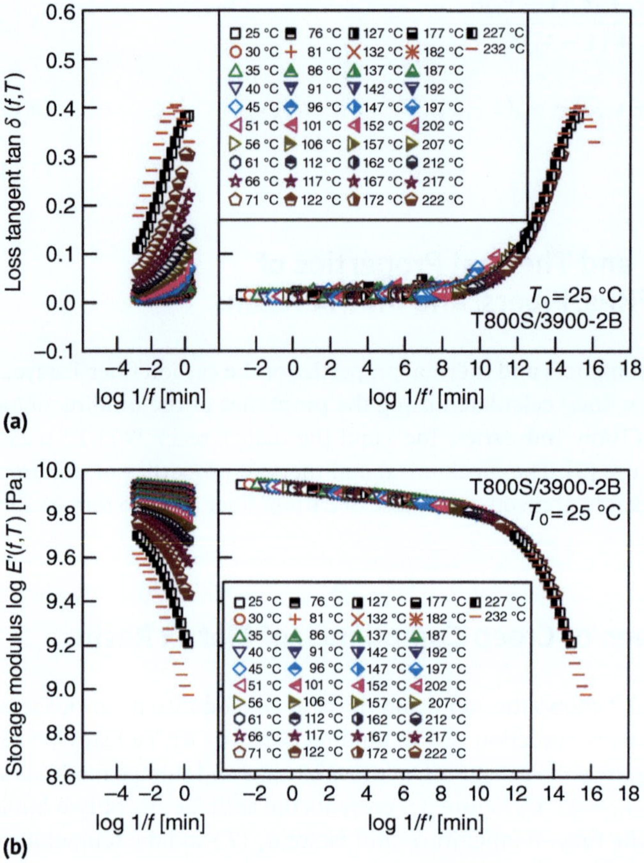

Figure 3.2 Master curves of tan δ and storage modulus for the transverse direction of a unidirectional CFRP.

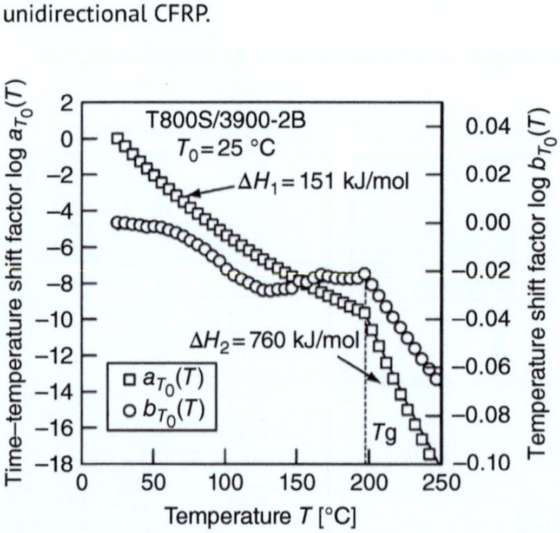

Figure 3.3 Two kinds of shift factors for a unidirectional CFRP.

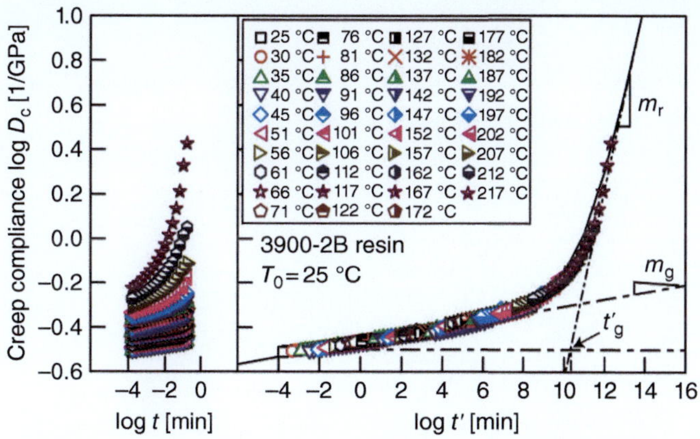

Figure 3.4 Master curve of creep compliance for the matrix resin.

Figure 3.4 shows the master curve of creep compliance of matrix resin 3900-2B calculated from the master curve of storage modulus for the transverse direction of the unidirectional CFRP T800S/3900-2B based on the approximate averaging method.

3.5 Conclusion

Carbon fibers show nonisotropic elastic behavior. Thermosetting resins show isotropic viscoelastic behavior. Therefore, a CFRP should be treated as a heterogeneous and nonisotropic viscoelastic material. This chapter introduced the rule of a mixture, showing the relationship between the mechanical and thermal properties of a unidirectional CFRP as well as those of the carbon fibers and the matrix resin. As an example, the mechanical and thermal properties of the unidirectional CFRP T800S/3900-2B, carbon fibers T800S, and matrix resin 3900-2B were introduced. The master curve of creep compliance of matrix resin 3900-2B was also introduced.

References

1 Christensen, R.M. (2005) *Mechanics of Composite Materials*, Dover Publications, Inc., New York.

2 Uemura, M. and Yamada, N. (1975) Elastic constants of carbon fiber reinforced plastic materials. *Journal of the Society of Material Science, Japan*, **24**, 156–163.

3 Tran, T., Simkins, D., Lim, S.H., Kelly, D., Pearce, G., Prusty, B.G., Gosse, J. and Christensen, S. (2012) *Application of a Scalar Strain-Based Damage Onset Theory to the Failure of a Complex Composite Specimens.* Proceedings of 28th International Congress of the Aeronautical Science, September 23–28, 2012, Brisbane, Australia.

4

Static and Fatigue Strengths of Fiber-reinforced Polymer

4.1 Introduction

The time–temperature dependence of the strengths for various loading directions of carbon-fiber-reinforced polymer (CFRP) laminates under a constant strain rate (CSR) and constant cyclic loadings has been studied by Miyano and Nakada [1]. The CSR loading and constant cyclic loading can be regarded, respectively, as typical types of static loading and fatigue loading. The fracture modes for these CFRP laminates are almost identical under static and fatigue loadings over a wide range of times and temperatures. The same time–temperature superposition principle (TTSP) as for the stress–strain relation of the matrix resin holds for these static and fatigue strengths of CFRP laminates. Results show that the master curves of static and fatigue strengths for CFRP laminates are obtainable from the corresponding parameters at a single loading rate under various temperatures using TTSP for the matrix resin.

In this chapter, we present time- and temperature-dependent static and fatigue strengths of unidirectional CFRPs for three loading directions within the plane of the unidirectional ply: the longitudinal tensile and compressive directions parallel to the fiber direction, and the transverse tensile direction, which is transverse to the fiber direction. The dependences of these static and fatigue strengths upon the loading rate and temperature are discussed based on TTSP [1].

4.2 Experimental Procedures

4.2.1 Preparation of Specimens

The tensile strength for the longitudinal direction of a unidirectional CFRP was measured from tension tests for the longitudinal direction of resin-impregnated carbon fiber strand (CFRP strand). The compressive strength for the longitudinal direction and the tensile strength for the transverse direction were measured using three-point bending tests for unidirectional CFRP laminates because the

Properties and Testing of Fiber-Reinforced Polymers, First Edition. Yasushi Miyano and Masayuki Nakada. © 2026 Wiley-VCH GmbH. All rights reserved, including rights for text and data mining and training of artificial intelligence technologies or similar technologies. Published 2026 by Wiley-VCH GmbH.

compressive fracture for the longitudinal direction and the tensile fracture for the transverse direction occur just under the loading point. Furthermore, three-point bending is easy and more stable for fatigue loading than either four-point bending or tests for axial tension or compression. Details of the three-point bending fracture are discussed later.

A CFRP strand was used as the tensile test specimen for the longitudinal direction of a unidirectional CFRP, which consists of the high-strength carbon fiber Torayca® T400-3K (Toray Industries, Inc.) and a general-purpose epoxy resin Epikote® 828 (Yuka Shell Epoxy Co.). These specimens were produced using the filament winding method. The glass transition temperature T_g of 828 is 145 °C. The CFRP strand diameter is approximately 1 mm.

Unidirectional CFRP laminates consisting of the high-strength carbon fiber Torayca T300-3K (Toray Industries, Inc.) and a general-purpose epoxy resin #2500 (Toray Industries, Inc.) were used as bending test specimens for longitudinal and transverse directions. These specimens were produced by hot pressing of prepreg sheets made from the fiber and resin. The glass transition temperature T_g of 2500 is 130 °C. The fiber volume fraction of CFRP laminates is approximately 55%.

4.2.2 Test Procedures

The tensile static tests for the CFRP strand were conducted under various constant temperatures at three loading rates using an Instron-type testing machine with a small temperature chamber, as shown in Figure 4.1. Loading rates (crosshead speeds) were 0.01, 1, and 100 mm/min. The tensile fatigue tests for the CFRP strand were conducted under several constant temperatures at two frequencies $f = 2$ and 0.02 Hz using an electro-hydraulic servo testing machine with a small temperature chamber. Stress ratio R (minimum stress/maximum stress) was 0.1. The tensile fracture of all specimens tested for both static and fatigue loadings occurred within the central 70 mm region of the specimen, as portrayed in Figure 4.1. The tensile static and fatigue strengths σ of the CFRP strand is defined as follows:

$$\sigma = P_{max} \frac{\rho}{t_e}, \tag{4.1}$$

where P_{max}, ρ, and t_e, respectively, stand for the maximum load of the CFRP strand, the density of the fiber, and the tex of the fiber strand (where the tex is defined by the mass (g)/length (1000 m)). Their respective units in this order are (N), (g/cm^3), and (g/10^6 m).

Three-point bending static tests for longitudinal and transverse directions of unidirectional CFRP laminates (abbreviated as longitudinal bending tests and transverse bending tests) were conducted using an Instron-type testing machine with a temperature chamber. The nominal dimensions of the test specimens were 80, 10, and 3 mm (length, width, and thickness) for the longitudinal bending tests and 65, 15, and 3 mm for the transverse bending tests. The test fixture span was 60 mm

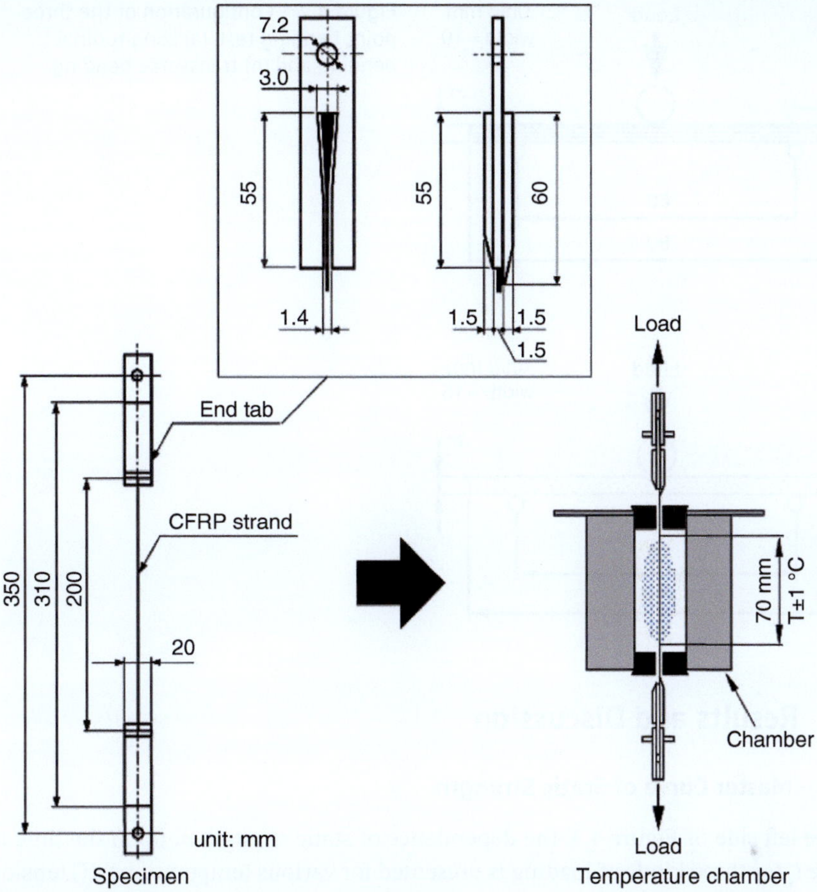

Figure 4.1 Tensile test method for CFRP strand.

for the longitudinal bending tests and 50 mm for the transverse bending tests. The three-point bending test fixture configuration is shown in Figure 4.2. The tests were conducted under various constant temperatures at three loading rates $V = 0.02$, 2, and 200 mm/min. Specimens having the same dimensions were used for three-point bending fatigue tests. The fatigue tests were conducted under several constant temperatures at $f = 2$ and 0.02 Hz using an electro-hydraulic servo testing machine equipped with a temperature chamber. Stress ratio R was 0.05. The flexural static and fatigue strengths σ of the CFRP laminates are defined as:

$$\sigma = \frac{3P_{\max}L}{2bh^2}, \tag{4.2}$$

where $P_{\max}$, L, b, and h, represent the maximum load, the span of the test fixture, width, and thickness of the specimen, respectively. The unit of the maximum load is (N), whereas those of the rest are (mm).

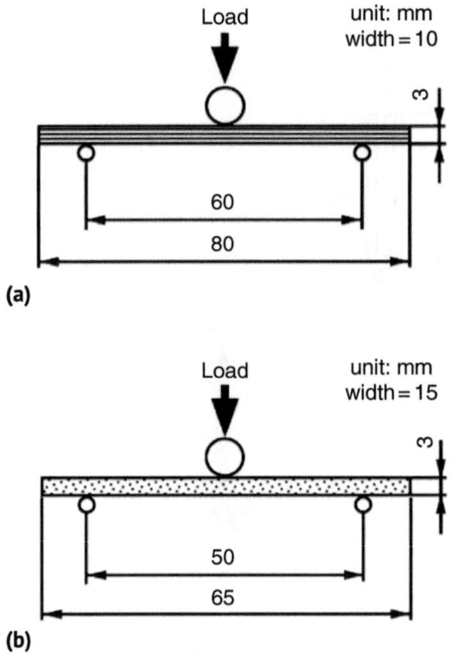

Figure 4.2 Configuration of the three-point bending test. (a) Longitudinal bending and (b) transverse bending.

4.3 Results and Discussion

4.3.1 Master Curve of Static Strength

On the left side of Figure 4.3, the dependence of static strength σ_s upon the time to failure t_s for three kinds of loading is presented for various temperatures: (i) tension in the longitudinal direction (longitudinal tension), (ii) bending in the longitudinal direction (longitudinal bending), (iii) bending in the transverse direction (transverse bending). Here, t_s is defined as the time period from the initial loading to the maximum load during the static test. The data are divided into three regions of t_s for a temperature correspond, respectively, to the data measured at one of the three loading rates. In each condition, three or five sets of test data (repeated test data) are included. It is clear from these figures that each σ_s for the three kinds of loading decreases with increasing t_s and temperature.

The master curve for each σ_s was constructed by shifting σ_s at constant temperatures other than the reference temperature T_0 along the log scale of t_s so that they overlap on σ_s at T_0 or on each other to form a single smooth curve. A smooth master curve for each σ_s is obtainable by curve fitting to these shifted data using a polynomial function, as shown on the right side of Figure 4.3. Therefore, the TTSP is applicable for each σ_s.

The time–temperature shift factors $a_{T_0}(T)$ for σ_s obtained experimentally in Figure 4.3 are shown, respectively, as circles of three kinds in Figure 4.4. The solid and dotted lines in this figure show $a_{T_0}(T)$ for the creep compliance of the matrix epoxy resin 828 and 2500, respectively. $a_{T_0}(T)$ for each σ_s agrees well with that for the creep compliance of the corresponding matrix resin, which is described

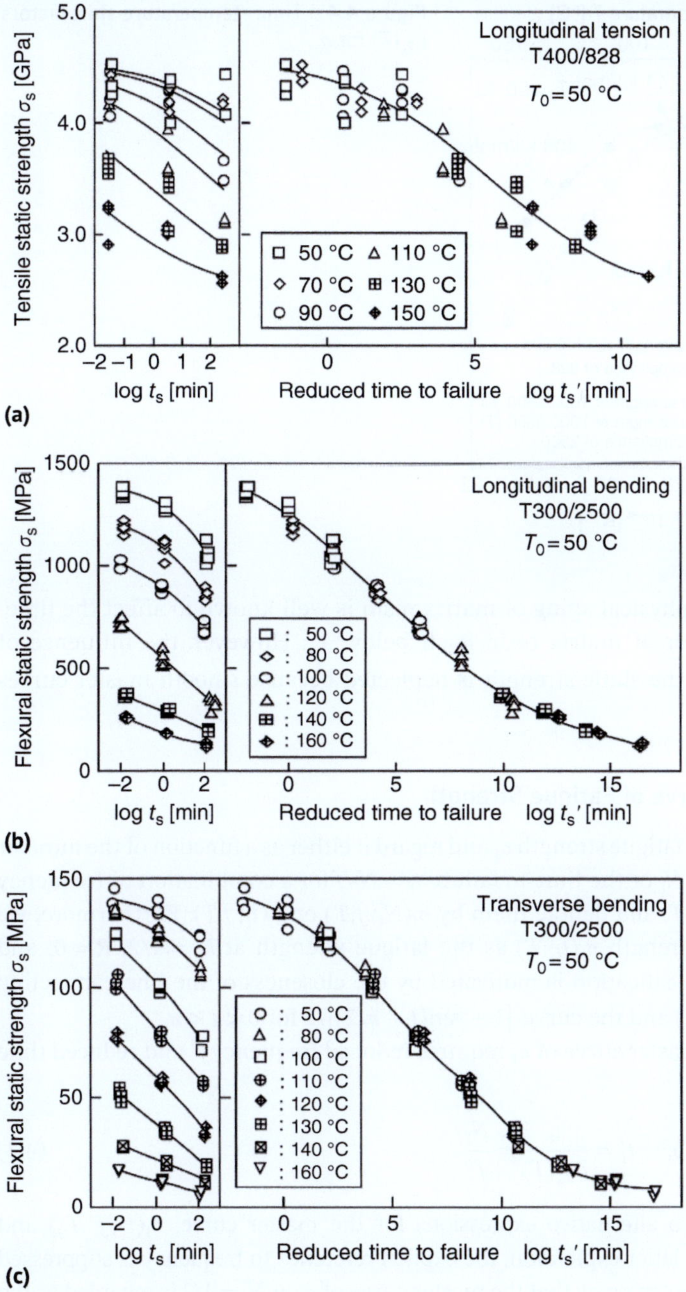

Figure 4.3 Master curves of σ_s for three kinds of loading. (a) Tensile σ_s for longitudinal direction, (b) flexural σ_s for longitudinal direction, and (c) flexural σ_s for transverse direction.

by Arrhenius equation (Eq. 2.6). The activation energy ΔH differs with the temperature range below and above the glass transition temperature T_g of the matrix resin. These facts clarify that the time–temperature dependence of static strength for three different loading types is controlled by the viscoelastic behavior of the

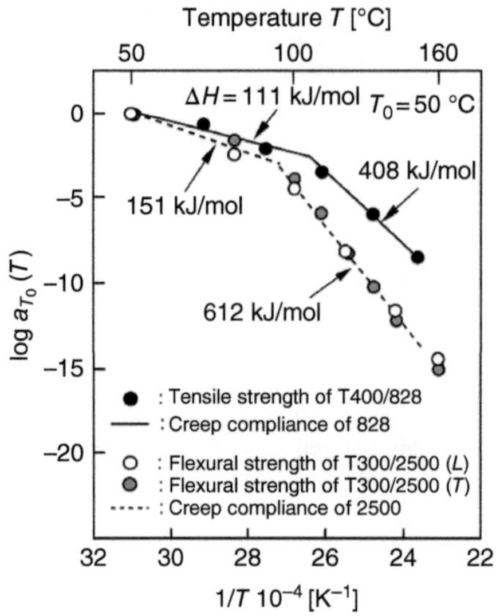

Figure 4.4 Time–temperature shift factors $a_{T_0}(T)$ for σ_s.

matrix resin. The physical aging of matrix resin is well known to affect the time-dependent behavior of matrix resin itself below T_g. However, the influence of physical aging on the static strength is neglected because smooth master curves are obtainable.

4.3.2 Master Curve of Fatigue Strength

We turn now to the fatigue strength σ_f and regard it either as a function of the number of cycles to failure N_f or the time to failure $t_f = N_f/f$ for a combination of frequency f and temperature T, and denote them by $\sigma_f(N_f; f,T)$ or $\sigma_f(t_f; f,T)$. Furthermore, we regard the static strength $\sigma_s(t_f; T)$ as the fatigue strength at $N_f = 1/2$, $R = 0$, and $t_f = 1/(2f)$. This classification is motivated by the closeness of the line connecting the origin and $(\pi, 1)$ and the curve $[1 + \sin(t - \pi/2)]/2$ for $0 < t < \pi$.

Describing the master curve of σ_f requires reduced frequency f' and reduced time t'_f defined as

$$f' = f \cdot a_{T_0}(T), \quad t'_f = \frac{t_f}{a_{T_0}(T)} = \frac{N_f}{f'}. \tag{4.3}$$

We introduce two alternative expressions for the master curve: $\sigma_f(t'_f; f', T_0)$ and $\sigma_f(t'_f; N_f, T_0)$. In the latter expression, the explicit reference to frequency is suppressed in favor of N_f. It is noteworthy that the master curve of σ_f at $N_f = 1/2$ is regarded as the master curve of σ_s. Eq. (4.3) enables us to construct the master curve for any arbitrary frequency from tests conducted at a single frequency under various temperatures.

The next four figures (Figures 4.5–4.8) display $\sigma_f - N_f$ curves or $\sigma_f - t'_f$ curves for loading of three kinds: (a) longitudinal tension, (b) longitudinal bending, and (c) transverse bending. These curves were obtained by curve fitting to the data using a polynomial function.

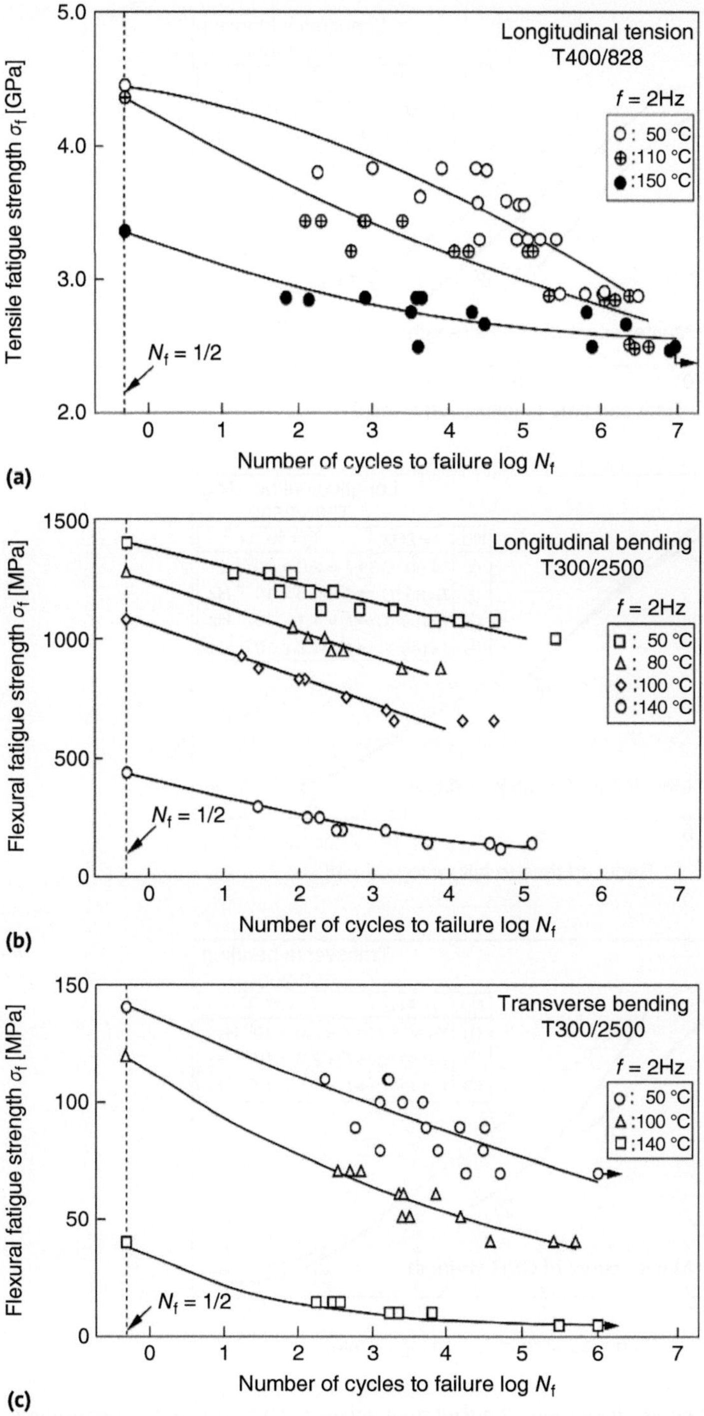

Figure 4.5 $\sigma_f - N_f$ curves for various temperatures T. (a) Tensile σ_f for longitudinal direction, (b) flexural σ_f for longitudinal direction, and (c) flexural σ_f for transverse direction.

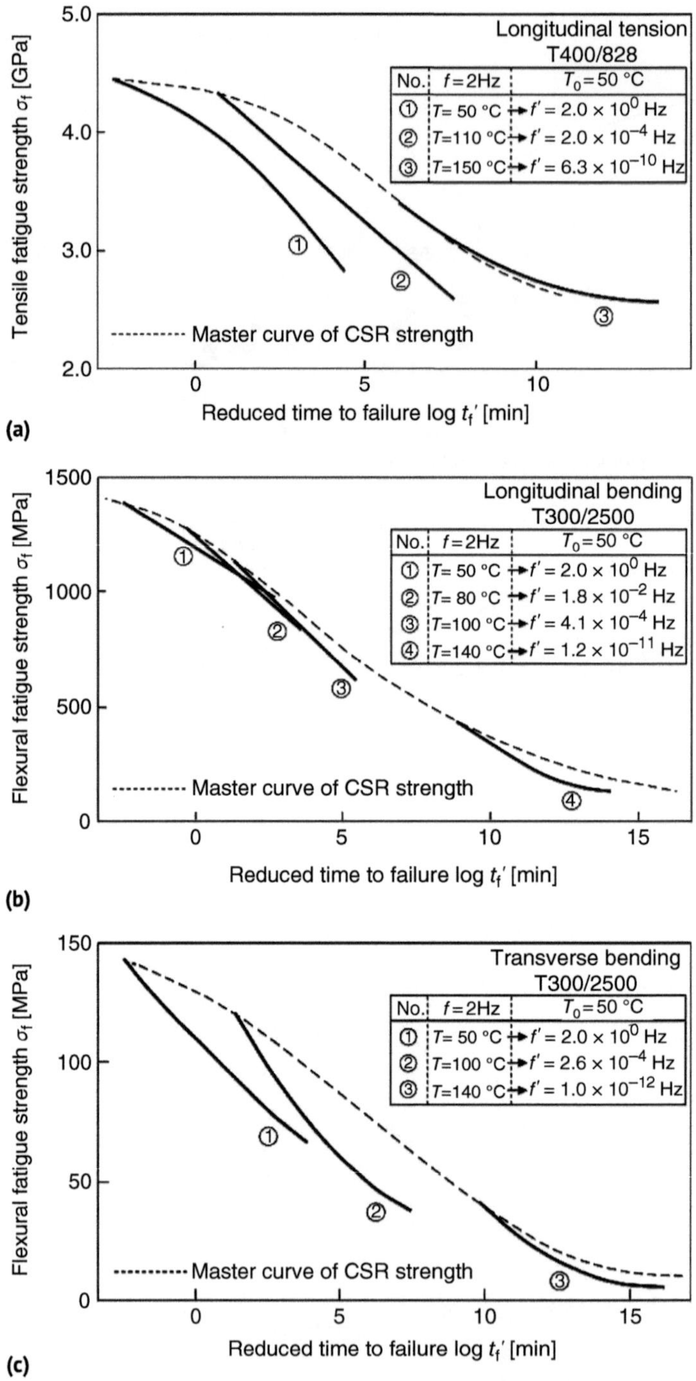

Figure 4.6 $\sigma_f - t'_f$ curves for various reduced frequencies f'. (a) Tensile σ_f for longitudinal direction, (b) flexural σ_f for longitudinal direction, and (c) flexural σ_f for transverse direction.

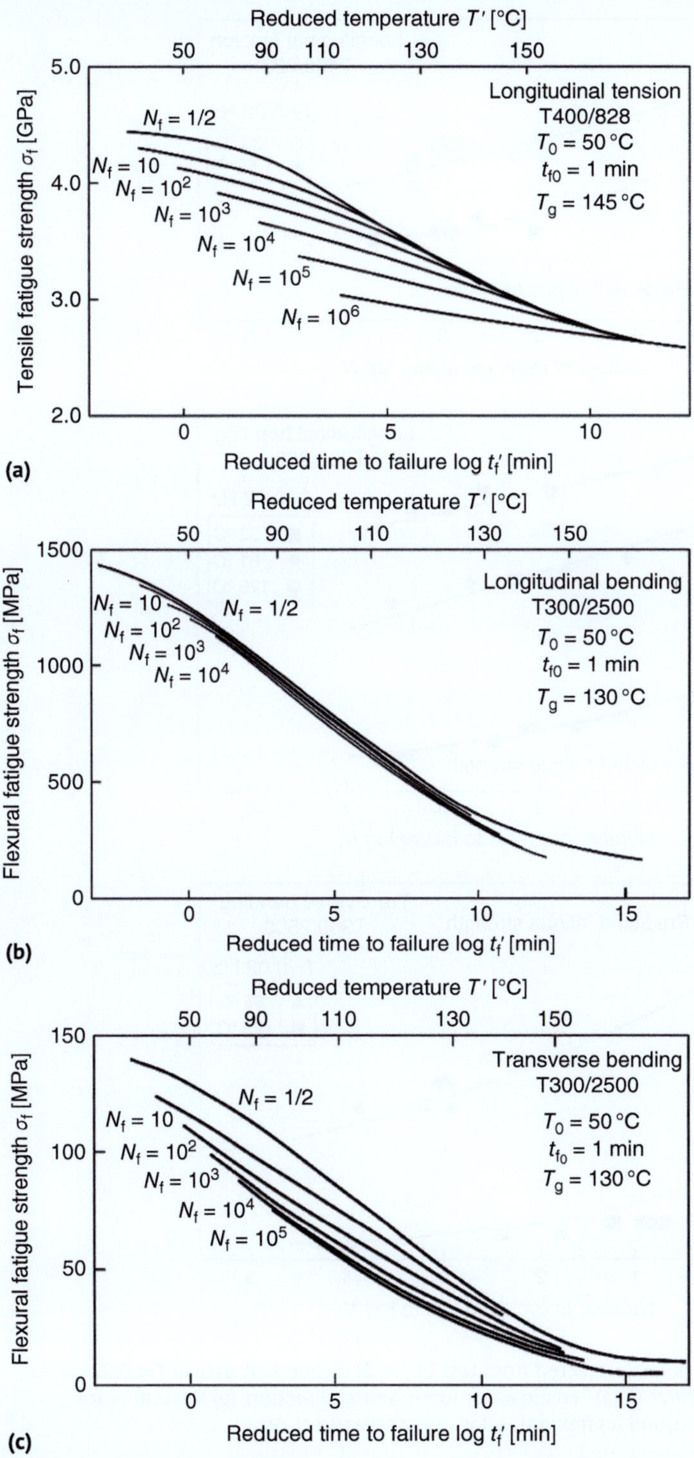

Figure 4.7 Master curves $\sigma_f - t_f'$ for various numbers of cycles to failure N_f. (a) Tensile σ_f for longitudinal direction, (b) flexural σ_f for longitudinal direction, and (c) flexural σ_f for transverse direction.

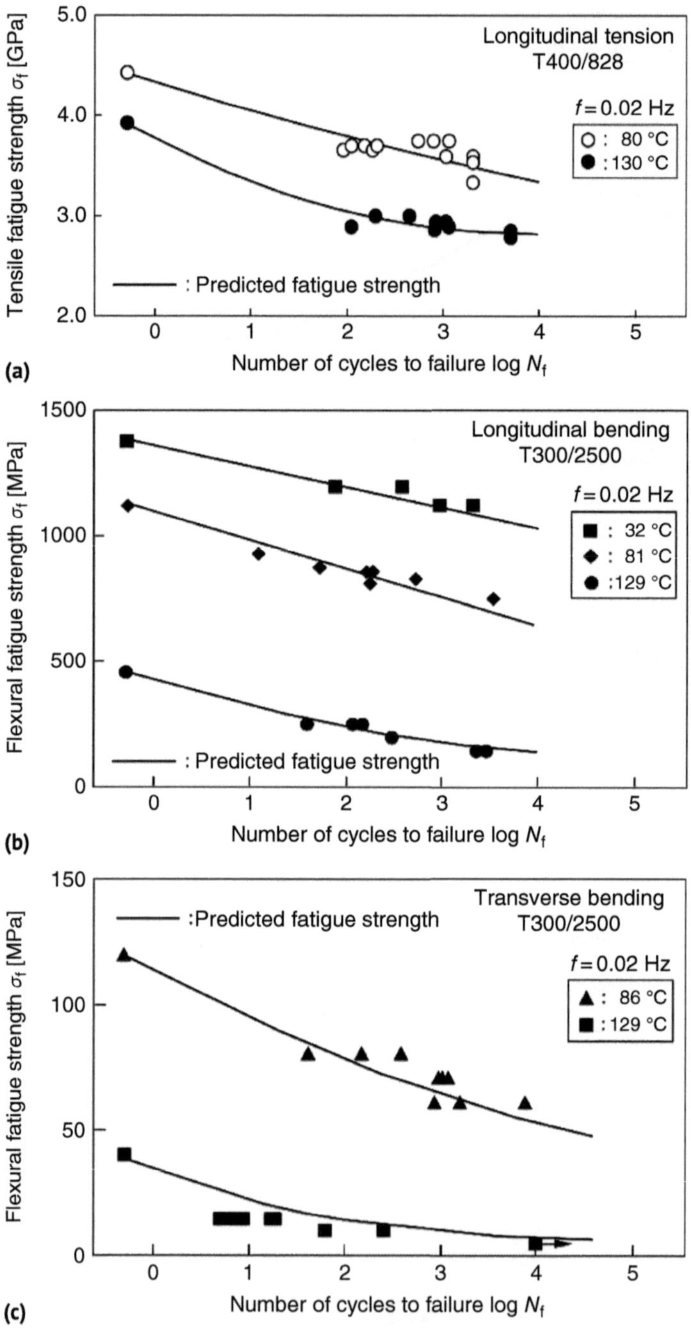

Figure 4.8 $\sigma_f - N_f$ curves (predicted from test at $f = 2$ Hz) and test data at $f = 0.02$ Hz for various temperatures T. (a) Tensile σ_f for longitudinal direction, (b) flexural σ_f for longitudinal direction, and (c) flexural σ_f for transverse direction.

Figure 4.5 portrays $\sigma_f - N_f$ curves at $f = 2$ Hz together with σ_s, which is regarded as σ_f at $N_f = 1/2$. In Figure 4.6, $\sigma_f - t_f'$ curves at $T_0 = 50$ °C are depicted as solid curves using $a_{T_0}(T)$ for σ_s; the master curve for σ_s is shown as a dashed curve. Figure 4.7 shows the master curves of σ_f versus t_f' for fixed N_f, which are constructed by connecting the points of the same N_f on the curves of each f'.

The σ_f–N_f curves at $f = 0.02$ Hz predicted from data for $f = 2$ Hz are displayed in Figure 4.8 together with test data. Because the $\sigma_f - N_f$ curves predicted on the basis of the shift factor for the creep compliance of the matrix resin capture fatigue test data satisfactorily; the TTSP for creep compliance also holds for the fatigue strength. Results confirm the validity of using the time–temperature shift factor for creep compliance for the construction of master curves of fatigue strength.

4.3.3 Characterization of Fatigue Strength for Loading Directions of Three Kinds

We return to Figure 4.7 and note that each of the three kinds of fatigue strength shows the characteristic time- and temperature-dependent behavior. The tensile fatigue strength for the longitudinal direction depends moderately on time and temperature, and on the number of cycles to failure N_f. The flexural fatigue strength for the longitudinal direction depends strongly on the time and temperature. However, this strength only slightly depends on N_f. The flexural fatigue strength for the transverse direction depends strongly on time, temperature, and on N_f.

The respective fracture modes in the tensile static and fatigue tests for the longitudinal direction are the fragmentation of carbon fiber at temperatures lower than T_g and the brush-type fiber delamination at temperatures higher than T_g. We have already confirmed that the time- and temperature-dependent tensile static strength is controlled by the viscoelastic behavior of the matrix resin based on Rosen's shear lag model [2]. From this result, it can be inferred that the time and temperature dependence of tensile fatigue strength is also controlled by the viscoelastic behavior of the matrix resin and that the dependence of N_f on the tensile fatigue strength results from the accumulation of fiber fragmentation at temperatures below T_g.

The fracture modes in the bending static and fatigue tests for the longitudinal direction are the compressive fracture on the compression side of the specimen just under the loading point in all ranges of time to failure, temperature, and N_f, as shown in Figure 4.9. A compressive fracture occurred instantaneously when the load reached the maximum load for static tests and the number of cycles reached N_f for fatigue tests at temperatures below T_g. At temperatures higher than T_g, compressive fracture occurs slightly before the load reaches the maximum load for static tests and the number of cycles reaches N_f for fatigue tests. In both cases, the compressive fracture is triggered by microbuckling of the carbon fiber. No other fracture such as delamination or crushing around the crack site was observed during static and fatigue tests at any temperature tested. The carbon fiber microbuckling is controlled by the viscoelastic behavior of the matrix resin based on Dow's microbuckling theory [3]. Therefore, it can be considered that the flexural fatigue strength for the longitudinal direction is controlled strongly by the viscoelastic behavior of the matrix resin and that it is not influenced by N_f.

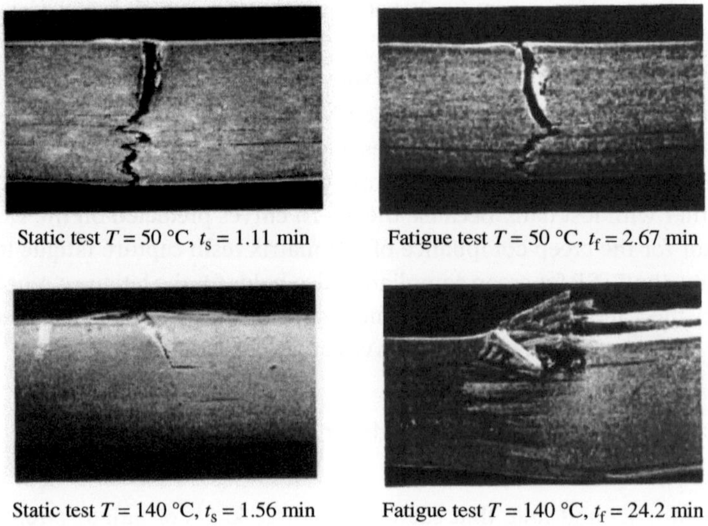

Figure 4.9 Side-view of specimens after the longitudinal bending tests. *Source:* Reproduced from [1], with permission of Springer Nature.

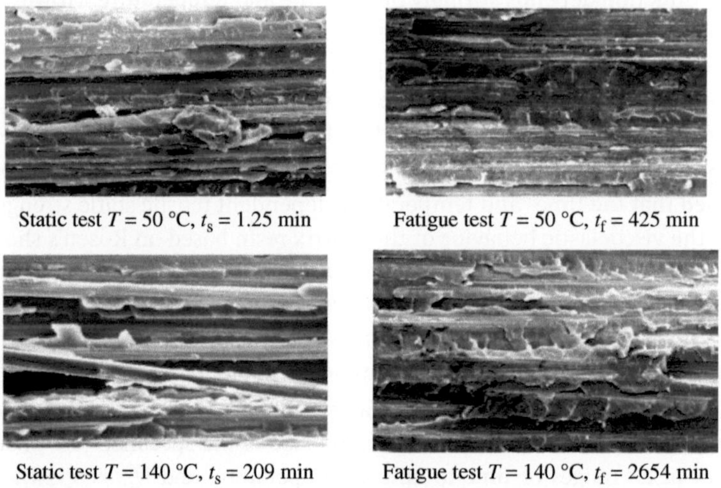

Figure 4.10 Fracture surface of specimens after the transverse bending tests. *Source:* Reproduced from [1], with permission of Springer Nature.

The fracture modes in the bending static and fatigue tests for the transverse direction are the tensile fracture on the tension side of the specimen just under the loading point in all ranges of time to failure, temperature, and N_f. Tensile fracture occurs instantaneously when the load reaches the maximum load for static tests, and the number of cycles reaches N_f for fatigue tests at all tested temperatures. The cohesive fracture of the matrix resin at temperatures below T_g and the interface fracture between the fiber and resin at temperatures above T_g were observed, respectively, on the fracture surface, as shown in Figure 4.10. It can be inferred from these results that the flexural fatigue strength for the transverse direction is controlled by the fracture of the matrix resin.

The fatigue fracture mechanisms of three kinds for loading direction differ, but they are controlled by the mechanical behavior of the matrix resin for which the TTSP holds as described previously. For that reason, the TTSP is applicable to fatigue strengths for three loading directions.

4.4 Applicability of TTSP

Table 4.1 shows the applicability of the accelerated testing methodology (ATM) to various FRP laminates with different combinations of fiber and matrix, weaving type, and loading direction. As the table shows, poly-acrylonitrile (PAN)-based CFRP meets the three conditions, irrespective of the structural configuration and loading style. Polyether ether ketone (PEEK) and pitch-based fibers do not meet the conditions.

The reasons why ATM is applicable (or not) are discussed. Combinations of the fiber and matrix when ATM is applicable are listed. Fibers are expected to have no time and temperature dependence on the mechanical properties; PAN-based carbon fibers have these properties. The matrix should follow TTSP for failure properties and stress–strain relations. Amorphous polymers have these properties. PEEK is liquid–crystal polymer that does not follow TTSP for the mechanical properties of the stress–strain relation and failure. Therefore, TTSP does not hold for strength under all loading conditions on CFRP with PEEK matrix. Pitch-based carbon fibers themselves show viscoelastic behavior, which differs from that of the matrix resin. TTSP does not hold for the longitudinal compressive loading of CFRP using pitch-based carbon fibers, which the viscoelastic behavior of fibers affects.

4.5 Conclusion

The time- and temperature-dependent static and fatigue strengths of a unidirectional CFRP for loading directions of three kinds were discussed based on TTSP. Master curves of static and fatigue strengths for three loading directions were obtained by the application of the same TTSP as that used for the viscoelastic behavior of the matrix resin. The static and fatigue strengths of each direction show characteristic behavior that differs from the static and fatigue strengths of others. (i) Tensile strength for the longitudinal direction strongly decreases concomitant with increasing time and temperature, and with an increased number of loading cycles. (ii) The compressive strength for the longitudinal direction strongly decreases, concomitant with increasing time and temperature, but it decreases very little with loading cycles. (iii) The tensile strength for the transverse direction strongly decreases concomitant with increasing time and temperature and with an increase in the number of loading cycles.

The applicability of TTSP to various FRPs combined with various fibers and matrix resins was assessed experimentally. That assessment clarified that TTSP is applicable to the strength of CFRPs combining PAN-based carbon fibers and an amorphous resin.

Table 4.1 Applicability of time–temperature superposition principle.

Fiber		Matrix	Deformation		Type	Fiber/matrix	Loading direction	Applicability		
			D_c	E^*				Static	Creep	Fatigue
Carbon	PAN	Epoxy	–	O	UD$^{\triangle}$	T400/828	LT	O	O	O
						HR40/828		O	–	–
						T300/828		O	O	–
							LT	O	O	–
			–	O	UD^{O}	T400/828	LT	O	O	O
			O	–	UD	Fortafil510/ Cape2002	LB	O	–	O
		PEEK	–	O	UD	T300/PEEK	LB	×	×	×
							TB	×	×	×
		Epoxy	–	O	SW	T400/3601	LB	O	O	O
			O	–	PW	T300/828	LB	O	–	O
			–	O	QIL	T800S/3900-2B	LB	O	O	O
						UT500/#135		O	O	O
						T800S/TR-A33		O	O	–
		PI	–	O	QIL	IM600/ PIXA-M	LB	O	–	–
			–	O		MR50K/ PETI-5		O	–	–
		BMI	–	O	QIL	G40-800/5260	LB	O	–	–
		Vinylester	O	O	PW	T300/VE	LB	O	O	O
			O	–	NCF	T700/VE		O	–	O
								O	–	O
	Pitch	Epoxy	–	O	UD$^{\triangle}$	XN05/828	LT	O	–	–
						XN50/828		O	–	–
			O	–	UD	XN40/25C	LB	×	×	×
						XN70/25C		×	–	×
						YS15/25P		O	–	O
						XN05/25P		O	–	O
Glass		Vinylester	–	O	PW	E-glass/VE	LB	O	O	O
			–	O		WE18W/VE		O	–	–

Deformation: D_c, creep compliance; E', storage modulus.

Type: UD, unidirectional plate; UD$^{\triangle}$, resin-impregnated strand; UD^{O}: hoop-wound ring; SW, satin woven fabric; PW, plain woven fabric; QIL, quasi-isotropic laminate; NCF, non-crimp fabric.

Loading direction: LT, longitudinal tension; LB, longitudinal bending; LC, longitudinal compression; TB, transverse bending.

References

1 Miyano, Y. and Nakada, M. (2006) Time and temperature dependent fatigue strengths for three directions of unidirectional CFRP. *Experimental Mechanics*, **46**, 155–162.

2 Nakada, M., Miyano, Y., Kinoshita, M., Koga, R., Okuya, T. and Muki, R. (2002) Time–temperature dependence of tensile strength of unidirectional CFRP. *Journal of Composite Materials*, **36**, 2567–2581.

3 Dow, F. and Gruntfest, I. J. (1960) Determination of most-needed, potentially possible improvements in materials for ballistic and space vehicles. *Space Sciences Laboratory Structures and Dynamics Operation*, T.I.S.R60SD389.

5

Application 1 of Accelerated Testing Methodology: Static and Fatigue Flexural Strengths of Various Fiber-reinforced Polymer Laminates Under Water Absorption Condition

5.1 Introduction

This chapter explains the prediction of the long-term fatigue life of fiber-reinforced polymer (FRP) laminates of five kinds designed for marine use under water absorption and temperature conditions. These FRP laminates were prepared under three water absorption conditions of dry, wet, and wet + dry after molding. The three-point bending static tests for FRP laminates of five kinds under three conditions of water absorption were conducted at various temperatures and strain rates. Furthermore, the three-point bending fatigue tests at zero stress ratio (minimum stress/maximum stress) for these specimens were conducted at various temperatures and frequencies. The characteristics of time, temperature, and water absorption dependences of flexural fatigue strength, in addition to flexural static strength, for these FRP laminates were examined based on the modified time–temperature superposition principle (modified TTSP) [1].

5.2 Specimen Preparation

The base material of the five FRP laminates used in this study was plain fabric carbon-fiber-reinforced polymer (CFRP) laminates of T300 carbon fibers/vinylester (VE) T300/VE (Toray Industries, Inc.). The first selection of FRP laminate to T300/VE was a combination of different fabrics: flat yarn plain fabric T700 carbon fibers/VE T700/VE-F (Toray Industries, Inc.), and multi-axial knitted T700 carbon fibers/VE T700/VE-K (Toray Industries, Inc.) for marine use. The second selection of FRP laminates to T300/VE was combinations with different fibers and matrix resin: plain fabric T300 carbon fibers/epoxy (EP) T300/EP (Toray Industries, Inc.), and plain fabric E-glass fibers/VE E-glass/VE (Toray Industries, Inc.), shown in Figure 5.1. These FRP laminates were formed by resin transfer molding (RTM), with the exception of T300/EP, which was formed by conventional hand lay-up. The laminate thickness was 2 mm.

These FRP laminates were prepared, as mentioned, under three water absorption conditions of dry, wet, and wet + dry after molding. Dry specimens were prepared by

Properties and Testing of Fiber-Reinforced Polymers, First Edition. Yasushi Miyano and Masayuki Nakada. © 2026 Wiley-VCH GmbH. All rights reserved, including rights for text and data mining and training of artificial intelligence technologies or similar technologies. Published 2026 by Wiley-VCH GmbH.

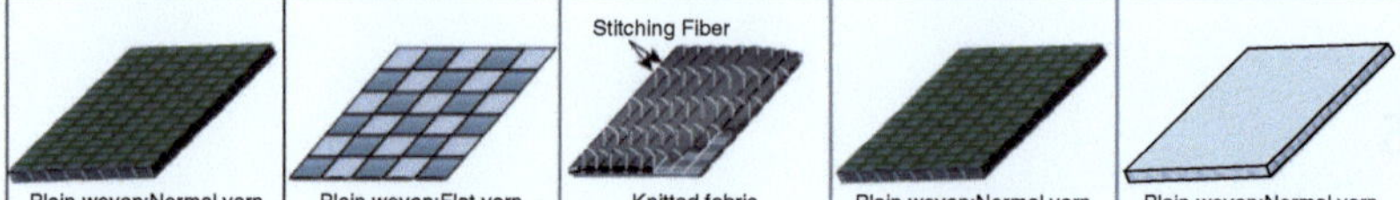

	T300/VE	T700/VE • F	T700/VE • K	T300/EP	E-glass/VE
Weave	Plain woven:Normal yarn (fiber bundle 2mm)	Plain woven:Flat yarn (fiber bundle 10mm)	Knitted fabric	Plain woven:Normal yarn (fiber bundle 2mm)	Plain woven:Normal yarn (fiber bundle 2mm)
Fiber	T300-3K-50B	T700SC-12K-50C		T300-3K-50B	WL230-104BS6
Matrix resin	DERAKANE MOMENTUM:100 Permek N:0.8 Cobalt naphthenate:0.06			Epikote 828:100 MHAC-P:103.6 2E4MZ:1	DERAKANE MOMENTUM:100 Permek N:0.8 Cobalt naphthenate:0.06
Number of layers	8	9	4	9	10
Volume fraction	49%	53%	52%	52%	33%
Forming	RTM (Resin transfer Molding)	RTM (Resin transfer Molding)	VARTM (Vacuum assisted RTM)	Hand lay up method	VARTM (Vacuum assisted RTM)
Cure schedule	25 °C × 48 h 150 °C × 2 h	25 °C × 48 h 150 °C × 2 h	25 °C × 48 h 150 °C × 2 h	70 °C × 12 h 150 °C × 4 h 190 °C × 2 h	25 °C × 48 h 150 °C × 2 h

Figure 5.1 The constitution of five kinds of FRP laminates combined with different resins and fibers. *Source:* Reproduced from [1], with permission of Elsevier.

Table 5.1 Conditions for dry, wet, and wet + dry specimens.

Specimen	In air	In water	In air
Dry	As cured		
Wet	As cured	+ 95 °C × 120 h	
Wet + dry	As cured	+ 95 °C × 120 h	+ 150 °C × 2 h

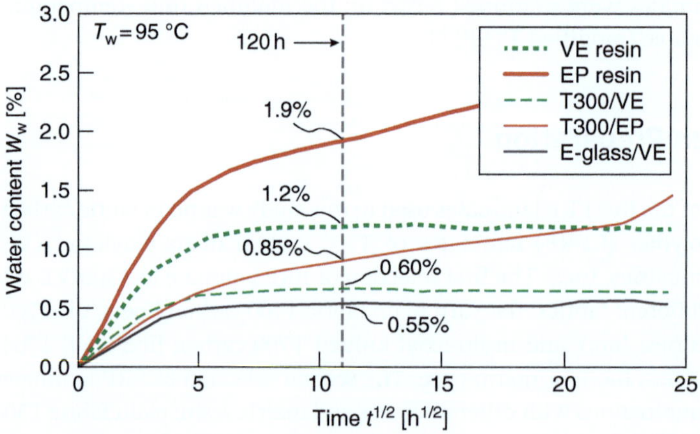

Figure 5.2 Water content versus soaking time.

holding the cured specimens at 150 °C for 2 hours in air; wet specimens were prepared by soaking dry specimens in hot water at 95 °C for 120 hours; and wet + dry specimens were prepared by dehydrating the wet specimens at 150 °C for 2 hours in air, as shown in Table 5.1. Figure 5.2 presents the water content versus soaking time at 95 °C. The soaking condition of 95 °C and 120 hours was applied to wet specimens. Figure 5.3 shows the water content in the resin and FRP for wet and wet + dry specimens. The water absorption of all FRP laminates increases with the wet condition of hot water

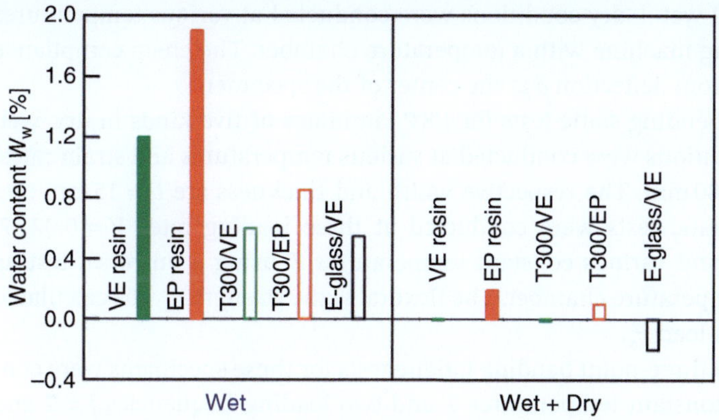

Figure 5.3 Water content in resin and FRP for wet and wet + dry conditions. *Source:* Reproduced from [1], with permission of Elsevier.

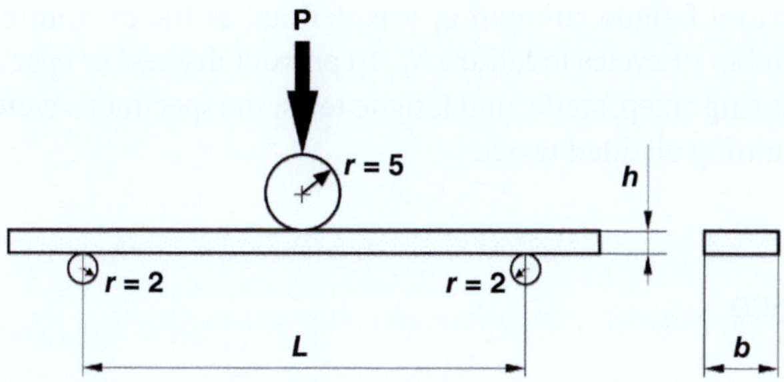

Figure 5.4 Configuration of three-point bending test.

Table 5.2 Test conditions.

Loading type	Deflection rate V(mm/min)	Frequency f(Hz)	Stress ratio $R(\sigma_{min}/\sigma_{max})$	Temperature T(°C)
Creep for neat resin	—	—	—	25–150
Static for FRP	0.02^a, 2, 200^a	—	—	25–160
Fatigue for FRP	—	0.02^a, 2	0.05	25–140

a. Test conditions for confirming the applicability of TTSP.

of 95 °C for 120 hours. The water absorption of neat VE resin and its CFRP laminates reverts to 0% with the wet + dry condition by re-drying; those of T300/EP and E-glass/VE do not return to 0%.

5.3 Experimental Procedures

Figure 5.4 shows the configuration of three-point bending tests. Table 5.2 presents the test conditions. To evaluate the viscoelastic behavior of VE and EP as the neat matrix resin, three-point bending creep tests for the neat VE and EP resins prepared under dry, wet, and wet + dry conditions were conducted at various temperatures using a creep testing machine with a temperature chamber. The creep compliance D_c was calculated from deflection δ at the center of the specimen.

The three-point bending static tests for FRP laminates of five kinds in dry, wet, and wet + dry conditions were conducted at various temperatures and strain rates. The span was $L = 80$ mm. The respective width and thickness are $b = 15$ mm and $h = 2.0$ mm. The static tests were conducted at three loading rates $V = 0.02$, 2, and 200 mm/min, and various constant temperatures T using a universal testing machine with a temperature chamber. The flexural static strength σ_s was calculated from the maximum load P_s.

Furthermore, the three-point bending fatigue tests for these specimens were conducted at various constant temperatures T and two loading frequencies $f = 2$ and 0.02 Hz using an electro-hydraulic servo testing machine with a temperature chamber. The stress ratio R (= minimum stress/maximum stress) was 0.05. The length, width, and thickness of the specimen, as well as the span, were the same as for flexural static tests. The flexural fatigue strength σ_f was defined as the maximum applied load P_{max} for the number of cycles to failure N_f. To prevent dryness of specimens in the wet condition during creep, static, and fatigue tests, the specimens were wrapped in a vinyl bag containing distilled water.

5.4 Creep Compliance

The left sides of the upper graphs in Figures 5.5 and 5.6 show the creep compliance D_c versus testing time t at various temperatures T for dry, wet, and wet + dry specimens of VE and EP resins. The master curves of D_c versus the reduced time t' were constructed by shifting D_c at various constant temperatures along the log scale of t and the log scale of D_c. Because the smooth master curve of D_c for each specimen is obtainable, as shown on the right sides of each graph, the modified TTSP is applicable for each D_c. These master curves clarify that D_c increases with water absorption, and that it reverts perfectly to that of the dry specimen by re-drying after water absorption.

The horizontal time–temperature shift factor $a_{T_0}(T)$ and the vertical temperature shift factor $b_{T_0}(T)$ at a reference temperature T_0 are shown in the lower graphs in these figures.

5.5 Flexural Static Strength

The left side of each graph in Figure 5.7 shows flexural static strength σ_s versus time to failure t_s at various temperatures T for dry, wet, and wet + dry specimens of five kinds of FRP laminates, where t_s is the time period from initial loading to the maximum load during testing. The master curves of σ_s versus the reduced time to failure t_s' were constructed by shifting σ_s at various constant temperatures along the log scale of t_s and the log scale of σ_s using the same time–temperature shift factors

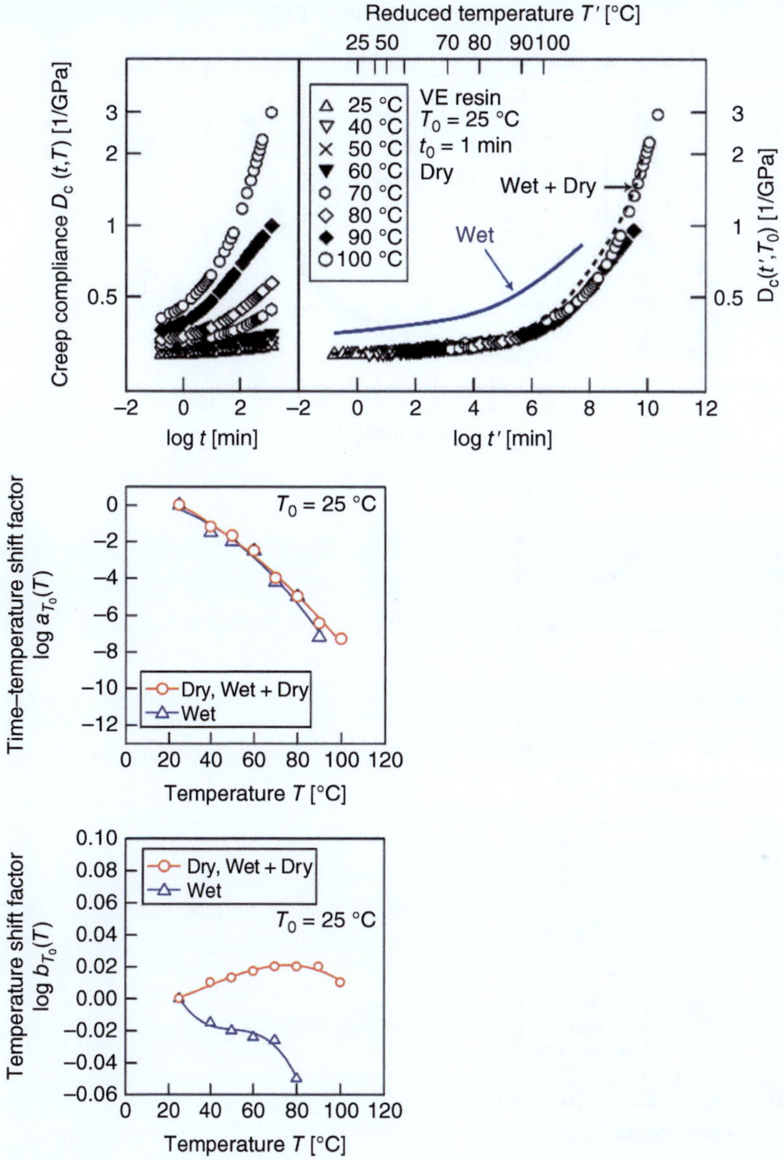

Figure 5.5 Master curve of creep compliance for neat vinylester (VE) resin and shift factors.

for D_c of the matrix resin shown in Figures 5.5 and 5.6. The smooth master curve of σ_s for each specimen is obtainable as shown on the right side of each graph. Therefore, the modified TTSP for D_c of the matrix resin is also applicable to the σ_s of corresponding FRP laminates. It is clear from Figure 5.7 that the σ_s for all five FRP laminates strongly decreases concomitantly with increasing time and temperature, and that these σ_s decrease with water absorption and revert to that of dry specimens by re-drying after water absorption, except that of E-glass/VE. The σ_s of wet + dry specimens of E-glass/VE does not revert to that of dry specimens.

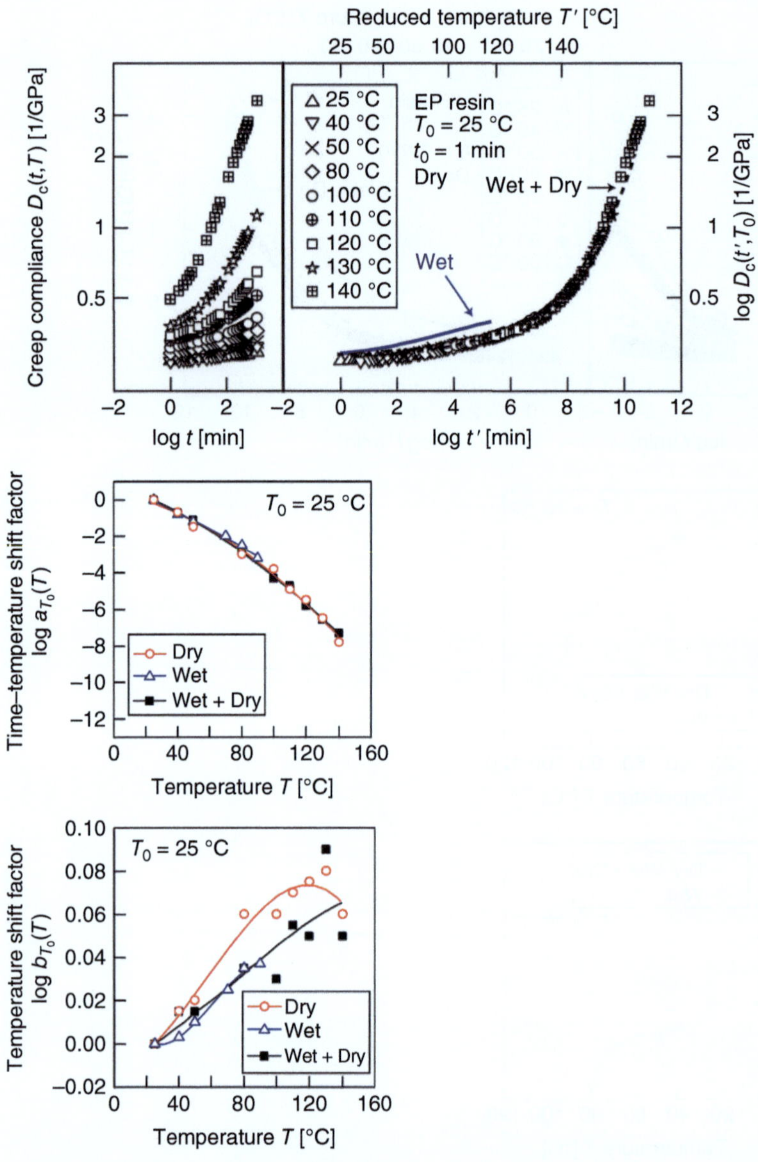

Figure 5.6 Master curve of creep compliance for neat epoxy (EP) resin and shift factors. *Source:* Reproduced from [1], with permission of Elsevier.

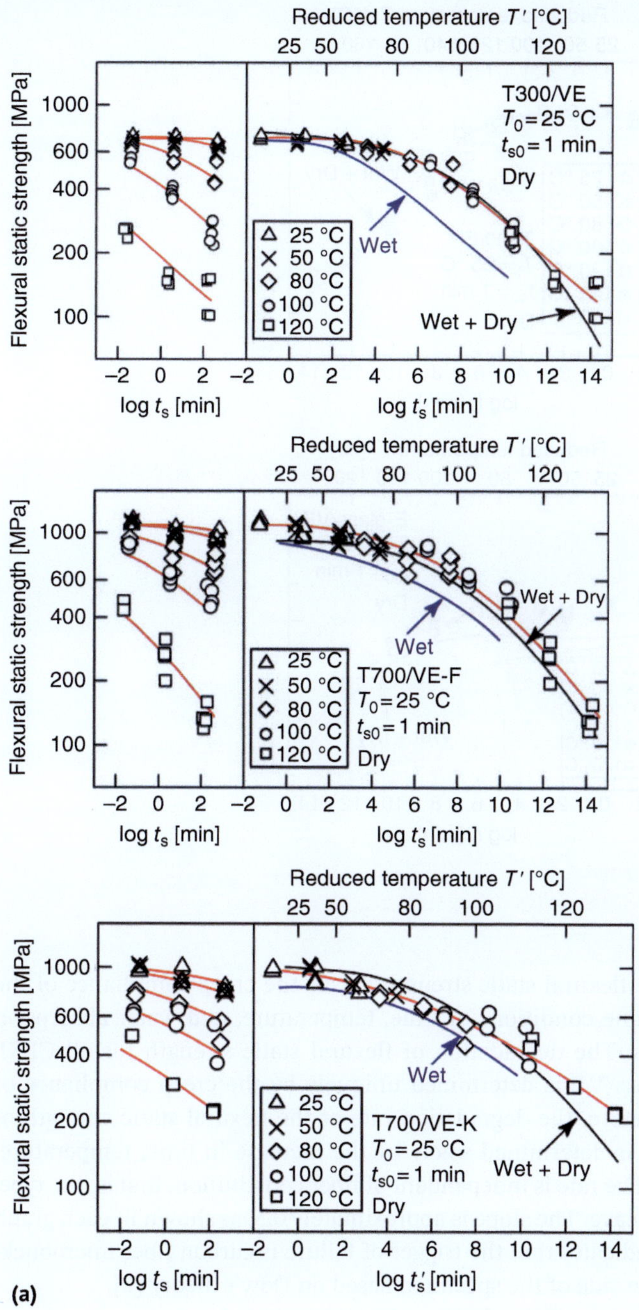

Figure 5.7 Master curves of flexural static strength in dry, wet, and wet + dry conditions.

(Continued)

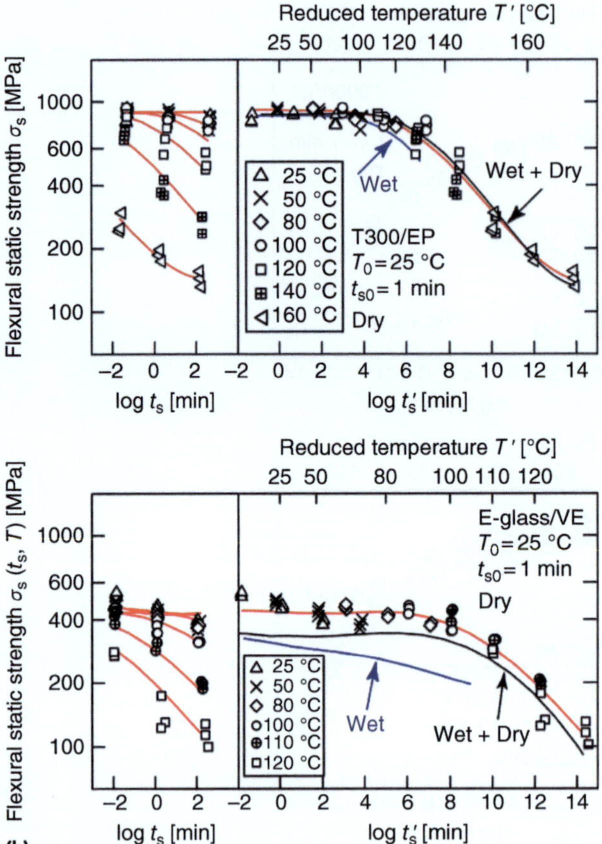

(b)

Figure 5.7 (Cont'd)

Figure 5.8 shows the flexural static strength versus the creep compliance of the matrix resin for the same conditions of time, temperature, and water absorption for five FRP laminates. The degradation of flexural static strength for all CFRP laminates except E-glass/VE is determined uniquely by the creep compliance of the matrix resin. Therefore, the degradation rate of the flexural static strength of these CFRP laminates is determined solely by the increase in time, temperature, and water absorption. The rate is independent of fiber constitution, that is, the type, volume fraction, and weave. The slope is approximately 0.5, as shown in each graph of this figure, which indicates that the trigger of failure is carbon fiber microbuckling on the compression side of the specimen based on Dow's theory [2].

Actually, the appearance of the fracture indicates that the fracture mode for these CFRP laminates is the compressive fracture of warp carbon fibers on the compression side of a specimen for all conditions tested, as shown in Figure 5.9. The fracture mode for T300/EP and E-glass/VE is tensile fracture on the tension side of the specimen at $T = 25$ °C. However, the fracture mode at high temperatures is the compressive fracture on the compression side of the specimen, which is the same as that for T300/VE.

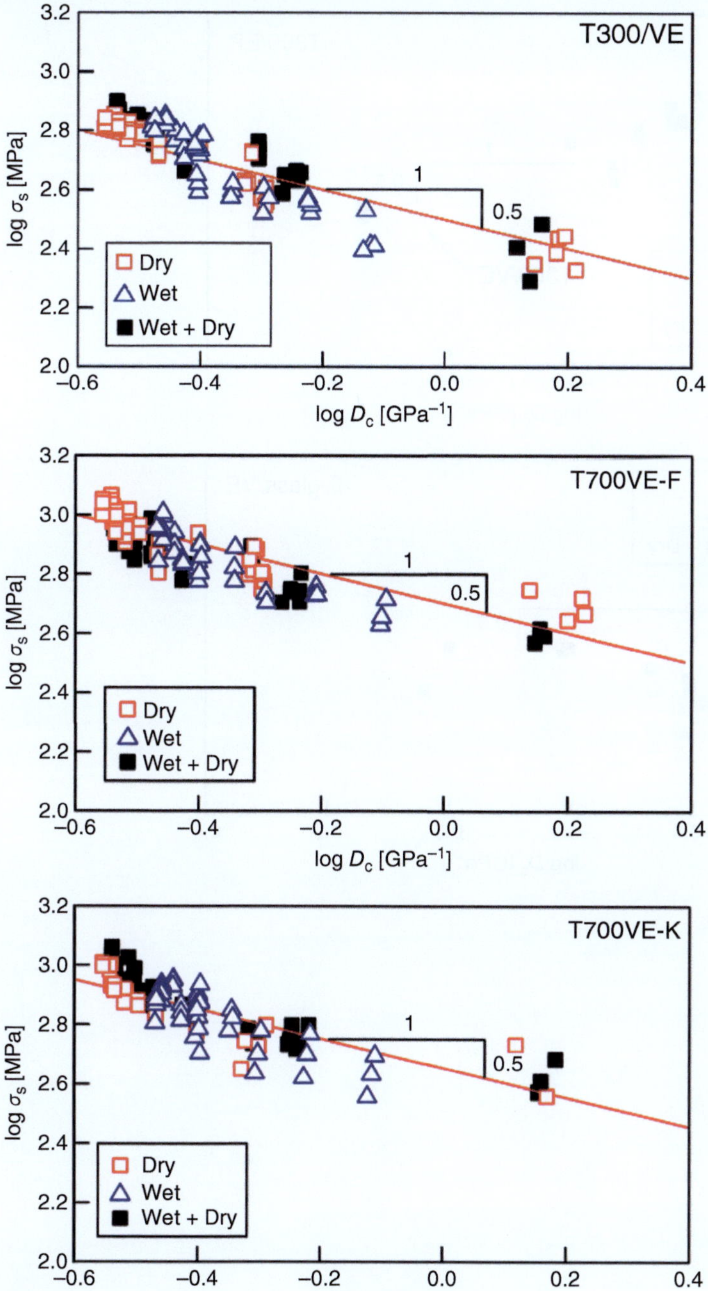

(a)

Figure 5.8 Flexural static strength versus creep compliance of matrix resin in dry, wet, and wet + dry conditions. *Source:* Reproduced from [1], with permission of Elsevier.

(Continued)

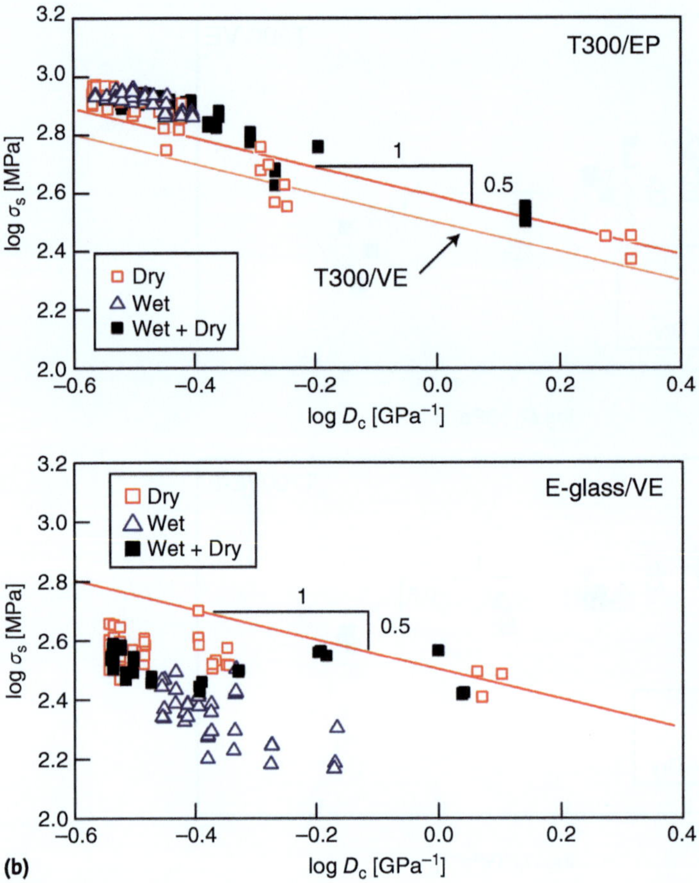

Figure 5.8 (Cont'd)

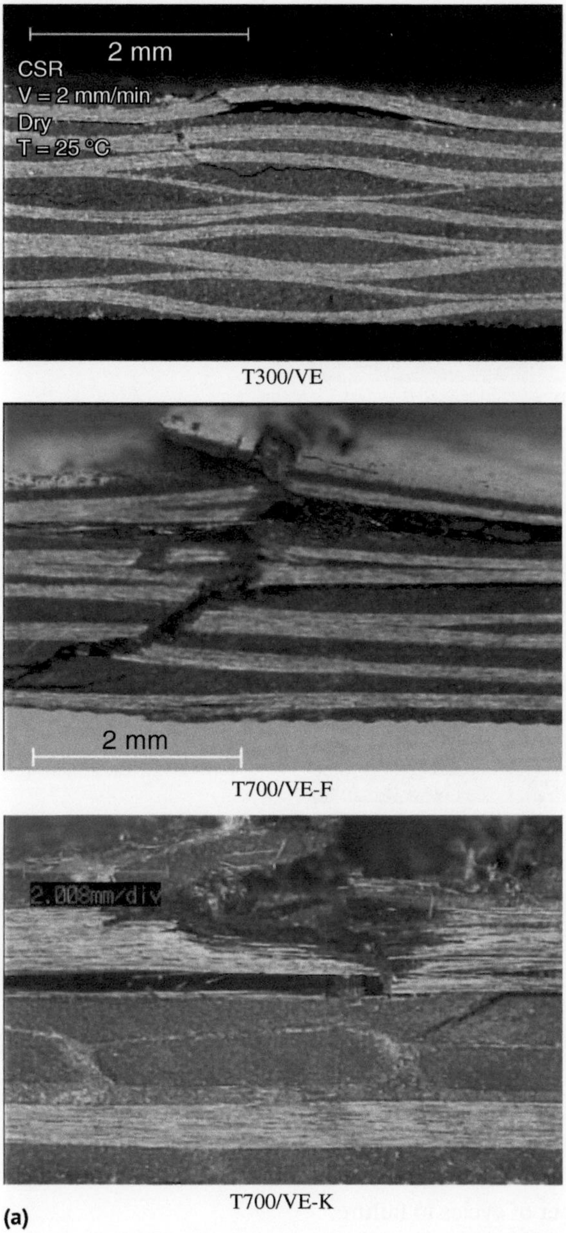

Figure 5.9 Fracture appearances of specimens after flexural static test at 25 °C in dry conditions. *Source:* Reproduced from [1], with permission of Elsevier.

(Continued)

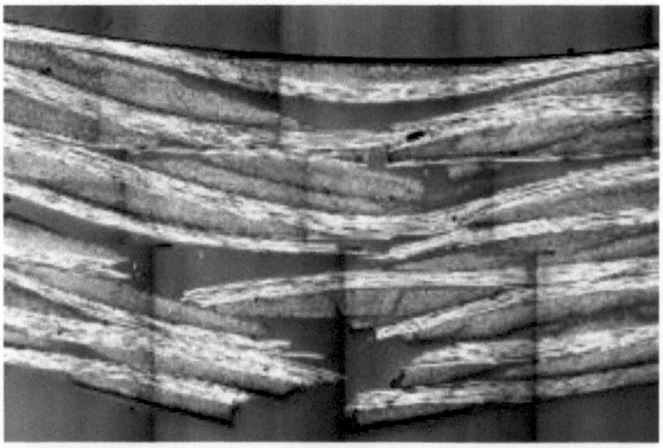

T300/EP

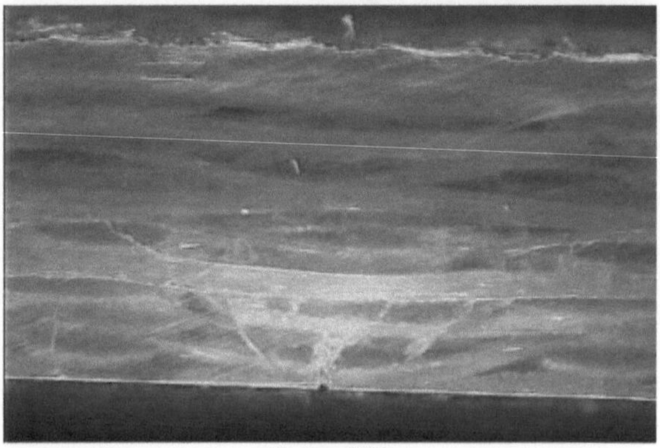

(b) E-glass/VE

Figure 5.9 (Cont'd)

5.6 Flexural Fatigue Strength

To construct the master curve of flexural fatigue strength σ_f, the reduced frequency f' is needed in addition to the reduced time to failure t_f', each of which is defined as

$$f' = f \cdot a_{T_0}(T), \quad t_f' = \frac{t_f}{a_{T_0}(T)} = \frac{N_f}{f'} \tag{5.1}$$

where N_f stands for the number of cycles to failure.

The σ_f versus N_f at frequency $f = 2$ Hz at various temperatures were measured for dry, wet, and wet + dry specimens of the FRP laminates of five kinds. For example, the σ_f versus N_f curves at various temperatures for the dry specimen are shown in Figure 5.10. By converting f and N_f into f' and t_f' using Eq. (5.1), the time–temperature shift factors $a_{T_0}(T)$ and temperature shift factors $b_{T_0}(T)$ of the creep compliance of the matrix resin for each specimen shown in Figures 5.5 and 5.6, the σ_f versus t_f' for each f' was constructed for dry, wet, and wet + dry specimens of the five FRP

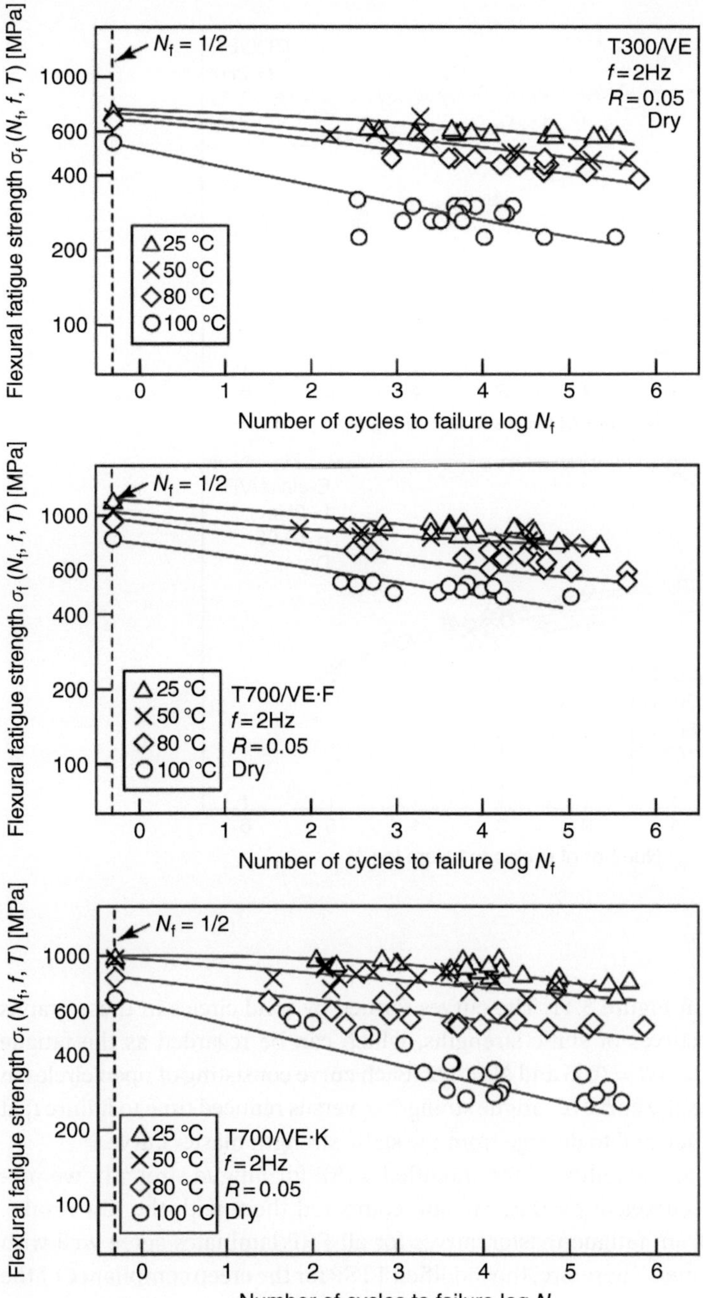

(a)

Figure 5.10 σ_f versus N_f curves at frequency 2 Hz for a dry specimen. *Source:* Reproduced from [1], with permission of Elsevier.

(Continued)

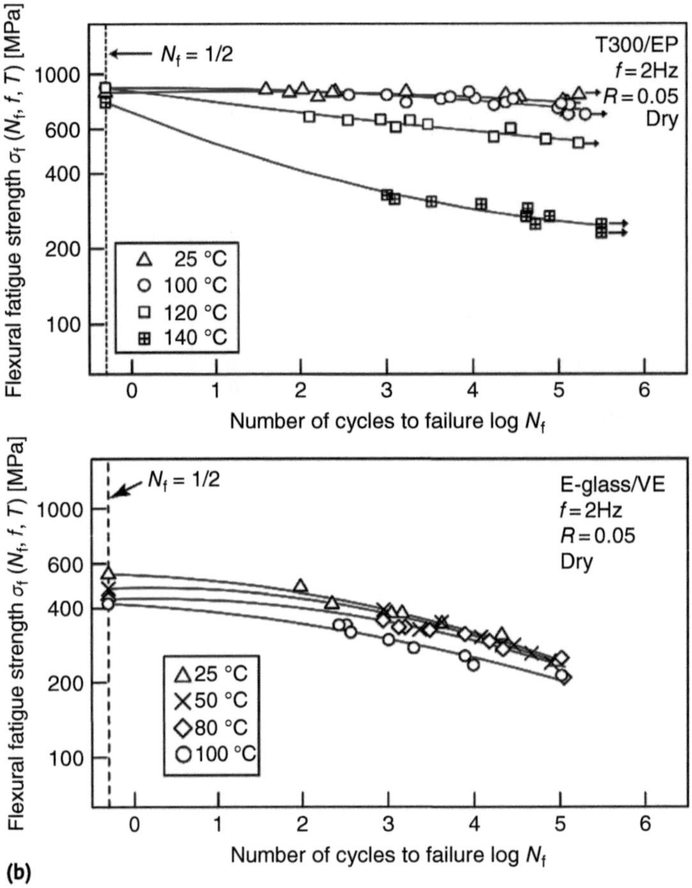

Figure 5.10 (Cont'd)

laminates shown in Figure 5.11. The curves formed by solid circles in these graphs show the master curves of static strengths, which can be regarded as the fatigue strength at stress ratio $R = 0.05$ and $N_f = 1/2$. Each curve consisting of open circles in these graphs shows the curve of fatigue strength σ_f versus reduced time to failure t'_f at each reduced frequency f' to diverge from the static strength master curve.

To confirm the applicability of the modified TTSP for fatigue strength, we predicted the $\sigma_f - N_f$ curves at $f = 0.02$ Hz and compared them with the test results. The predicted σ_f from fatigue master curves for all FRP laminates agree well with the experimental ones. Therefore, the modified TTSP for the creep compliance of the matrix resin also holds for the fatigue strength of the corresponding FRP laminates.

It is clear from Figure 5.11 that the σ_f of all five FRP laminates strongly decreases with time to failure, temperature, and water absorption. They also clarify that the σ_f of four kinds of FRP laminates, except E-glass/VE, decreases only slightly with N_f, although that of E-glass/VE decreases strongly with N_f. The degradation rate to time and temperature for the fatigue strength σ_f of these CFRP laminates closely resembles that for static strength. The σ_f of all FRP laminates also decreases with water absorption and returns to that of the dry specimens by re-drying after water absorption, except that of T300/EP in the long time range and that of E-glass/VE for all ranges of time examined. The σ_f of wet + dry specimens of T300/EP and E-glass/VE does not revert to that of dry specimens: it shows irreversible behavior.

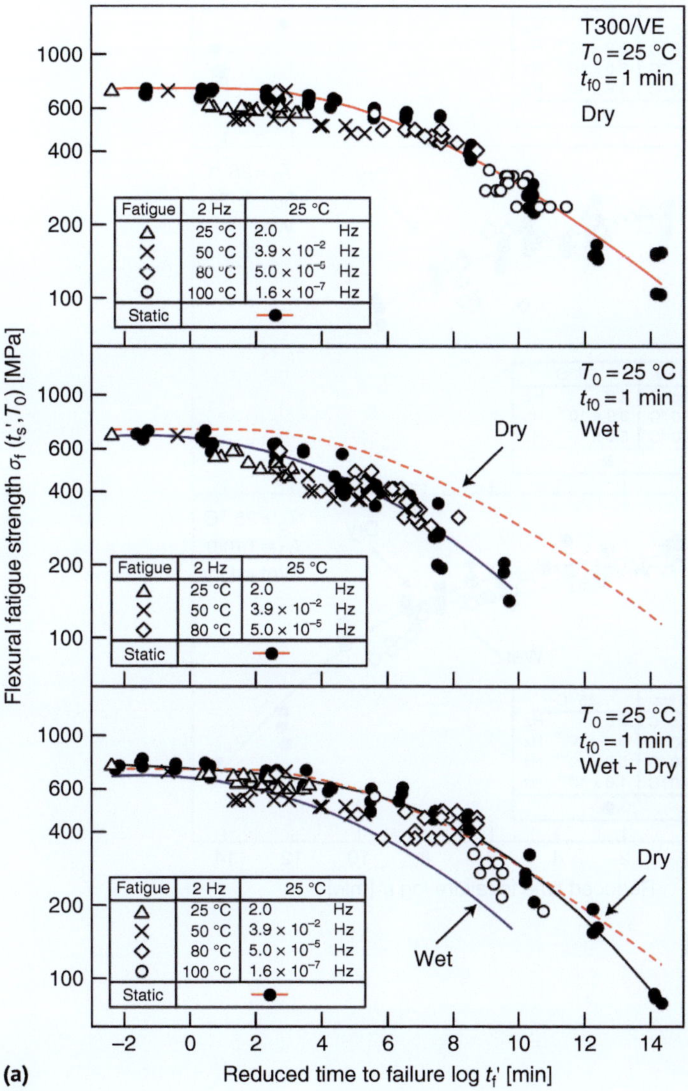

(a)

Figure 5.11 Master curves of flexural fatigue strength for (a) T300/VE, (b) T700/VE-F, (c) T700/VE-K, (d) T300/EP, and (e) E-glass/VE. *Source:* Reproduced from [1], with permission of Elsevier.

(Continued)

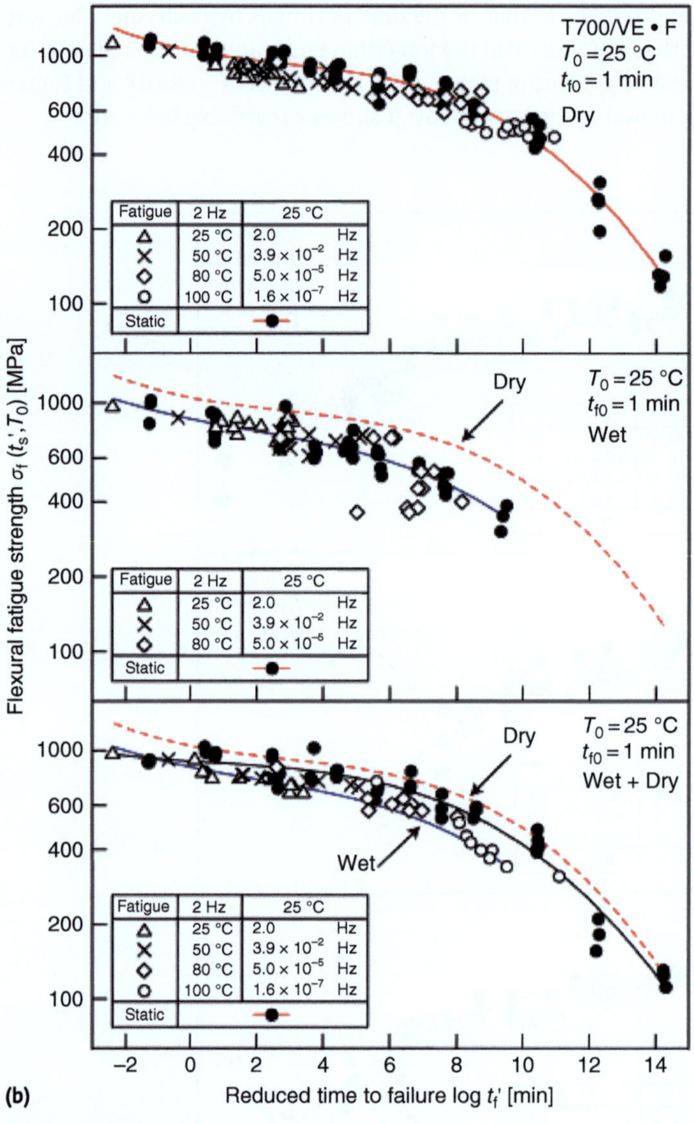

(b)

Figure 5.11 (Cont'd)

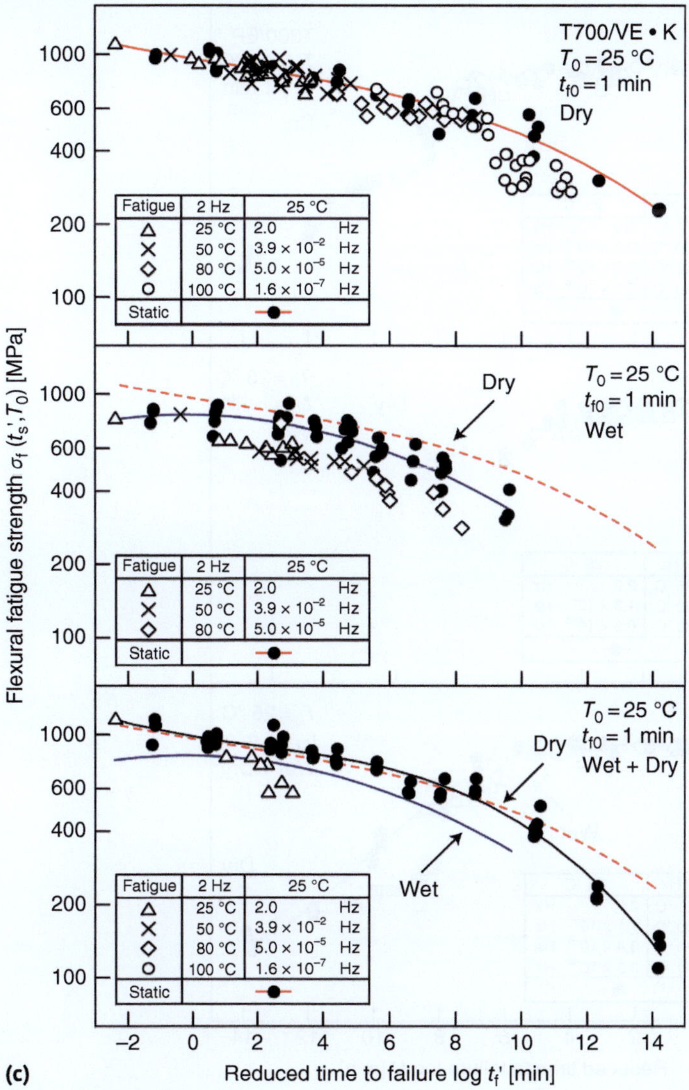

(c)

Figure 5.11 (Cont'd)

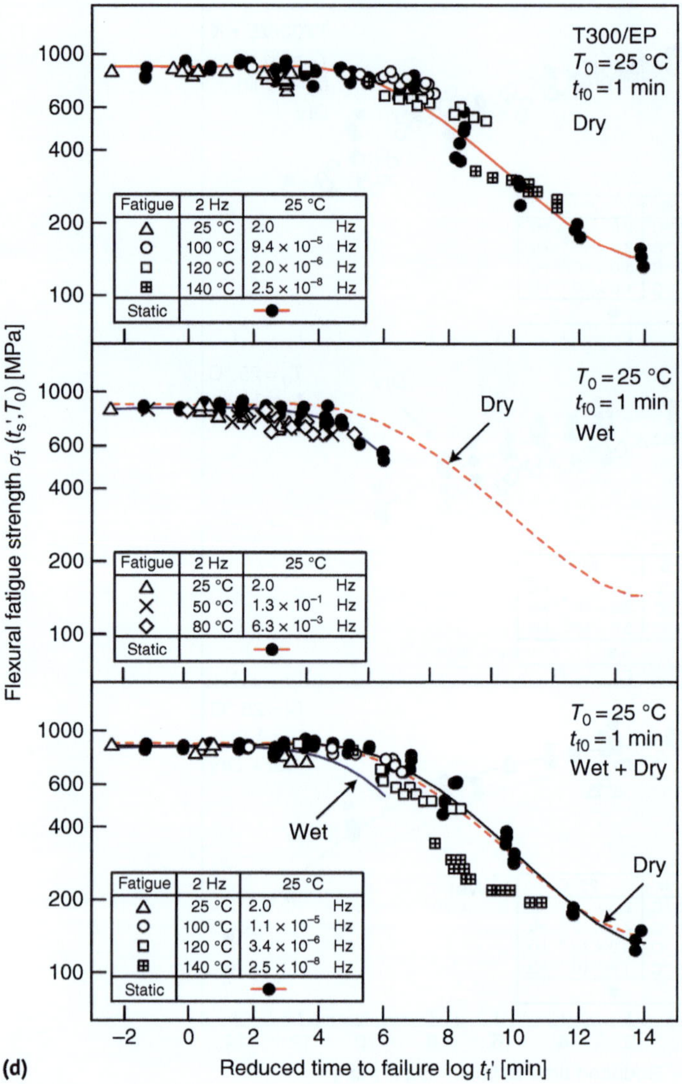

(d)

Figure 5.11 (Cont'd)

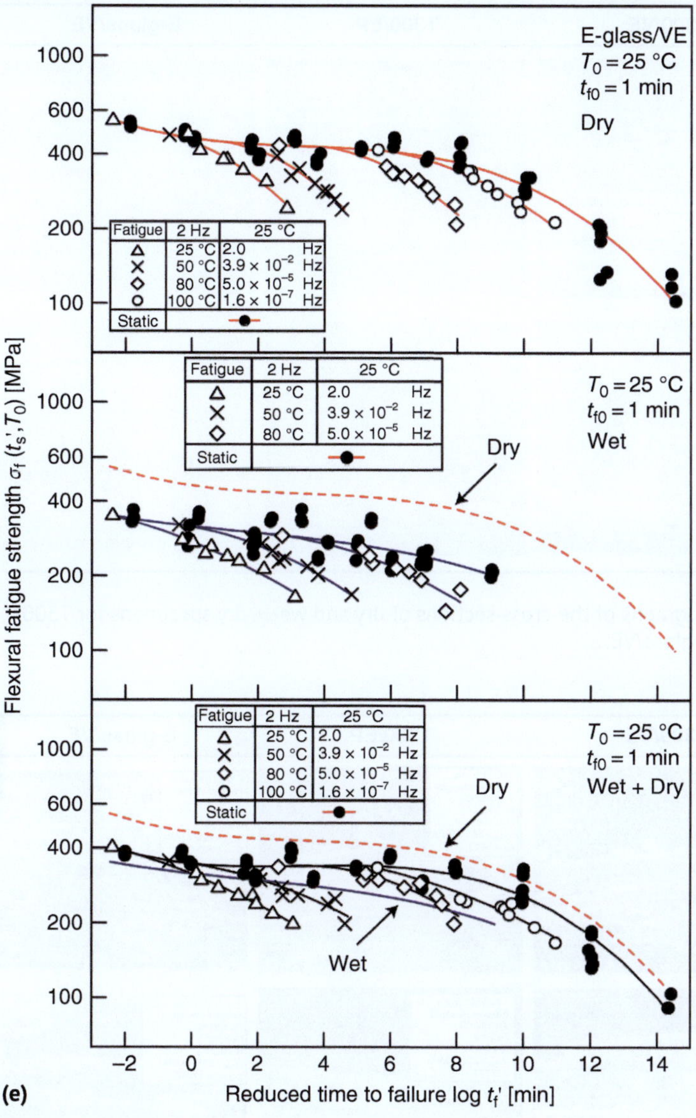

(e)

Figure 5.11 (Cont'd)

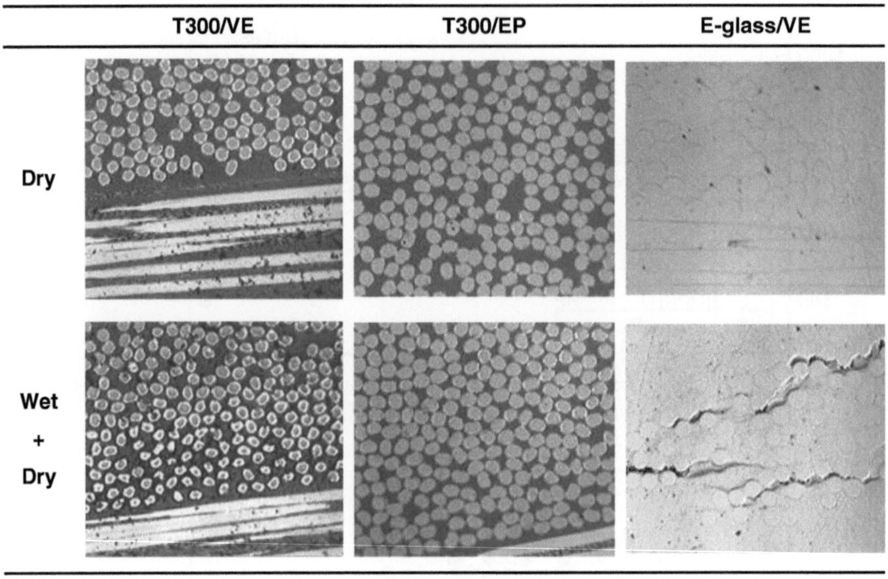

Figure 5.12 Photographs of the cross sections of dry and wet + dry specimens for T300/VE, T300/EP, and E-glass/VE.

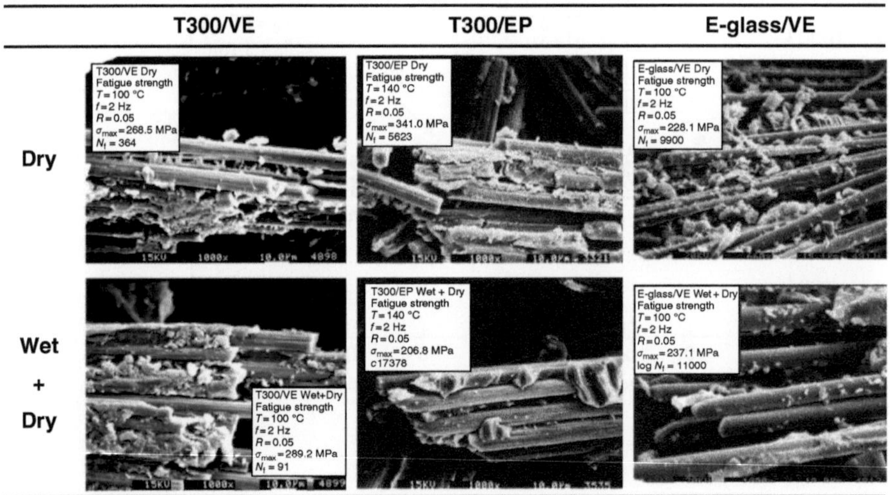

Figure 5.13 Fractographs of dry and wet + dry specimens for T300/VE, T300/EP, and E-glass/VE after fatigue tests.

Figure 5.12 presents photographs depicting cross sections of dry and wet + dry specimens for T300/VE, T300/EP, and E-glass/VE. Figure 5.13 shows fractographs of these specimens after fatigue tests. It is clear from these pictures that the interfaces of carbon fibers and VE resin for T300/VE adhere to each other after the wet + dry treatment because the resin adhesion on carbon fibers is observed on the wet + dry specimen after fatigue tests. The adhesion of carbon fibers and EP resin for T300/EP decreases with the wet + dry treatment because the resin adhesion on carbon fibers is not observed on wet + dry specimen after fatigue tests. The gaps in the interfaces of glass fibers and VE resin for E-glass/VE are generated by the wet + dry treatment, as shown in Figure 5.12. Presumably, for that the reason, the fatigue strength is not reversed by the wet + dry treatment for T300/EP and E-glass/VE.

5.7 Conclusion

Prediction of the long-term fatigue life of five kinds of FRP laminates combined with matrix resin, fiber, and fabric for marine use was performed under temperature and water environments using our developed ATM based on modified TTSP. The three-point bending static and fatigue tests for five kinds of FRP laminates at three conditions of water absorption were conducted at various temperatures and loading rates.

Results show that the flexural fatigue strengths of three kinds of CFRP laminates with VE resin as matrix depend strongly on the water absorption as well as on the time and temperature, but they depend only weakly on the number of cycles to failure. The master curves of fatigue strength for these CFRP laminates were constructed using the test data based on modified TTSP. The fatigue strength of these CFRP laminates decreases with water absorption. It reverts to the initial fatigue strength by re-drying after water absorption. Furthermore, the degradation rate of the fatigue strength of these CFRP laminates is clearly determined only by the increase in the time, temperature, and water absorption. It is independent of fiber constitution: that is, the type, volume fraction, and weave.

Results also show that CFRP laminates with EP resin as matrix and glass fiber-reinforced polymer (GFRP) laminates with VE resin as matrix chemically change according to the process of water absorption and re-drying, and that the flexural fatigue strength of these FRP laminates decreases with chemical degradation through this process.

References

1 Nakada, M. and Miyano, Y. (2009) Accelerated testing for long-term fatigue strength of various FRP laminates for marine use. *Composites Science and Technology*, **69**, 805–813.

2 Dow, N.F. and Gruntfest, I.J. (1960) *Space Sciences Laboratory, Structures and Dynamics Operation*, T.I.S.R60SD389.

6

Application 2 of Accelerated Testing Methodology: Life Prediction of Carbon-fiber-reinforced Polymer/Metal Bolted Joint

6.1 Introduction

This chapter specifically examines the life prediction of a carbon-fiber-reinforced polymer (CFRP)/metal bolted joint under water absorption conditions based on the accelerated testing methodology (ATM). The CFRP/metal bolted joints were prepared under three conditions of dry, wet, and wet + dry. The tensile static and fatigue tests of CFRP/metal bolted joints under these three conditions are performed at various loading rates and temperatures. From these tests, the master curves of static and fatigue failure loads of these joints for the long-term life prediction were obtained. The results also clarify the influence of water absorption on the long-term life of these joints [1].

6.2 Experimental Procedures

6.2.1 Preparation of CFRP/Metal Bolted Joints

Figures 6.1 and 6.2 show the CFRP/metal bolted joints constructed from a CFRP pipe, steel rods (C45), and bolts. A CFRP pipe is joined to the steel rods by two 1/4-20UNC bolts with small washers. The CFRP pipe thickness is 2.6 mm. CFRP prepreg sheets are made using unidirectional carbon fibers Fortafil 510 and epoxy resin Cape 2002. The CFRP pipes are produced by sheet-winding molding. The laminate construction and cure condition of CFRP pipes are $[0/45/90/\text{-}45]_3$ and 120 °C × 2 hours + 150 °C × 2 hours. The fracture of these bolted joints occurs at the corner of the CFRP pipe around the bolts under tensile axial loading to the steel rods, as described later.

Details of specimen conditions for CFRP pipes are presented in Table 6.1. Dry specimens were prepared by holding the cured specimens in an oven at 150 °C for 2 hours. Wet specimens were prepared by soaking the dry specimens in hot water at 80 °C for 48 hours. Wet + dry specimens were prepared by dehydrating the wet specimens in the oven at 150 °C for 2 hours. The water content of specimens was found by weighing the CFRP pipe specimens before and after water absorption.

Properties and Testing of Fiber-Reinforced Polymers, First Edition. Yasushi Miyano and Masayuki Nakada. © 2026 Wiley-VCH GmbH. All rights reserved, including rights for text and data mining and training of artificial intelligence technologies or similar technologies.
Published 2026 by Wiley-VCH GmbH.

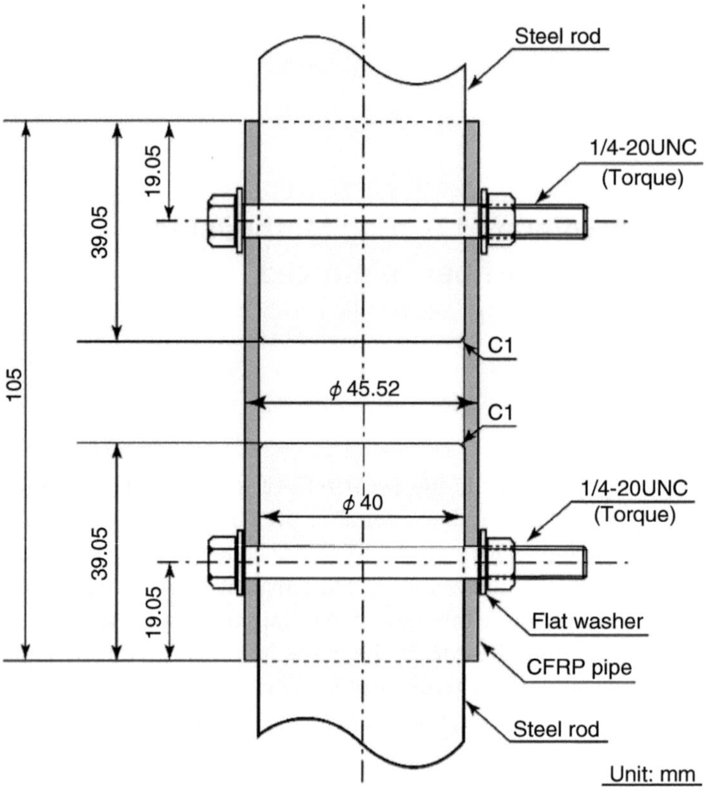

Figure 6.1 Configuration of CFRP/metal bolted joint system.

Figure 6.2 Photograph of CFRP/metal bolted joint system.

Table 6.1 Specimen conditions for dry, wet, and wet + dry.

	Water content (wt%)	In oven	→	In water	→	In oven
Dry	0	150 °C × 2 h		—		—
Wet	1.0	150 °C × 2 h	→	80 °C × 48 h		—
Wet + dry	0	150 °C × 2 h	→	80 °C × 48 h	→	150 °C × 2 h

Table 6.2 Test conditions.

	Test temperature T (°C)	Loading rate V (mm/min)	
(a) Static test			
Dry	25, 50, 80, 100, 120	0.01, 1, 100	
Wet (W = 1.0 wt%)	25, 50, 80, 95	0.01, 1, 100	
Wet + dry	25, 50, 80, 100, 120	0.01, 1, 100	
	Test temperature T (°C)	**Frequency f (Hz)**	**Load ratio R (P_{min}/P_{max})**
(b) Fatigue test			
Dry	25, 80, 100, 120	5, 0.05	0.05
Wet (W = 1.0 wt%)	25, 50, 80, 90	5, 0.05	0.05
Wet + dry	25, 80, 120	5	0.05

6.2.2 Tensile Static and Fatigue Tests

Tensile static tests for specimens prepared under three conditions, namely dry, wet, and wet + dry, were conducted under various conditions of three constant loading rates of 0.01, 1, and 100 mm/min and five constant temperatures from 25 °C to 120 °C, as shown in Table 6.2. At least three specimens were tested for each condition. Tensile fatigue tests for specimens of these three kinds were conducted under various conditions of two constant frequencies of 5 and 0.05 Hz and four constant temperatures of 25–120 °C, shown also in Table 6.2. At least five specimens were tested for each condition. Figure 6.3 shows that the tensile tests for wet specimens were conducted under the device while maintaining constant water absorption during testing.

Additionally, three-point bending creep tests for the transverse direction of unidirectional CFRP laminates, which consist of the same fiber and resin used in CFRP pipes, prepared under three conditions of dry, wet, and wet + dry were conducted to elucidate the time–temperature shift factors for creep compliances of these specimens. These time–temperature shift factors were used for constructing the master curves of static and fatigue failure loads for the CFRP/metal bolted joint.

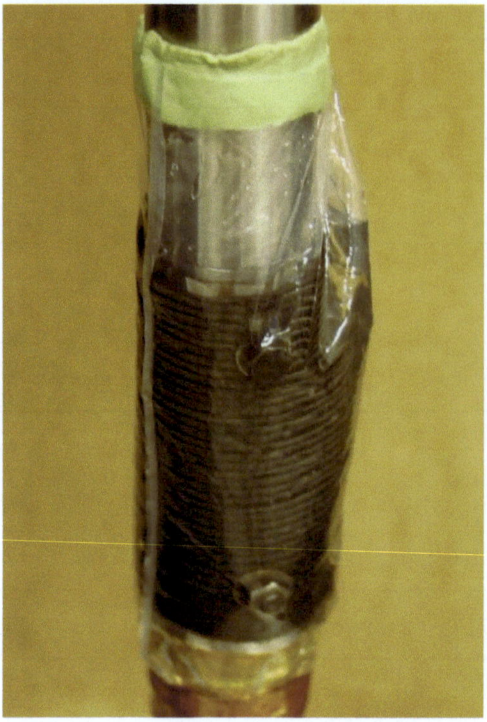

Figure 6.3 Photograph of CFRP bolted joint system for "wet" specimen.

6.3 Results and Discussion

6.3.1 Master Curves of Creep Compliance for Transverse Direction of Unidirectional CFRP Laminates

The left sides of Figures 6.4–6.6, respectively, show creep compliance D_c versus loading time t at various temperatures T for the transverse direction of unidirectional CFRP laminates prepared under the three conditions of dry, wet, and wet + dry. These master curves of D_c are constructed by shifting D_c horizontally at various temperatures along the logarithmic scale of time t so that they overlap smoothly on D_c versus the reduced time t' at the reference temperature T_0, as shown on the right side of each graph. The upper horizontal axis on the right side of each graph shows the reduced temperature T' at reference time t_0. Because the smooth master curves of D_c for dry, wet, and wet + dry specimens are obtainable, the time–temperature superposition principle (TTSP) is applicable for D_c of dry, wet, and wet + dry specimens. These figures show that D_c increases with water absorption and that it returns perfectly to that of the dry specimen by re-drying after water absorption.

The time–temperature shift factor $a_{T_0}(T)$ at T_0 for each D_c obtained experimentally in Figures 6.4–6.6 is shown in Figure 6.7. The $a_{T_0}(T)$ for each D_c are described by two Arrhenius equations with different activation energies ΔH shown in Chapter 2. Figure 6.7 clarifies that the knee point temperature of $a_{T_0}(T)$ decreases with water

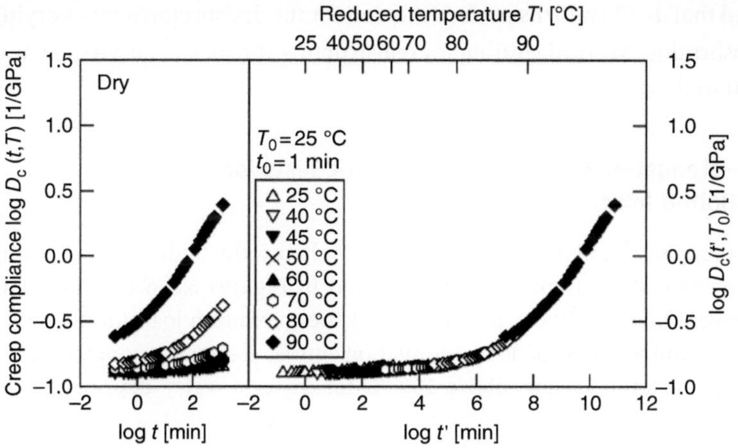

Figure 6.4 Master curve of creep compliance for "dry" specimen.

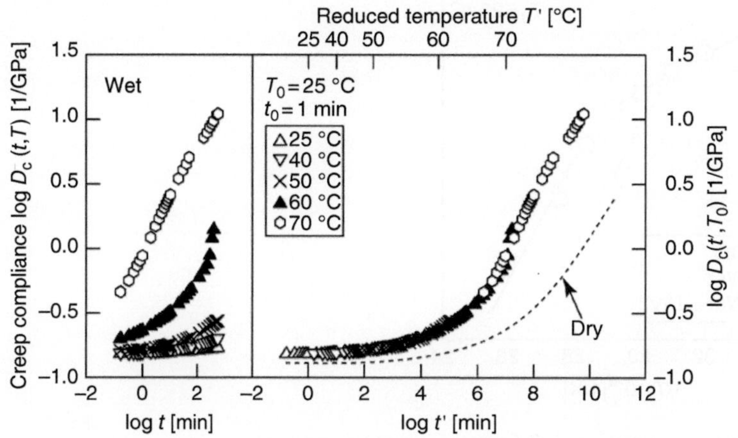

Figure 6.5 Master curve of creep compliance for "wet" specimen.

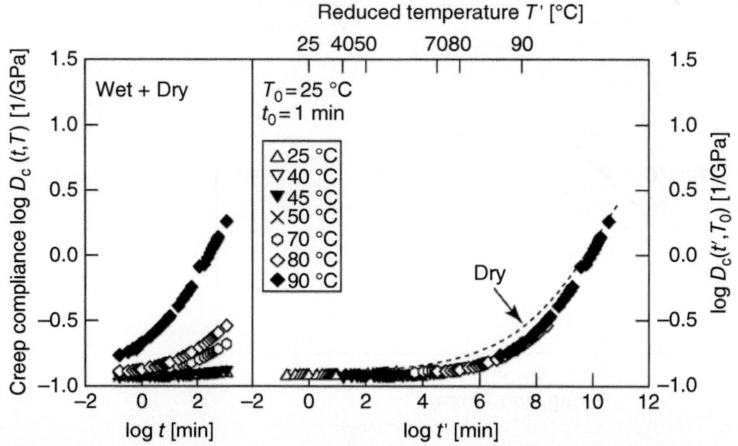

Figure 6.6 Master curve of creep compliance for "wet + dry" specimen.

absorption and that $a_{T_0}(T)$ returns perfectly to that for the dry specimen by re-drying after water absorption. Vertical shifting for constructing the master curve of D_c was not performed in this process.

6.3.2 Load–elongation Curves at Tensile Static Tests for CFRP/Metal Bolted Joint

Typical load–elongation curves for CFRP/metal bolted joint under static loading with elongation rate $V = 1$ mm/min are presented in Figures 6.8–6.10. The load–elongation curves show nonlinear behavior until the maximum load for all conditions tested. The initial slope of load–elongation curves decreases concomitantly with increasing test temperatures above 100 °C for dry and wet + dry specimens,

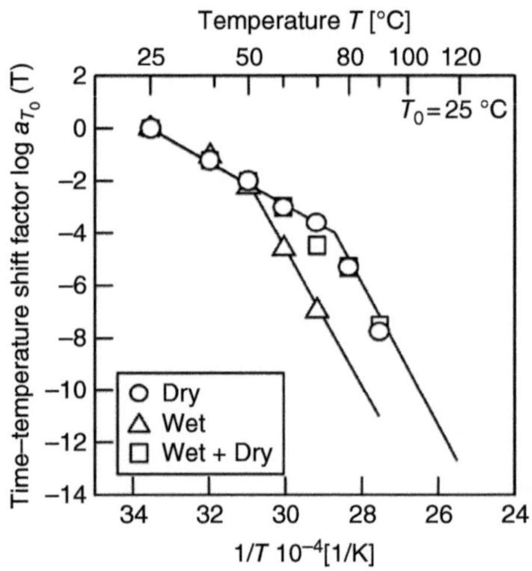

Figure 6.7 Time–temperature shift factors for creep compliance.

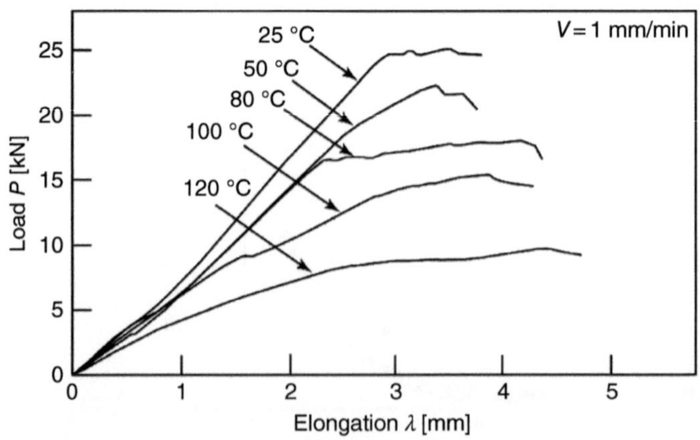

Figure 6.8 Load–elongation curves under static loading for "dry" specimen.

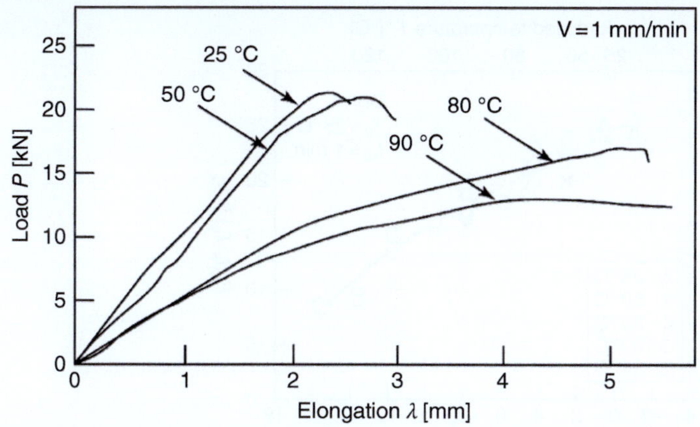

Figure 6.9 Load–elongation curves under static loading for "wet" specimen.

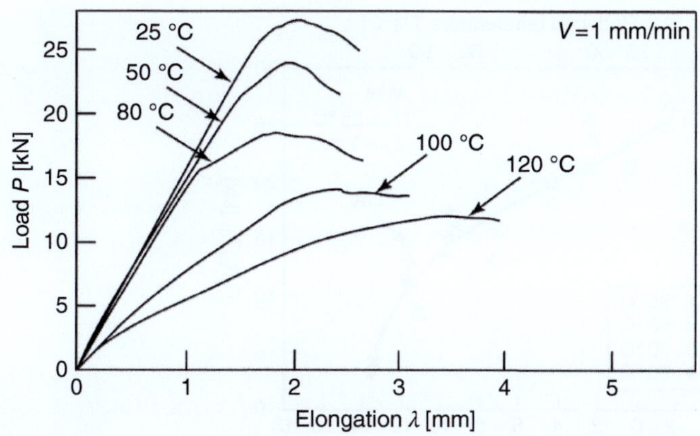

Figure 6.10 Load–elongation curves under static loading for "wet + dry" specimen.

whereas it decreases concomitantly with the increase of the test temperature above 80 °C for wet specimens. The static failure load P_s is defined as the maximum load for all tested conditions.

6.3.3 Master Curves of Static Failure Load for CFRP/Metal Bolted Joint

The left sides of Figures 6.11–6.13 show the tensile static failure load P_s versus time to failure t_s at various constant temperatures T, where t_s is the time period from the initial load to the maximum load. The master curves of P_s versus the reduced time to failure t'_s shown at the right sides of these figures are constructed by shifting P_s at various constant temperatures T along the logarithmic scale of t_s using the $a_{T_0}(T)$ for D_c shown in Figure 6.7. The upper horizontal axes in these figures show the reduced temperature T' at the reference time to failure t_{s0}. Because the smooth master curves are obtainable, the same TTSP for D_c is applicable for P_s of dry, wet, and wet + dry

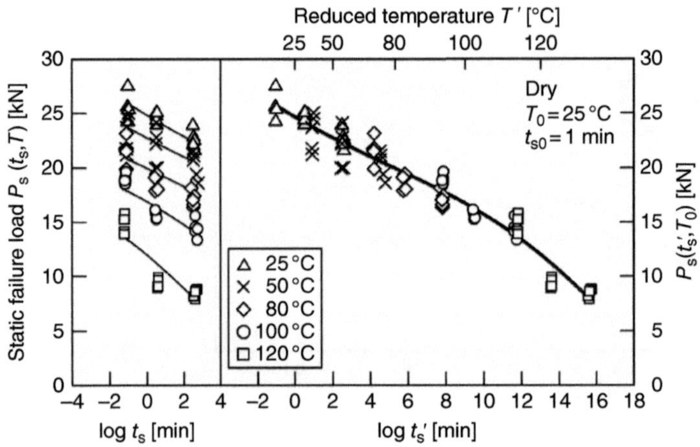

Figure 6.11 Master curve of static failure load for "dry" specimen.

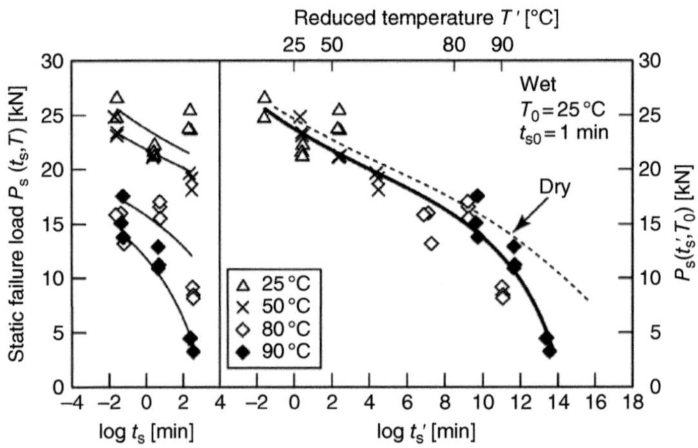

Figure 6.12 Master curve of static failure load for "wet" specimen.

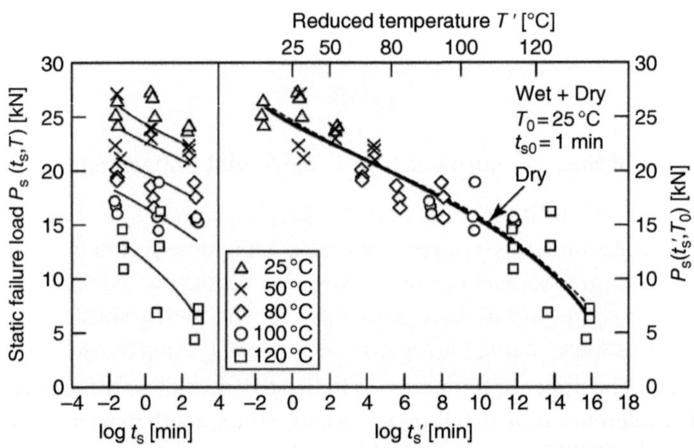

Figure 6.13 Master curve of static failure load for "wet + dry" specimen.

specimens. The TTSP for viscoelastic behavior is applicable for failure load of the bolted joint for reasons that are discussed later. From these figures, it is clarified that P_s decreases with water absorption and that it returns perfectly to that of the dry specimen by re-drying after water absorption.

6.3.4 Master Curves of Fatigue Failure Load for CFRP/Metal Bolted Joint

We regard the tensile fatigue failure load P_f as a function of either the number of cycles to failure N_f or the time to failure $t_f = N_f/f$ for a combination of frequency f, load ratio R, and temperature T and denote them by $P_f(N_f; f, R, T)$ or $P_f(t_f; f, R, T)$. Furthermore, we consider that the static failure load $P_s(t_f; T)$ is equal to the fatigue failure load at $N_f = 1/2$ and $R = 0$ by choosing $t_f = 1/(2f)$. Figures 6.14–6.16, respectively, show P_f

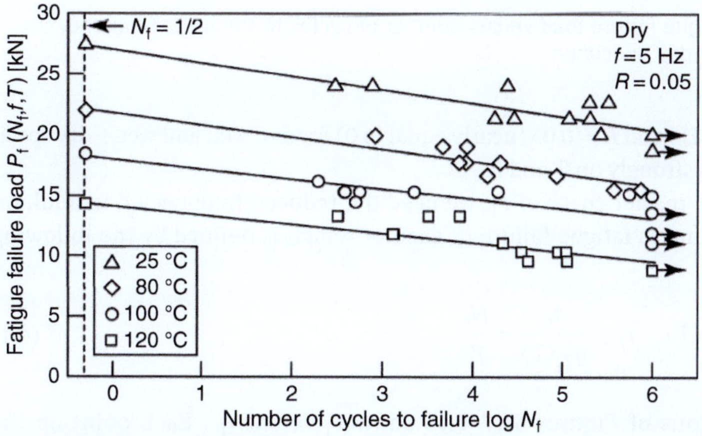

Figure 6.14 Fatigue failure load versus number of cycles to failure at frequency f = 5 Hz for "dry" specimen.

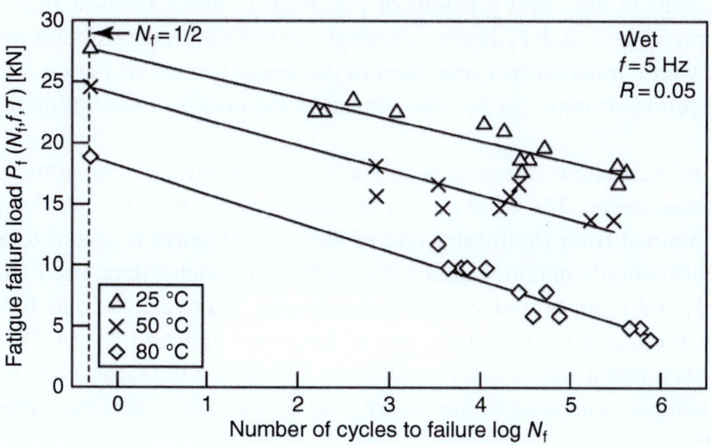

Figure 6.15 Fatigue failure load versus number of cycles to failure at frequency f = 5 Hz for "wet" specimen.

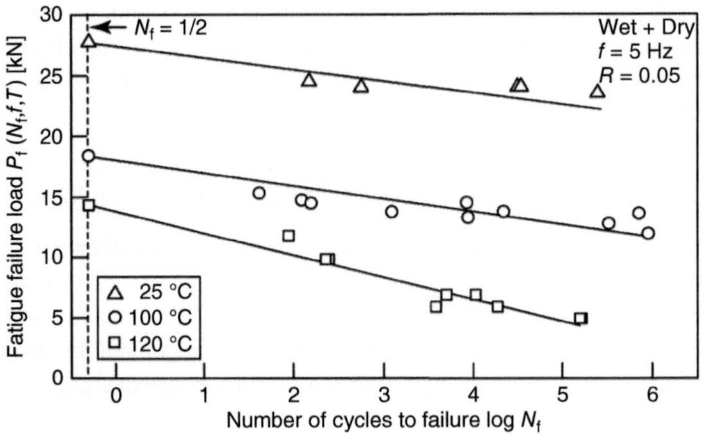

Figure 6.16 Fatigue failure load versus number of cycles to failure at frequency $f = 5$ Hz for "wet + dry" specimen.

versus N_f at $f = 5$ Hz and $R = 0.05$ (nearly equal to 0) for dry, wet, and wet + dry specimens. P_f depends strongly on T and N_f.

To describe the master curve of P_f, we need the reduced frequency f' in addition to the reduced time to fatigue failure t_f', each of which is defined by the following equations:

$$f' = f \cdot a_{T_0}(T), \quad t_f' = \frac{t_f}{a_{T_0}(T)} = \frac{N_f}{f'} \tag{6.1}$$

The upper portions of Figures 6.17–6.19 show P_f versus t_f'. Each point on the master curve is at a constant f' represents N_f. By connecting the points of the same N_f on these curves, the master curves of P_f at a constant N_f are constructed as shown in the lower portions of these figures. The master curves for dry and wet + dry specimens in the lower portions of Figures 6.17 and 6.19 show that P_f hardly depends on N_f, although P_f depends strongly on the time and temperature. However, the master curves for wet specimen in the lower portion of Figure 6.18 show that P_f depends strongly on N_f. The reason is explained in the following section.

Figures 6.20 and 6.21 show P_f versus N_f at $f = 0.05$ Hz at various temperatures for dry and wet specimens. The solid curves in these figures show the predicted $P_f - N_f$ curves obtained from the master curves shown in Figures 6.17 and 6.18. Because the experimentally obtained data agree with the predicted data, the TTSP for D_c holds for P_f and P_s for both dry and wet specimens. Therefore, we conclude that the ATM is applicable for the prediction of long-term fatigue life of CFRP/metal bolted joints under a wet or a dry conditions. The reason the same TTSP for viscoelastic behavior is applicable to the failure load of a bolted joint is examined specifically later.

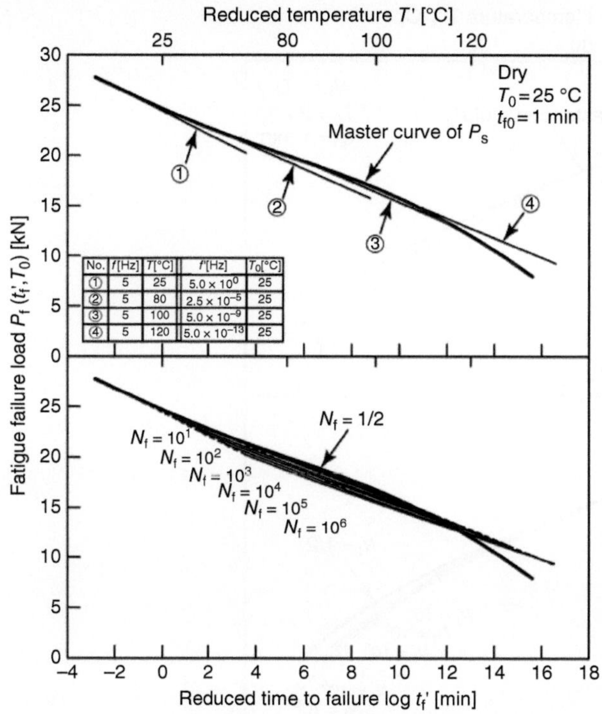

Figure 6.17 Master curves of fatigue failure load for "dry" specimen.

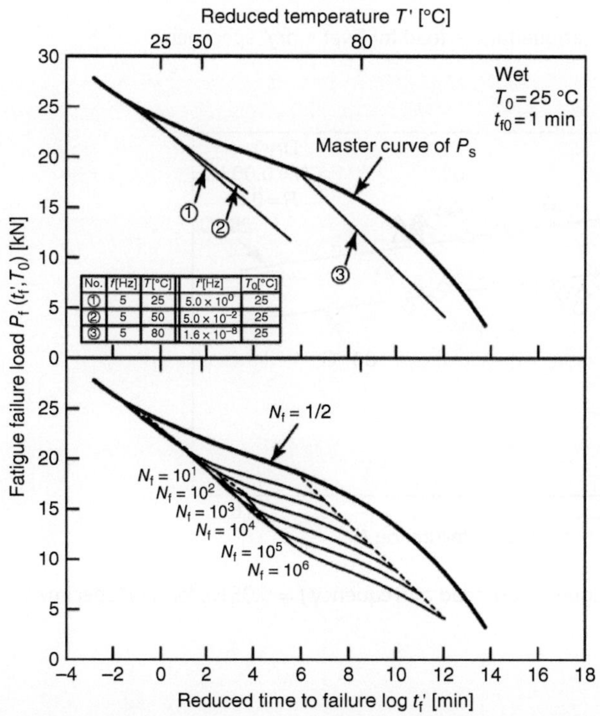

Figure 6.18 Master curves of fatigue failure load for "wet" specimen.

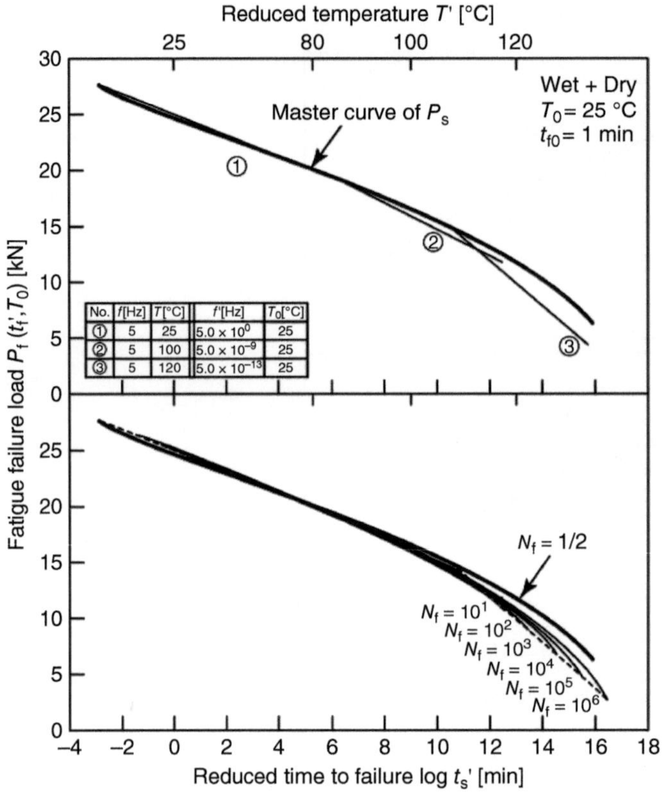

Figure 6.19 Master curves of fatigue failure load for "wet + dry" specimen.

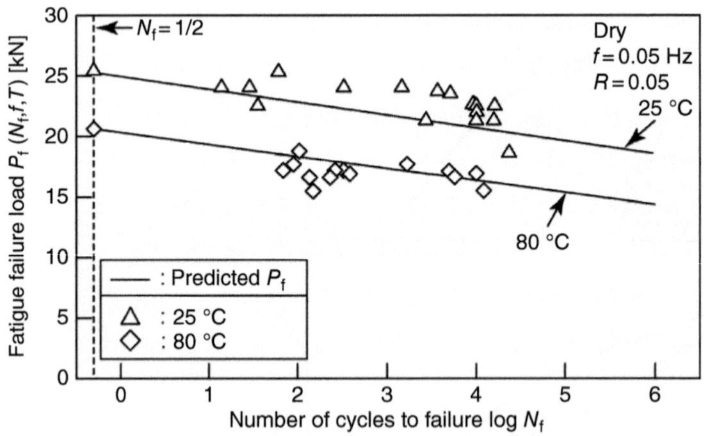

Figure 6.20 Prediction of fatigue failure load at frequency $f = 0.05$ Hz for "dry" specimen.

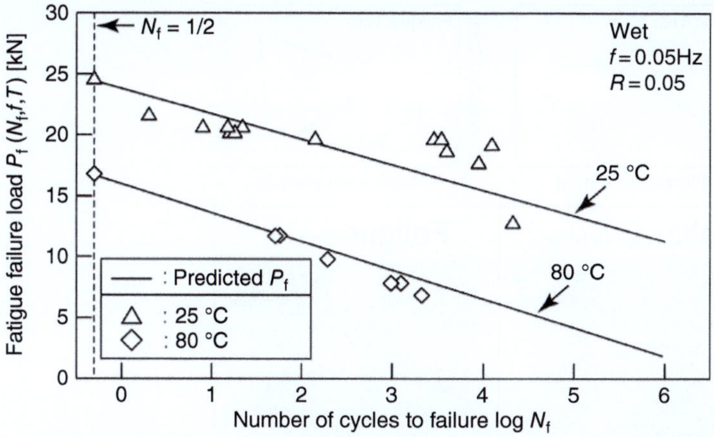

Figure 6.21 Prediction of fatigue failure load at frequency $f = 0.05$ Hz for "wet" specimen.

6.3.5 Fracture Appearance of CFRP/Metal Bolted Joints Under Static and Fatigue Loadings

Figure 6.22 portrays the fracture appearance of CFRP/metal bolted joint after static and fatigue tests at 25 °C and 120 °C under the dry conditions. The fracture of the bolted joint occurs at the corner of a hole in the CFRP pipe under both static and fatigue loadings at all tested temperatures, as shown in Figure 6.22. Bearing failures of the CFRP pipe were observed originally. Then the cracks occurred along the circumference of the pipe from both edges of the hole.

Figure 6.23 shows that the cross-section of CFRP/metal bolted joint observed after static and fatigue loadings. The fracture appearances at the cross-section are shown in Figure 6.24. It is clear from the upper portion of this figure that the fracture is triggered by carbon fiber microbuckling in 0° layers by the compressive stress. The critical load by the carbon fiber microbuckling is controlled by the viscoelastic stress–strain relation of the matrix resin with the TTSP [2]. For this reason, the same TTSP for the viscoelastic behavior of the matrix resin is applicable for the failure load of the bolted joint. The microbuckling fractures are observed under static and fatigue loadings in all test conditions for dry, wet, and wet + dry specimens, except during fatigue loading for the wet specimen, in which delamination is observed, as shown in the lower portion of Figure 6.24. This behavior corresponds closely with the fatigue failure loading characteristics of master curves, which only slightly depend on the number of cycles to failure for dry and wet + dry specimens but strongly depend on the number of cycles to failure for the wet specimens. The viscoelastic behavior of the matrix resin also controls the critical load by delamination [3] as well as by carbon fiber microbuckling.

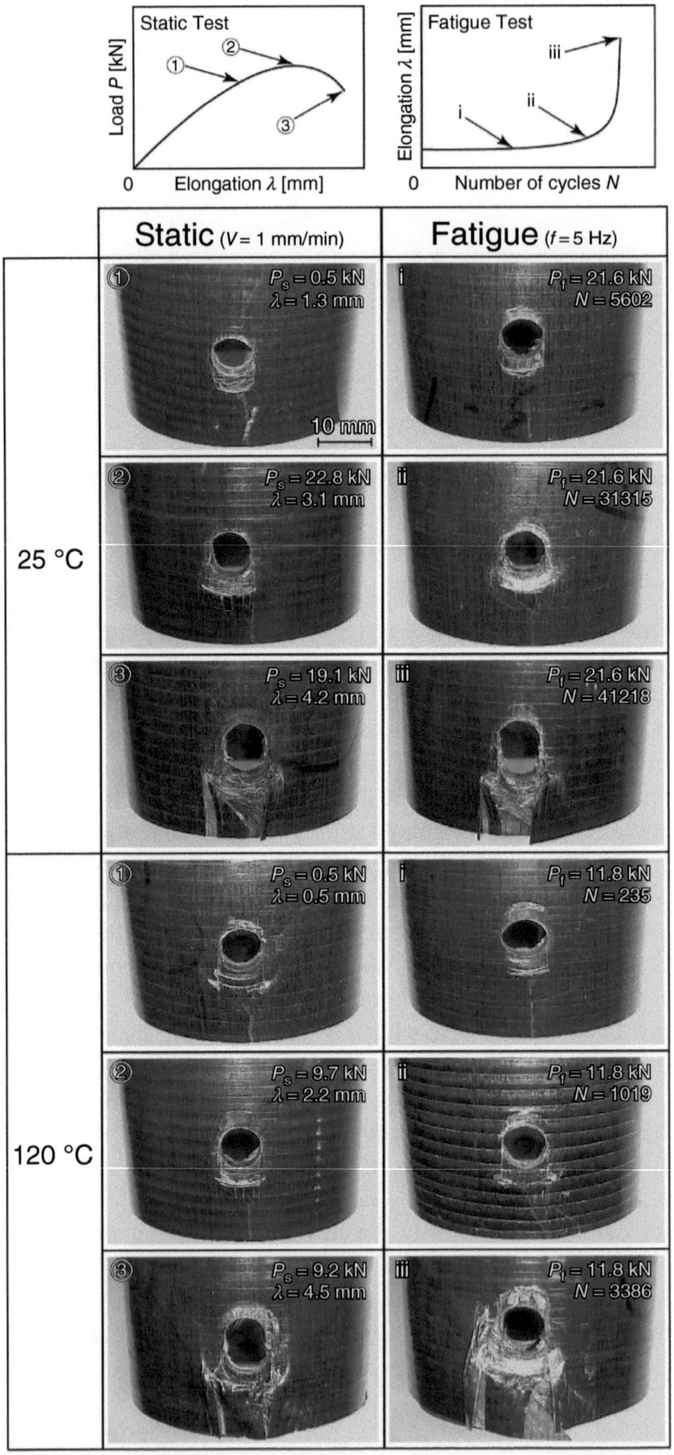

Figure 6.22 Fracture appearances of CFRP/metal bolted joint during static and fatigue tests.

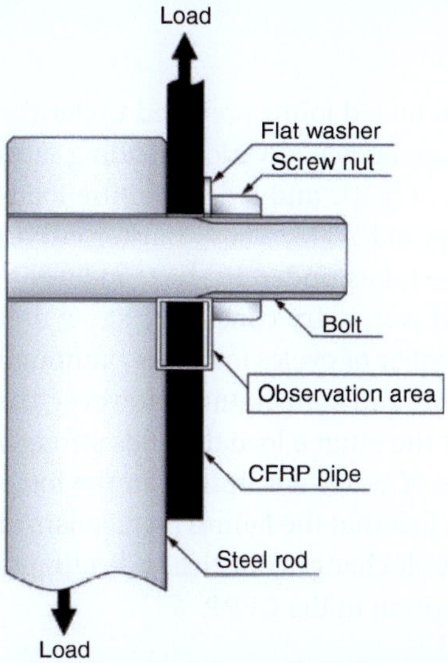

Figure 6.23 Observation area of CFRP/metal bolted joint.

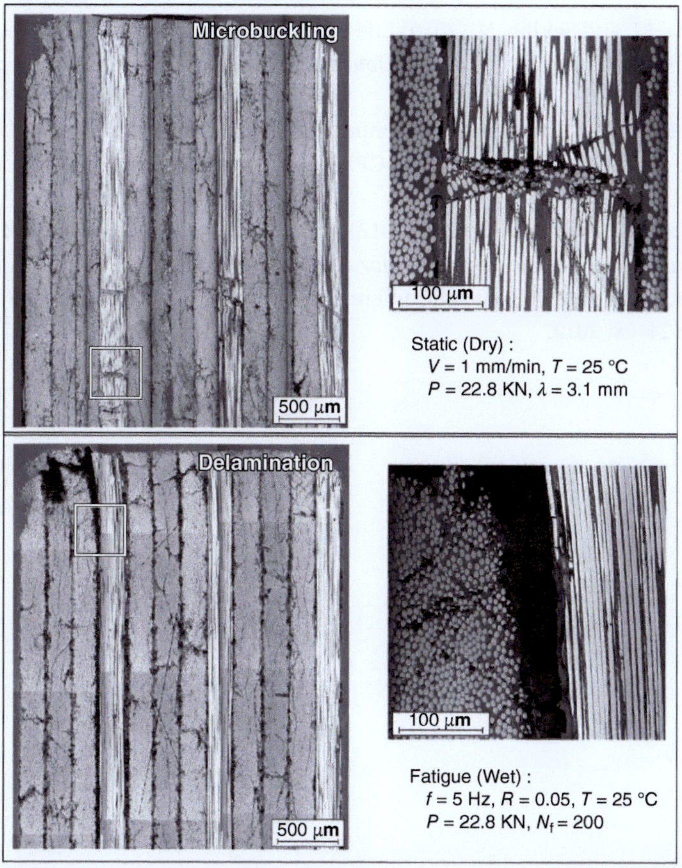

Figure 6.24 Fracture appearance in the cross section of CFRP/metal bolted joint under static and fatigue loadings.

6.4 Conclusion

The tensile static and fatigue tests of CFRP/metal bolted joints prepared under the three conditions of dry, wet, and wet + dry were conducted at various loading rates and temperatures. As a result, the master curves of static and fatigue failure loads of these joints were obtained based on our proposed ATM. These master curves enable the prediction of the long-term life of these joints under dry, wet, and wet + dry conditions. The master curves under dry and wet + dry conditions show that the fatigue failure load hardly depends on the number of cycles to failure, although the fatigue failure load depends strongly on the time and temperature. However, the master curves under the wet condition show that the fatigue load depends strongly on the number of cycles to failure. The influence of water absorption on the long-term life of these joints corresponds well with the fact that the failure mechanism at the edge of the hole in CFRP in contact with the bolt changes drastically, leading to delamination from microbuckling by water absorption in the CFRP.

References

1 Miyano, Y., Nakada, M. and Sekine, N. (2010) Life prediction of CFRP/metal bolted joint under water absorption condition. *Journal of Composite Materials*, **44**, 2393–2411.

2 Miyano, Y. and Nakada, M. (2006) Time and temperature dependent fatigue strengths for three directions of unidirectional CFRP. *Experimental Mechanics*, **46**, 155–162.

3 Moromizato, T., Nakada, M. and Miyano, Y. (2012) *Applicability of Time–temperature Superposition Principle to the Mode I Interlaminar Fracture Behavior of CFRP Laminates*. Proceedings of 15th European Conference on Composite Materials, Venice, Italy, June 24–28, 2012.

Part 2

Advanced Accelerated Testing Methodology

Introduction

In this part, we aim to evaluate the long-term life under static, fatigue, and creep loads for carbon-fiber-reinforced polymers (CFRP), which is a combination of carbon fiber, whose mechanical properties such as elastic modulus and strength do not change even when exposed to any load for a long period of time under temperature and water environments, and a matrix resin that exhibits so-called viscoelastic behavior, whose mechanical properties change significantly depending on time, temperature, and water.

Based on Christensen's crack kinetic theory for viscoelastic bodies, we propose a formula, advanced accelerated testing methodology (Advanced ATM), that statistically expresses the long-term life when static, fatigue, and creep loads are applied to CFRP. We then introduce four examples of the application of this formula.

In Chapter 7, the formulation of static strengths of polymer and CFRP is introduced based on the modified time-temperature superposition principle (modified TTSP) and Christensen's viscoelastic crack kinetic theory for matrix resin. Then this formulation is applied to the time and temperature dependent static strengths for the three typical directions of unidirectional CFRP.

In Chapter 8, the formulation of long-term fatigue strength of CFRP is accomplished based on the formulation of static strength of CFRP demonstrated in Chapter 7. This formulation is completed by predicting the long-term static and fatigue strengths of four typical directions of unidirectional CFRP laminates.

In Chapter 9, the formulation of the long-term creep strength of CFRP is performed based on the formulation of static strength of CFRP. This formulation is demonstrated by statistical prediction of the long-term creep failure time under longitudinal tension loading for unidirectional CFRP.

Chapter 10 presents the Advanced ATM evaluation of the tensile and compressive static strengths in the longitudinal and transverse directions of two kinds of unidirectional CFRP under wet conditions.

The stress-based micromechanics of failure (MMF) have been proposed by Sung-Kyu Ha and others for FRP. In Chapter 11, the procedure of MMF/ATM method combined with Advanced ATM and MMF is proposed for fatigue life prediction of structures made of CFRP laminates. The MMF/ATM method validity was confirmed for the open hole compression fatigue strengths of quasi-isotropic CFRP laminates, as one example of CFRP structures.

In Chapter 12, two kinds of resin impregnated carbon fiber strands (CFRP strands) as the tensile test specimens for unidirectional CFRP molded by different cure conditions as molding condition are prepared. The statistical static tensile strengths for both CFRP strands are measured at various temperatures and the statistical creep failure times for both CFRP strands are predicted by the same procedure of Chapter 9. The effect of molding condition on the statistical static and creep strengths of CFRP strands are discussed theoretically and experimentally.

In Chapter 13, the statistical static strengths under tension loading of CFRP strands with carbon fibers of various kinds are measured at various temperatures and are discussed in terms of the role of the mechanical properties of carbon fiber as well as the matrix resin viscoelasticity.

7

Formulation of Static Strength of Fiber-reinforced Polymers

7.1 Introduction

The time and temperature dependence of the static and fatigue strengths for various directions of carbon-fiber-reinforced polymer (CFRP) laminates with various combinations of fiber and matrix were measured in the previous part. The master curves of these static and fatigue strengths of CFRP laminates were constructed using measured data based on the modified time–temperature superposition principle (modified TTSP), which holds for the viscoelastic behavior of the matrix resin. Results clarified experimentally that the long-term static and fatigue strengths of CFRP laminates can be predicted using measured short-term strength based on modified TTSP for the viscoelastic behavior of the matrix resin.

Christensen and Miyano [1, 2] developed a lifetime prediction methodology based upon kinetic crack growth in polymers and fiber-reinforced polymers (FRPs) showing viscoelastic behavior. Their formulations were performed theoretically and statistically based on the fact that the fracture behavior of the polymer and FRP can be characterized statistically.

In this chapter, the formulation of static strengths of the polymer and FRP as the advanced accelerated testing methodology (Advanced ATM) is introduced based on modified TTSP and Christensen's viscoelastic crack kinetic theory as the first step. Then, this formulation is applied to the time- and temperature-dependent static strengths for the three typical directions of a unidirectional CFRP measured at various deformation rates and temperatures. The three typical directions selected are tension and compression for the longitudinal direction of the unidirectional CFRP, and tension for the transverse direction of the unidirectional CFRP. To construct the master curves of these static strengths based on modified TTSP to hold for the viscoelastic behavior of the matrix resin, the relations between these static strengths and the viscoelastic behavior of the matrix resin are evaluated from the viewpoints of the failure mechanism. The quantitative characteristics of these master curves of the static strength are examined on the basis of the formulation related to their failure mechanisms [3].

Properties and Testing of Fiber-Reinforced Polymers, First Edition. Yasushi Miyano and Masayuki Nakada. © 2026 Wiley-VCH GmbH. All rights reserved, including rights for text and data mining and training of artificial intelligence technologies or similar technologies. Published 2026 by Wiley-VCH GmbH.

7.2 Formulation of Static Strength

The long-term life related to the static strength of an FRP can be shown by the following equation and Figure 7.1 based on the conditions A and B:

Condition A: The failure probability is independent of the temperature and load history.

Condition B: Time and temperature dependence of strength is controlled by the viscoelasticity of the matrix resin.

$$\sigma_s = \sigma_0 \cdot f_A \cdot f_B \quad \log \sigma_s = \log \sigma_0 + \log f_A + \log f_B \tag{7.1}$$

where σ_0 represents the static strength at the reference time and temperature determined by the type of fiber and the type of weave, the volume fraction, the load direction, and other factors. Also, f_A is the scatter of strength as a function of the failure probability P_f, which is determined by the type of fiber and the type of weave, the volume fraction, the load direction, and other factors. Finally, f_B is the time- and temperature-dependent strength determined by viscoelastic compliance $D^*(t,T)$ of the matrix resin.

The static strength in the case of exposing to constant strain rate loading can be shown by the following equation based on conditions A and B:

$$\log \sigma_s(P_f, t, T) = \log \sigma_0(t_0, T_0) + \frac{1}{\alpha}\log[-\ln(1-P_f)] - n_R \log\left[\frac{D^*(t,T)}{D_c(t_0,T_0)}\right] \tag{7.2}$$

where P_f signifies the failure probability, t denotes the failure time, t_0 represents the reference time, T is the temperature, T_0 stands for the reference temperature, σ_0 and α, respectively, denote the scale parameter and the shape parameter on Weibull distribution of static strength, n_R is the viscoelastic parameter, and D_c and D^*, respectively, represent the creep and viscoelastic compliances of the matrix resin. The viscoelastic compliance D^* for the static load with a constant strain rate is shown by the following equation:

$$D^*(t,T) = D_c(t/2,T) \tag{7.3}$$

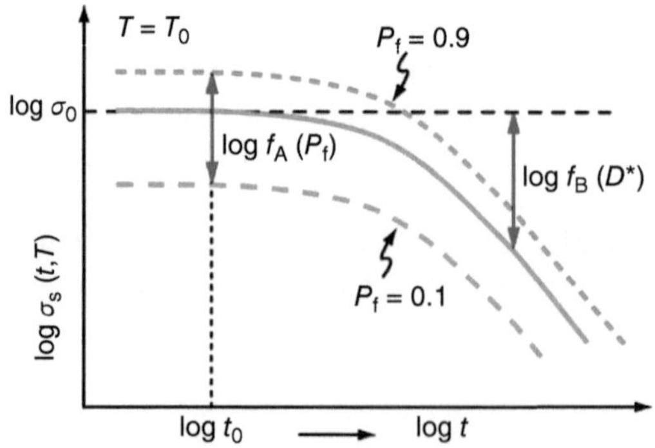

Figure 7.1 Static strength against failure time at a constant temperature T_0.

The statistical static strength σ_s is shown by the following equation by substituting Eq. (7.3) into Eq. (7.2):

$$\log\sigma_s(P_f,t,T) = \log\sigma_0(t_0,T_0) + \frac{1}{\alpha}\log[-\ln(1-P_f)] - n_R \log\left[\frac{D_c(t/2,T)}{D_c(t_0,T_0)}\right] \quad (7.4)$$

The scale parameter σ_0, the shape parameter α, and the viscoelastic parameter n_R should be determined through measurement of the static strengths at various temperatures.

7.3 Application of Formulation

The formulation of static strengths of polymer and FRP developed in the previous section is applied to the tensile strength of neat resin as the matrix of CFRP and the static strengths of three kinds of load directions of a unidirectional CFRP.

7.3.1 Experimental Procedures

The tensile static strength for the neat resin used for the matrix resin of CFRP was measured using tension tests. The tensile static strength for the longitudinal direction of the unidirectional CFRP was measured from tension tests of a resin-impregnated carbon fiber strand (CFRP strand). The CFRP strand has a small cross-sectional area. It is possible to heat a specimen in the region of the gage length so that the tensile failure occurs in the region of the gage length for all tested temperatures. The compressive static strength for the longitudinal direction and the tensile static strength for the transverse direction were measured using three-point bending tests of CFRP laminates because the compressive fracture for the longitudinal direction and the tensile fracture for the transverse direction occur just under the loading point. Furthermore, three-point bending is easier and more stable than either four-point bending or axial tension and compression.

7.3.2 Preparation of Specimens

The compositions and cure conditions for the epoxy (EP) resin used as the matrix and the unidirectional CFRP are presented in Table 7.1. The EP resin plates were made using the infusion technique. Unidirectional CFRP of two kinds was prepared. One is the CFRP strand made using the filament winding technique. The other is the unidirectional CFRP laminate made using the autoclave technique. The CFRP strand is used for the evaluation of tensile strength in the longitudinal direction. The fiber volume fraction and the cross-sectional area of the CFRP strand are not constant. However, the number of filaments in a CFRP strand is perfectly constant. Therefore, we evaluated the tensile strength of a carbon fiber bundle as the tensile strength of a CFRP strand. CFRP laminates were used for the evaluation of flexural strengths in the longitudinal and transverse directions. The respective thickness and fiber volume fraction of CFRP laminates are 3 mm and 56%.

Table 7.1 Compositions and cure conditions for EP resin and unidirectional CFRP.

Composition	
Resin (weight ratio)	CFRP
Epoxy: Epikote 828 (100)	Matrix: epoxy resin
Hardener: Methyl himic anhydride (104)	Carbon fiber: Torayca T300-3k
Cure accelerator: 2-ethyl-4-methyl imidazol (1)	Volume fraction of fiber for laminates: 56%
Curing condition 70 °C for 12 h + 150 °C for 4 h + 190 °C for 2 h	

7.3.3 Test Procedures

Bending creep tests for the EP resin were conducted under various temperatures to obtain the creep compliance as a viscoelastic coefficient. The tensile static tests for the EP resin were conducted under various deformation rates and temperatures to obtain the tensile static strength.

The tensile static tests for the longitudinal direction and the bending static tests for the longitudinal and transverse directions of the unidirectional CFRP were conducted with various deformation rates and temperatures to obtain the tensile and flexural static strengths. These testing methods and conditions are shown, respectively, in Figure 7.2 and Table 7.2. The flexural static strengths in the longitudinal and transverse directions of unidirectional CFRP can be regarded, respectively, as compressive and tensile static strengths because the flexural failure mode for the longitudinal direction is compressive and that for the transverse direction is tensile over all ranges of the deformation rate and temperature. In addition, CFRP strands are used in tests to determine the longitudinal tensile strength of unidirectional CFRP, and the details of the molding method for test specimens and the loading method for test specimens are described in Section 12.2.

Creep compliance D_c of the EP resin was calculated from the deflection δ at the center of the specimen using Eq. (7.5), as

$$D_c = \frac{4bh^3\delta}{P_0 L^3} \tag{7.5}$$

where P_0 stands for the dead load, L signifies the span, and b and h, respectively, represent the width and thickness of the specimen.

The tensile static strength σ_s of the EP resin was calculated from the maximum load P_s using Eq. (7.6), as

$$\sigma_s = \frac{P_s}{bh}. \tag{7.6}$$

The longitudinal tensile static strength σ_s of a CFRP strand is defined by Eq. (7.7):

$$\sigma_s = P_s \frac{\rho}{t_e}. \tag{7.7}$$

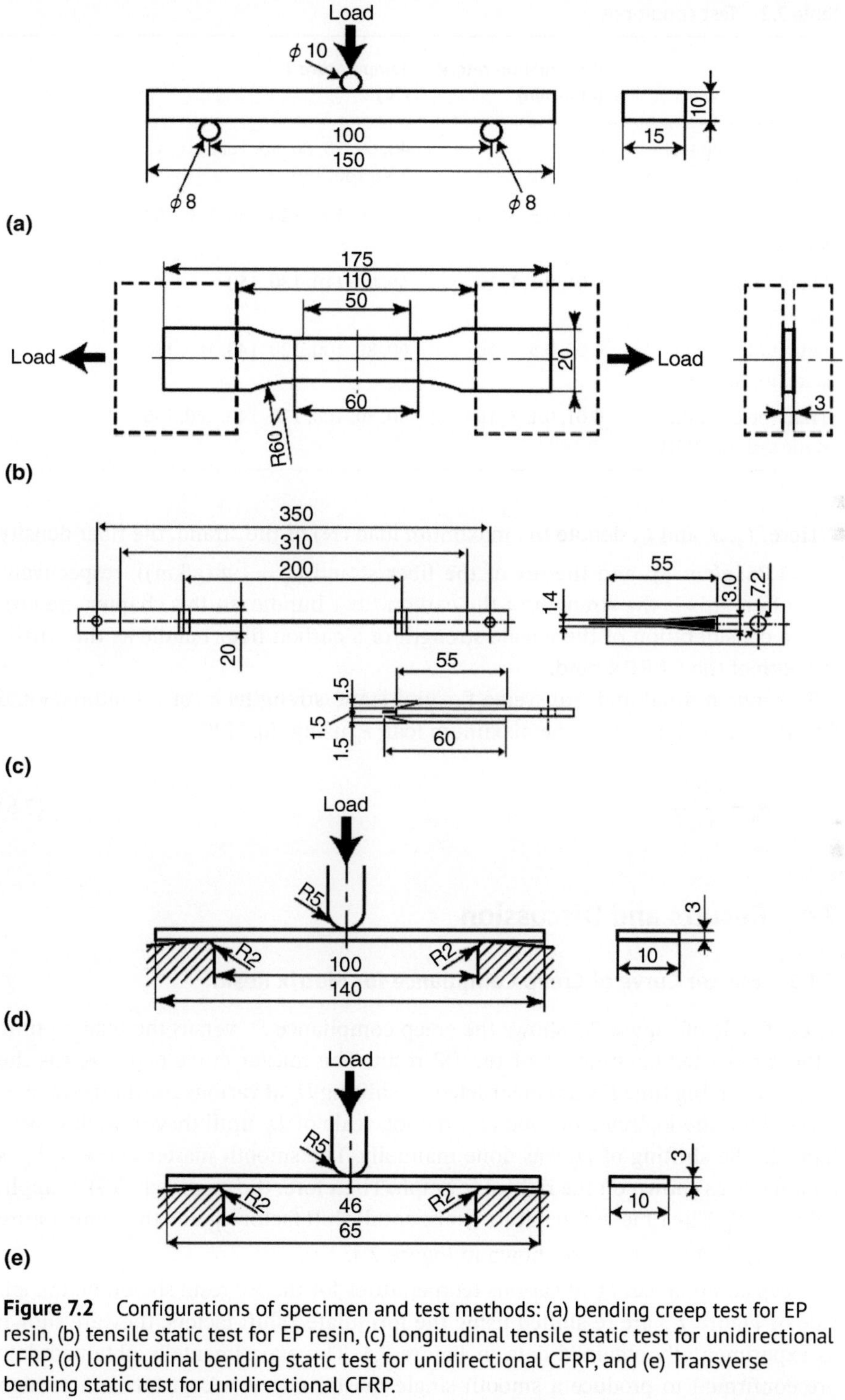

Figure 7.2 Configurations of specimen and test methods: (a) bending creep test for EP resin, (b) tensile static test for EP resin, (c) longitudinal tensile static test for unidirectional CFRP, (d) longitudinal bending static test for unidirectional CFRP, and (e) Transverse bending static test for unidirectional CFRP.

Table 7.2 Test conditions.

	Deformation rate V (mm/min)	Temperature T (°C)
Bending creep test for EP resin	–	40, 60, 70, 80, 90, 100, 110, 120, 130, 140, 150, 160, 180
Tensile static test for EP resin	0.05, 0.5, 5, 50	60, 80, 100, 120, 140, 160, 180
Longitudinal tensile static test for CFRP	0.01, 0.1, 1, 10	50, 90, 110, 130, 150
Longitudinal bending static test for CFRP	0.05, 0.5, 5, 50	60, 80, 100, 120, 140, 160, 180
Transverse bending static test for CFRP	0.01, 0.1, 1, 10	60, 80, 100, 120, 140, 160, 180

Here, P_s, ρ, and t_e, denote the maximum load (N) of the strand, the fiber density $(\rho = 1.76\,(\mathrm{g/cm^3}))$, and the tex of the fiber strand ($t_e = 198\,(\mathrm{g/km})$), respectively. Therefore, this is the strength of the carbon fiber bundle. In this chapter, we present a consideration of the tensile strength of a carbon fiber bundle as the tensile strength of the CFRP strand.

The longitudinal and transverse flexural static strengths σ_s of a unidirectional CFRP are calculated from the maximum load P_s using Eq. (7.8):

$$\sigma_s = \frac{3P_s L}{2bh^2}. \tag{7.8}$$

7.4 Results and Discussion

7.4.1 Master Curve of Creep Compliance for Matrix Resin

The left side of Figure 7.3 shows the creep compliance D_c versus the loading time t for various temperatures T of the EP resin. The master curve of D_c versus the reduced loading time t' was constructed by shifting D_c at various constant temperatures along the log scale of time and the log scale of D_c until they mutually overlapped. The shifting of D_c was done manually. The smooth master curve of D_c is obtainable as shown on the right side graph. Therefore, the modified TTSP is applicable for D_c. The time–temperature (horizontal) shift factor a_{T_0} and the temperature (vertical) shift factor b_{T_0} are shown in Figure 7.4.

Creep compliances D_c at various temperatures for the EP resin shown on the left side of Figure 7.3 are re-shifted using the formulated shift factors; they are shown as experimentally obtained data in Figure 7.5. The experimentally obtained data are confirmed to produce a smooth single master curve, except for data at high temperatures, that is, $T = 160$ and 180 °C, at which CFRP structures are usually not used. The master curve of D_c for the EP resin is shown as a solid curve in Figure 7.5.

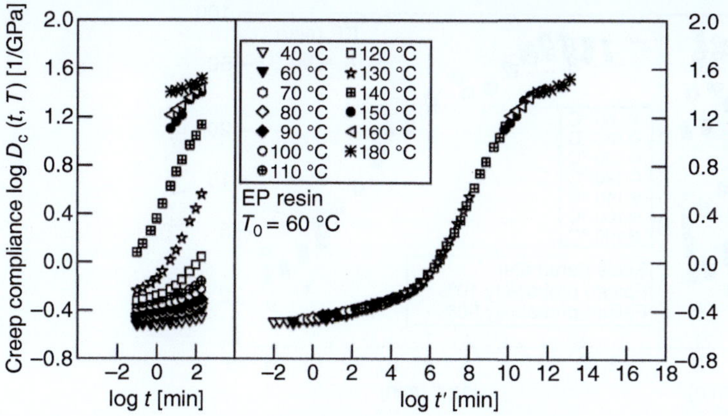

Figure 7.3 Master curve of creep compliance for EP resin.

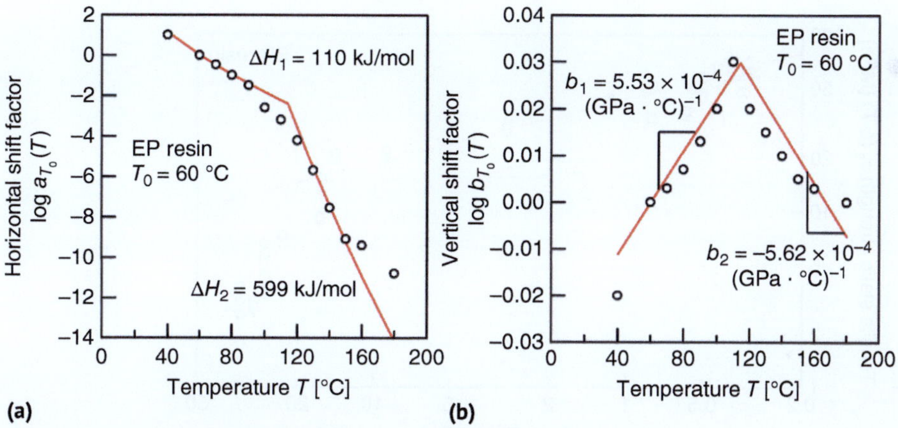

Figure 7.4 Horizontal and vertical shift factors for EP resin.

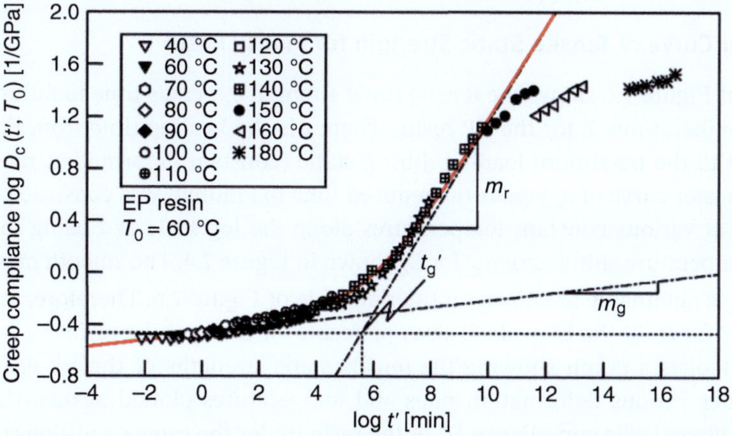

Figure 7.5 Formulation of the master curve of creep compliance for EP resin.

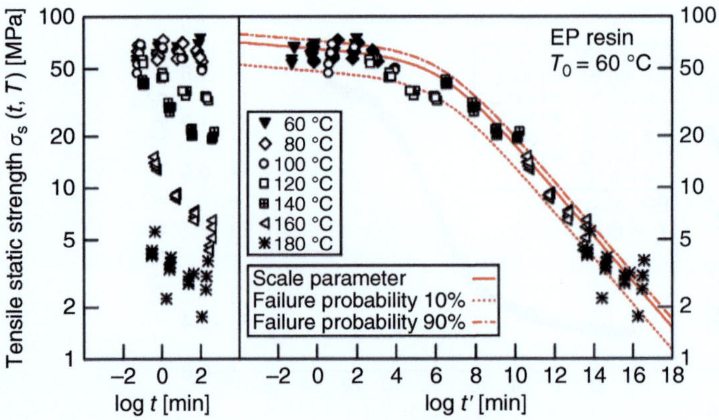

Figure 7.6 Master curve of tensile static strength for EP resin.

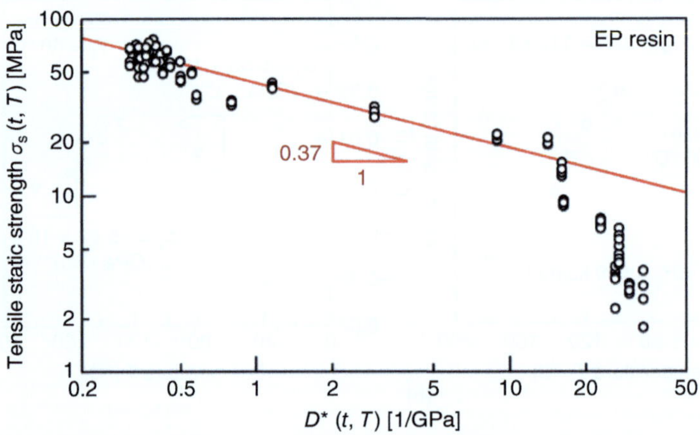

Figure 7.7 Tensile static strength versus viscoelastic compliance for EP resin.

7.4.2 Master Curve of Tensile Static Strength for Matrix Resin

The left side of Figure 7.6 shows the tensile static strength σ_s versus time to failure t at various temperatures T for the EP resin, where t is the loading time from the initial loading to the maximum loading during static (constant deformation rate) testing. The master curve of σ_s versus the reduced time to failure t' was constructed by shifting σ_s at various constant temperatures along the log scale of t using the same time–temperature shift factor a_{T_0} for D_c shown in Figure 7.4. The smooth master curve of σ_s is obtainable as shown on the right side of Figure 7.6. Therefore, the modified TTSP for D_c of the EP resin is also applicable for the σ_s of the resin.

Figure 7.7 depicts a graph showing the tensile static strengths of the EP resin measured under various deformation rates and temperatures plotted against the inverse of the viscoelastic compliance D^* of the resin under the same conditions of time and temperature. D^* under static loading is obtained using Eq. (7.3) for creep

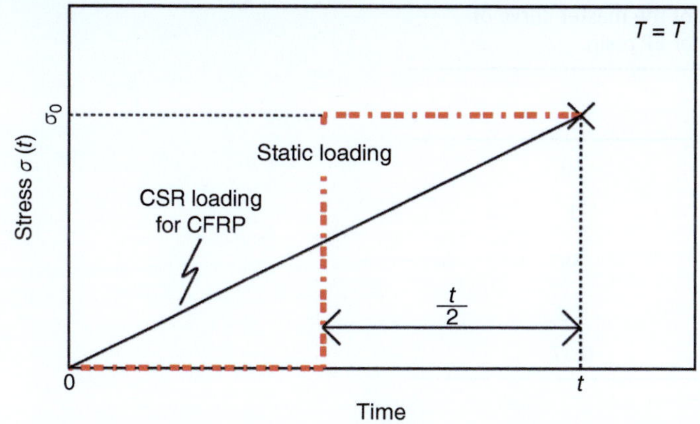

Figure 7.8 Stress history under constant deformation rate loading (static loading) for simplified determination of the viscoelastic compliance D^*.

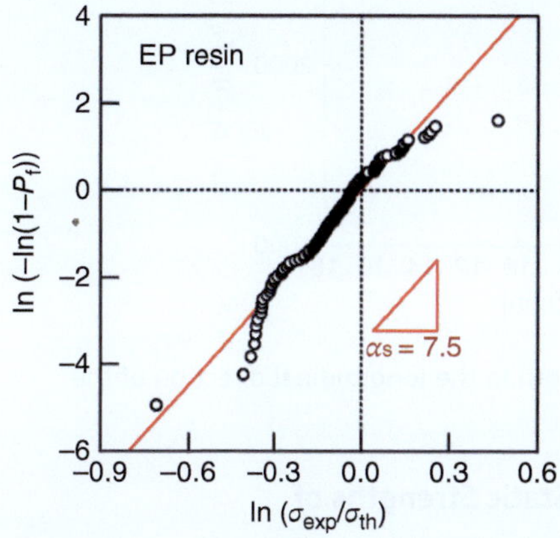

Figure 7.9 Weibull distribution of tensile static strength for EP resin.

compliance D_c by assuming the stress history under constant deformation rate loading as the step loading indicated by a dotted line in Figure 7.8.

From Figure 7.7, the time and temperature dependence of tensile static strength for the EP resin is determined uniquely by D^* for the resin. The relation between the tensile static strength and the viscoelastic compliance for the EP resin produces a straight line except in the range of high temperatures and long times. Therefore, the master curve of tensile static strength for the EP resin is also formulated according to Eq. (7.4).

The Weibull distribution obtained through the formulation of the master curve of tensile static strength for the EP resin assuming the shape parameter is independent of time and temperature is shown in Figure 7.9. Three levels of probability are shown as solid and dotted curves in Figure 7.6. The parameters in this equation are shown in Table 7.3.

Table 7.3 Parameters of the master curve of tensile static strength for EP resin.

EP resin	
T_0(°C)	60
t'(min)	1
$\sigma_0(t'_0, T_0)$(MPa)	65
α	7.5
n_R	0.37

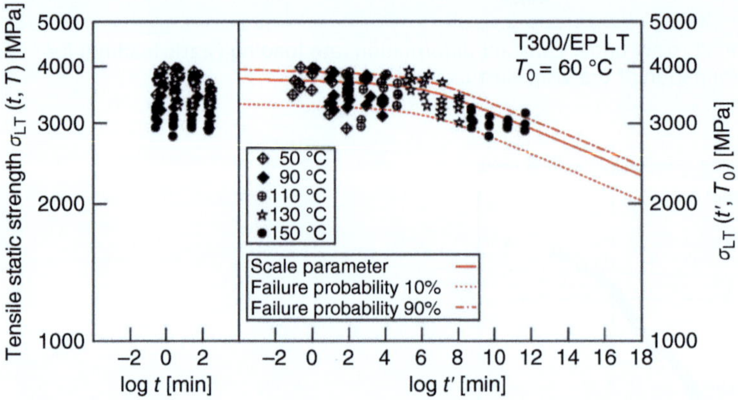

Figure 7.10 Master curve of tensile static strength in the longitudinal direction of the unidirectional CFRP.

7.4.3 Master Curves of Three Kinds of Static Strengths of Unidirectional CFRP

The left sides of Figures 7.10–7.12, respectively, show the tensile and flexural static strengths in the longitudinal direction and the flexural static strength in the transverse direction of the unidirectional CFRP versus time to failure t at various temperatures. The master curve of these strengths σ_s versus reduced time to failure t' was constructed by shifting σ_s at various constant temperatures along the log scale of t using the same time–temperature shift factors a_{T_0} for D_c shown in Figure 7.4. Because the smooth master curves of σ_s are obtainable as shown on the right sides of Figures 7.10–7.12, respectively, the modified TTSP for D_c of EP resin is also applicable to these σ_s of the unidirectional CFRP.

Figures 7.13–7.15 show the static strengths of various load directions measured under various deformation rates and temperatures, shown against the inverse of viscoelastic compliance D^* of the matrix resin (EP resin) under the same time and temperature conditions. The time and temperature dependence for these static strengths is determined uniquely by D^* for the matrix resin because each of the relations between the static strength for the unidirectional CFRP and D^* for the resin produces a straight line except the range of high D^* for the case of static strength for

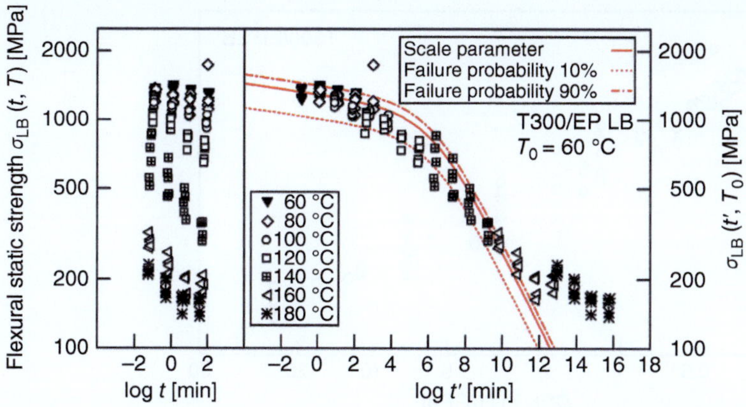

Figure 7.11 Master curve of flexural static strength in the longitudinal direction of the unidirectional CFRP.

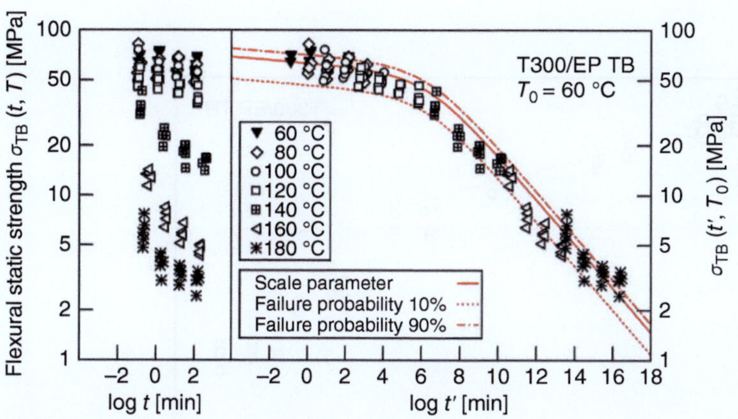

Figure 7.12 Master curve of flexural static strength in the transverse direction of the unidirectional CFRP.

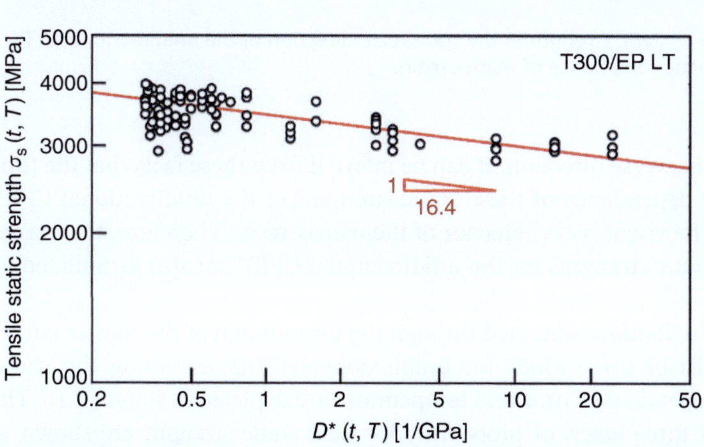

Figure 7.13 Tensile static strength in the longitudinal direction of the unidirectional CFRP versus the viscoelastic compliance of matrix resin.

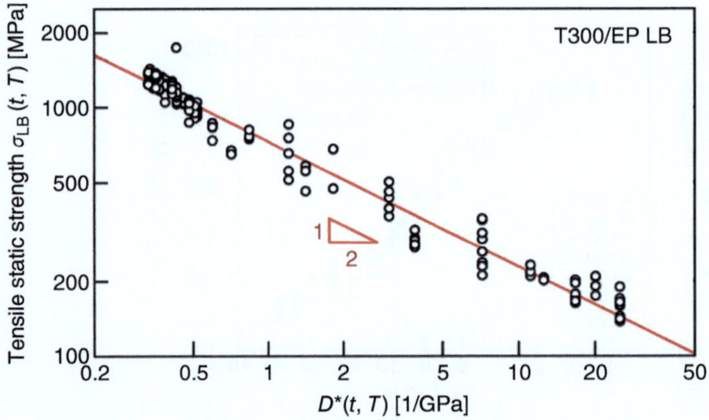

Figure 7.14 Flexural static strength in the longitudinal direction of the unidirectional CFRP versus the viscoelastic compliance of matrix resin.

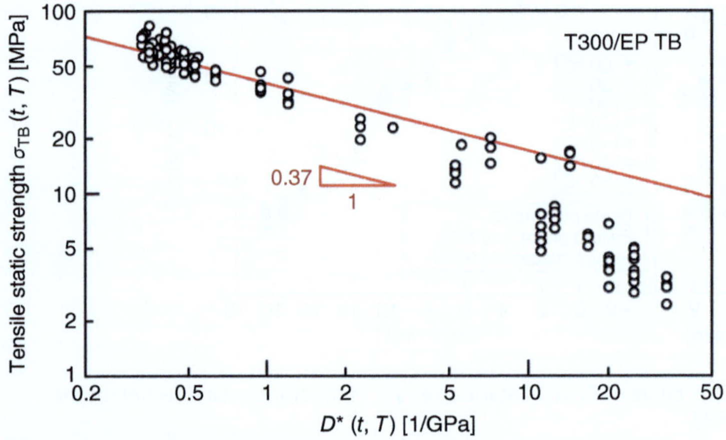

Figure 7.15 Flexural static strength in the transverse direction of the unidirectional CFRP versus the viscoelastic compliance of matrix resin.

bending in the transverse direction. It can be inferred from these facts that the time and temperature dependence of these static strengths of the unidirectional CFRP is controlled by the viscoelastic behavior of the matrix resin. Therefore, the master curves of these static strengths for the unidirectional CFRP are also formulated by Eq. (7.4).

The Weibull distributions obtained through the formulation of the master curves of static strengths of three kinds for unidirectional CFRP assuming the shape parameter is independent of time and temperature are depicted in Figure 7.16. The master curves of three levels of probability for each static strength are shown as solid and dotted curves, respectively, in Figures 7.10–7.12. The parameters in this equation are shown in Table 7.4.

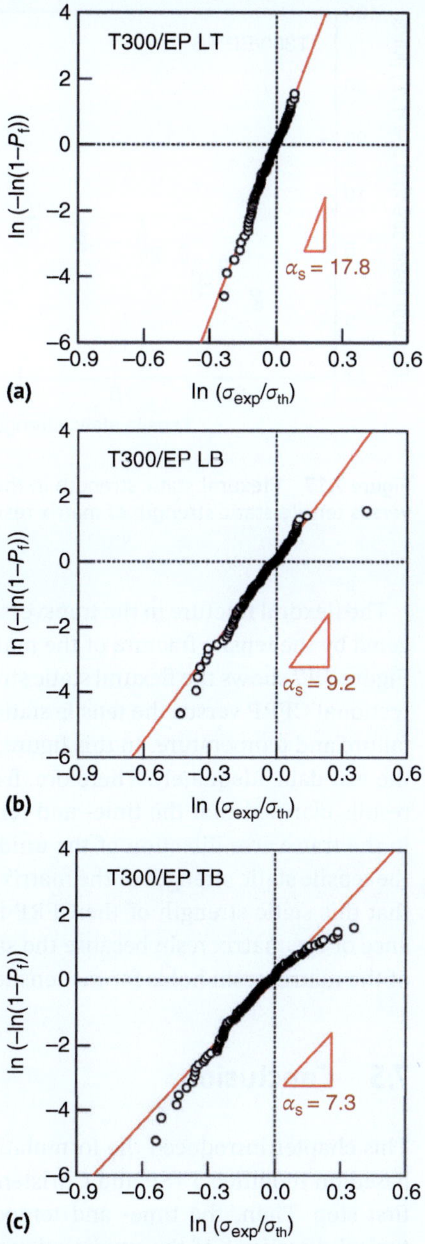

Figure 7.16 Weibull distribution of three kinds of static strength for the unidirectional CFRP.

Table 7.4 Parameters of master curves of three kinds of static strengths for unidirectional CFRP.

T300/EP	**LT**	**LB**	**TB**
$\sigma_0(t'_0, T_0)$ (MPa)	3710	1280	63
α	17.8	9.2	7.3
n_R	0.061	0.50	0.37

LT, longitudinal tension; LB, longitudinal bending; TB, transverse bending.

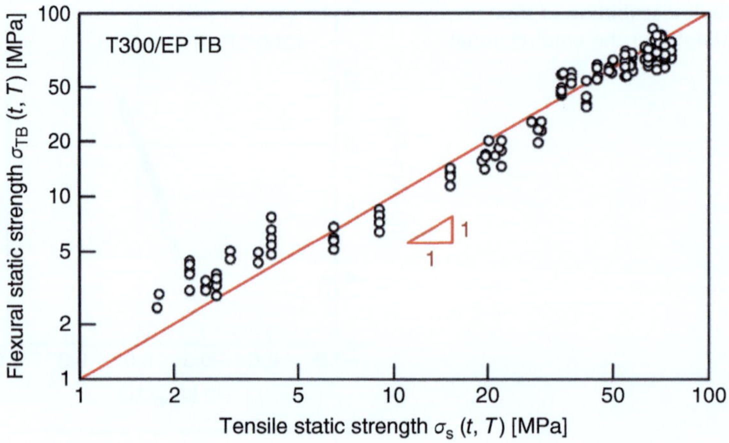

Figure 7.17 Flexural static strength in the transverse direction of the unidirectional CFRP versus tensile static strength of matrix resin.

The flexural fracture in the transverse direction of the unidirectional CFRP is triggered by the tensile fracture of the matrix resin in the tension side of the specimen. Figure 7.17 shows the flexural static strength in the transverse direction of the unidirectional CFRP versus the tensile static strength of the EP resin at the same time of failure and temperature. In this figure, the straight line with a slope of 1/1 captures the test data adequately. Therefore, from the viewpoint of the failure mechanism, results clarified that the time- and temperature-dependent flexural static strength in the transverse direction of the unidirectional CFRP is most strongly affected by the tensile static strength of the matrix resin. It can be inferred on a theoretical basis that this static strength of the CFRP is determined uniquely by the creep compliance of the matrix resin because the same modified TTSP for the creep compliance of the matrix resin holds for the tensile static strength of epoxy resin.

7.5 Conclusion

This chapter introduced the formulation of static strengths of polymers and FRPs based on modified TTSP and Christensen's viscoelastic crack kinetic theory as the first step. Then, the time- and temperature-dependent static strengths for three typical directions of the unidirectional CFRP, which are the tensile and compressive static strengths for the longitudinal direction and the tensile static strength for the transverse direction, were measured at various deformation rates and temperatures. Using these measured data, the master curves of these static strengths were constructed based on modified TTSP, which holds for the creep compliance of the matrix resin. Furthermore, the relations between the viscoelastic behavior of the matrix resin and these static strengths were evaluated from the viewpoint of the failure mechanism. The quantitative characteristics of these master curves of the static strength were discussed using the formulation based on their failure mechanisms.

The results show that the strength master curve produced using our formulation based on their statistics, including the failure mechanism, is available and practical for the design of long-term CFRP durability.

References

1 Christensen, R. and Miyano, Y. (2006) Stress intensity controlled kinetic crack growth and stress history dependent life prediction with statistical variability. *International Journal of Fracture*, **137**, 77–87.

2 Christensen, R. and Miyano, Y. (2007) Deterministic and probabilistic lifetimes from kinetic crack growth-generalized forms. *International Journal of Fracture*, **143**, 35–39.

3 Nakada, M. and Miyano, Y. (2012) Formulation of time–temperature dependent strength of unidirectional CFRP. *Journal of Composite Materials*, **47**, 1897–1906.

8

Formulation of Fatigue Strength of Fiber-reinforced Polymer

8.1 Introduction

The formulation of the long-term fatigue strength of fiber-reinforced polymer (FRP) as the advanced accelerated testing methodology (Advanced ATM) is accomplished on the basis of the formulation of the static strength of the FRP demonstrated in the previous chapter. This formulation is completed by predicting the long-term static and fatigue strengths of four typical directions of unidirectional carbon-fiber-reinforced polymer (CFRP) laminates [1].

8.2 Formulation

The formulation of static strength σ_s established in the previous chapter is presented again as follows:

$$\sigma_s = \sigma_0 \cdot f_A \cdot f_B \quad \log \sigma_s = \log \sigma_0 + \log f_A + \log f_B$$

The formulation of fatigue strength σ_f can be shown by the following equation and Figure 8.1 based on the following conditions of A, B, and the additional condition of C.

Condition A: The failure probability is independent of the temperature and load history.

Condition B: The time and temperature dependence of the strengths is controlled by the viscoelasticity of the matrix resin.

Condition C: Strength decreases with cumulative damage of cyclic loading.

$$\sigma_f = \sigma_0 \cdot f_A \cdot f_B \cdot f_C \quad \log \sigma_f = \log \sigma_0 + \log f_A + \log f_B + \log f_C \tag{8.1}$$

where σ_0 is the static strength at the reference time and temperature determined by the type of fiber, type of weave, volume fraction, load direction, and other factors. f_A is the scatter of strength as a function of the failure probability P_f determined by the type of fiber, type of weave, the volume fraction, the load direction, and other factors. f_B is the time- and temperature-dependent strength degradation determined by the viscoelastic compliance $D^*(t,T)$ of the matrix resin. Additionally, f_C is the

Properties and Testing of Fiber-Reinforced Polymers, First Edition. Yasushi Miyano and Masayuki Nakada. © 2026 Wiley-VCH GmbH. All rights reserved, including rights for text and data mining and training of artificial intelligence technologies or similar technologies.
Published 2026 by Wiley-VCH GmbH.

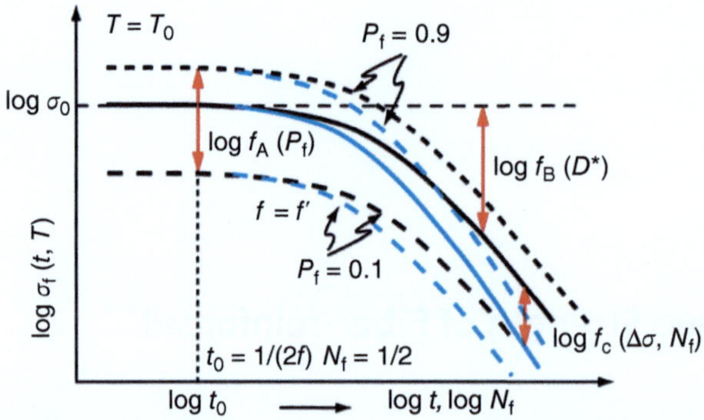

Figure 8.1 Explanation of the formulation of fatigue strength.

strength degradation caused by the cumulative damage of cyclic loading. The details of Eq. (8.1) are illustrated by the following equation:

$$\begin{aligned}\log \sigma_f(P_f, t, T, N_f) &= \log \sigma_0(t_0, T_0) + \frac{1}{\alpha}\log[-\ln(1 - P_f)] \\ &\quad - n_R \log\left[\frac{D^*(t,T)}{D_c(t_0,T_0)}\right] - \frac{1}{2} n_F \log(2N_f)\end{aligned} \tag{8.2}$$

Here, D^* is expressed by the following equation for a sinusoidal cyclic load pattern:

$$D^*(t,T) \approx \frac{1}{2}D_c(t,T) + \frac{1}{2}D_c\left(\frac{1}{4f},T\right),\ f = \frac{N_f}{t} \tag{8.3}$$

where f stands for the frequency and N_f represents the number of cycles to failure. The degradation of fatigue strength with increasing time to failure t and the number of cycles to failure N_f is shown in Figure 8.1.

8.3 Application of Formulation

The formulations of fatigue strength and static strength of FRP developed in the previous section are applied to the static and fatigue strengths of the four load directions of unidirectional CFRP laminates.

8.3.1 Specimens and Test Methods

The test specimens were fabricated from unidirectional CFRP laminates of MR60H/1053 (Mitsubishi Rayon Co., Ltd.), which consist of the carbon fiber MR60H and epoxy resin 1053. All laminates were produced using the autoclave technique. The curing procedure includes heating at 180 °C for 2 hours, followed by post-curing at 160 °C for 70 hours. The volume fraction of the fiber is approximately 55%. The laminates were cut to specific sizes for tests using a diamond-grit wheel.

Dynamic mechanical analysis (DMA) tests were performed for various frequencies and temperatures for the transverse direction of unidirectional CFRP laminates. The master curve of creep compliance and the time–temperature shift factor for matrix resin are obtainable indirectly from data obtained from these DMA tests.

Static and fatigue tests were conducted under various temperatures for four typical directions of unidirectional CFRP laminates. Those four directions are the longitudinal tension, longitudinal bending, transverse bending, and transverse compression. From the results of these tests, we evaluated the static and fatigue strengths in longitudinal tension (X), longitudinal compression (X′), transverse tension (Y), and transverse compression (Y′). Details related to test methods and test results presented here were explained in our earlier report [2]. Therefore, the explanation presented here addresses the formulations of Advanced ATM using the measured data.

8.3.2 Creep Compliance of Matrix Resin

The master curve of creep compliance and the time–temperature shift factor of the matrix resin can be ascertained practically by the storage modulus for the transverse direction of unidirectional CFRP laminates measured at various frequencies and

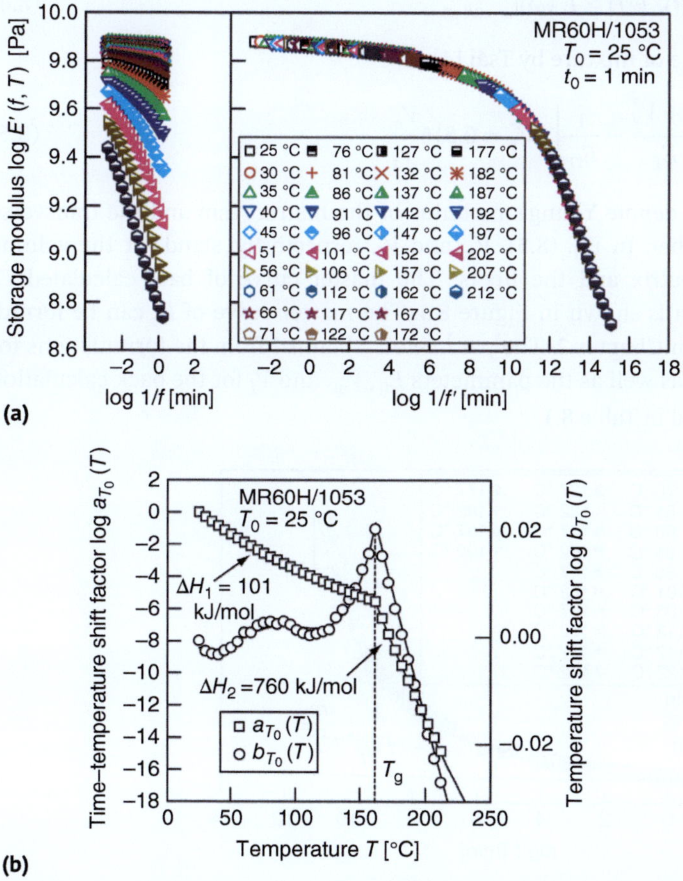

Figure 8.2 (a) Master curve and (b) shift factors for storage modulus in the transverse direction of unidirectional CFRP laminates [2]. *Source:* Reproduced from [2], with permission of Sage Publications.

temperatures with DMA tests. The left side of Figure 8.2a shows the storage modulus E' for the transverse direction of unidirectional CFRP laminates versus time t, where time t is the inverse of frequency. The right side shows the master curve of E', which is constructed by shifting E' at various constant temperatures along the logarithmic scale of t and a logarithmic scale of E' until they overlap, for reduced time t' at the reference temperature $T_0 = 25$ °C. Furthermore, E' at various constant temperatures can be superimposed so that a smooth curve is obtained. Therefore, the modified TTSP is applicable for the storage modulus for the transverse direction of unidirectional CFRP laminates.

The time–temperature shift factor $a_{T_0}(T)$, which is the horizontal shift amount, shown by square symbols in Figure 8.2b, can be formulated using Eq. (2.6) in Chapter 2. The temperature shift factor $b_{T_0}(T)$, which is the amount of vertical shift, shown by circular symbols in Figure 8.2b, can be fit with Eq. (2.7) in Chapter 2.

The creep compliance D_c of the matrix resin was back-calculated from the storage modulus E' for the transverse direction of unidirectional CFRP laminates using the following relations [3]:

$$D_c(t) \sim 1/E(t), \ E(t) \cong E'(\omega)|_{\omega \to 2/(\pi t)} \tag{8.4}$$

and a modified rule of mixture by Tsai [4] as

$$\frac{1}{E_m} = \frac{1}{V_y^*}\left[\frac{1+V_y^*}{E_T} - \frac{1}{E_{fT}}\right], \ V_y^* = 0.516\frac{V_m}{V_f} \tag{8.5}$$

where E_m and E_{fT} denote Young's modulus of the matrix resin and the transverse modulus of the fiber. In Eq. (8.5), V_m and V_f, respectively, stand for the volume fraction of the matrix and the fibers. The master curve of back-calculated D_c of the matrix resin is shown in Figure 8.3. The master curve of D_c can be formulated as Eq. (2.5) in Chapter 2. The parameters obtained from the formulations for $a_{T_0}(T)$, $b_{T_0}(T)$, D_c, as well as the parameters E_{ft}, V_m, and V_f for the back calculation of D_c, are portrayed in Table 8.1.

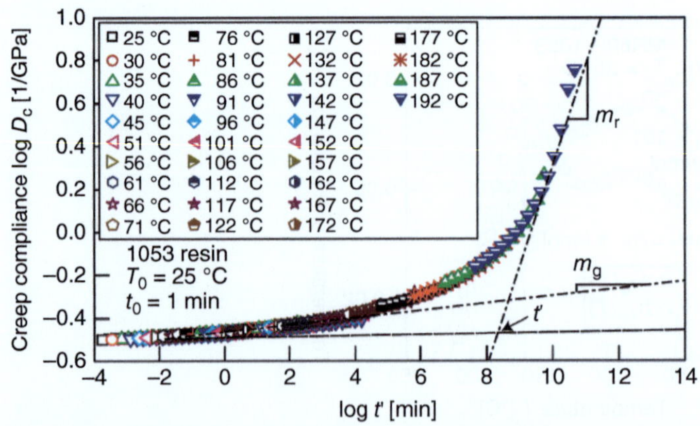

Figure 8.3 Master curve of creep compliance for the matrix resin calculated from the storage modulus for the transverse direction of unidirectional CFRP laminates [2]. *Source:* Reproduced from [2], with permission of Sage Publications.

Table 8.1 Parameters for the master curve of creep compliance for matrix resin.

T_0 (°C)	25
D_{c0} (GPa^{-1}) at t_0 and T_0	0.330
t_0 (min)	1
m_g	0.00950
m_r	0.260
t_g(min)	1.12×10^8
ΔH_1 (kJ/mol)	101
ΔH_2 (kJ/mol)	760
T_g (°C)	162
b_0	1.13×10^{-2}
b_1	-9.859×10^{-4}
b_2	2.43×10^{-5}
b_3	-2.23×10^{-7}
b_4	6.98×10^{-10}
E_{ft}(GPa)	29.3
V_m	0.45
V_f	0.55

8.3.3 Master Curves of Static and Fatigue Strengths for Unidirectional CFRP Laminates

Figure 8.4a shows the Weibull distribution of the static strength of CFRP laminates at $T = 25$ °C for the four directions X, X′, Y, and Y′. From these results, the scale and shape parameters σ_{0s} and α_s for the static strength at $T = 25$ °C can be inferred, as given in Table 8.2. Figure 8.5 shows the temperature dependence of the static strength of CFRP laminates for the four directions of X, X′, Y, and Y′. These strengths clearly depend on the temperature. Figure 8.6 shows the relationship between the static strength of CFRP laminates and the viscoelastic compliance of the matrix resin corresponding to the same time and temperature. The time and temperature dependence of static strengths for CFRP laminates against the viscoelastic compliance of the matrix resin is determined uniquely. Their slopes are constant. The material parameter n_R can be determined as the slope of this relation. Figure 8.4b shows the Weibull distribution of static strength of CFRP laminates for all tested temperatures. From these results, the scale and shape parameters σ_{0r} and α_r for the static strength for all tested temperatures can be determined, as shown in Table 8.2. The parameters σ_{0r} and α_r for the static strength at all temperatures are almost equal to those at room

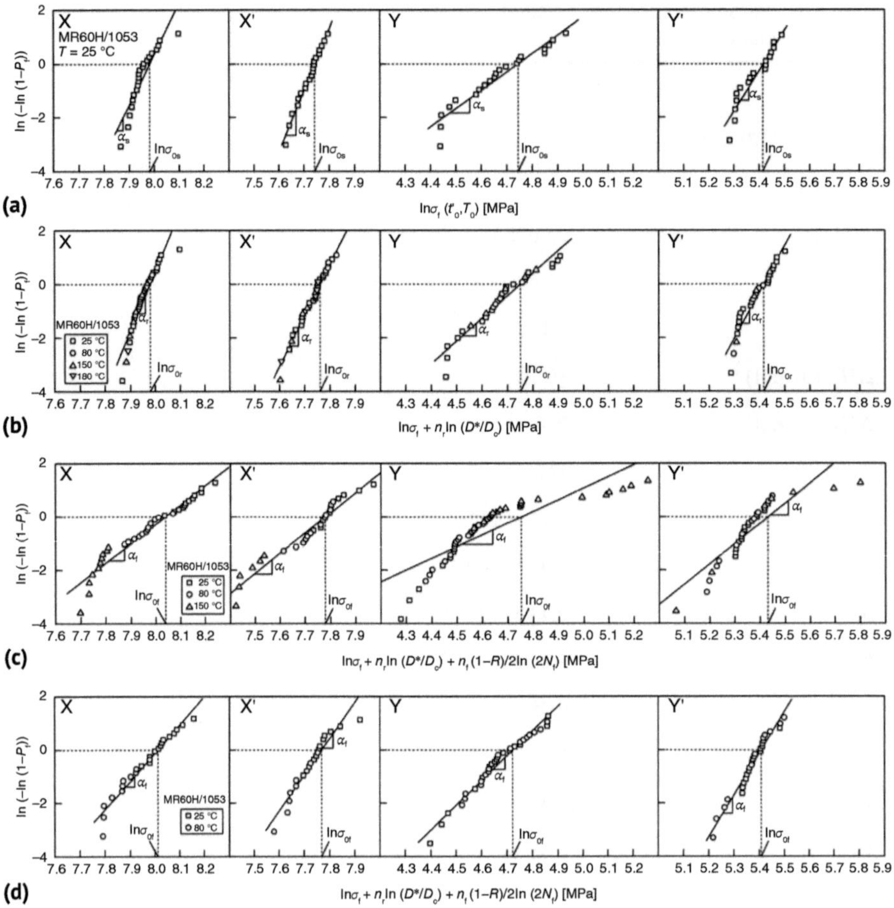

Figure 8.4 Weibull distributions for the strength of CFRP laminates: (a) static strength of CFRP laminates at $T = 25\,°C$, (b) static strength of CFRP laminates for all temperatures tested, (c) fatigue strength of CFRP laminates for all temperatures tested, and (d) fatigue strength of CFRP laminates at $T = 25$ and $80\,°C$.

temperature for each direction. Therefore, the failure probability of static strength of CFRP laminates in each direction is independent of temperature.

Figure 8.7 shows the fatigue strength of CFRP laminates versus the number of cycles to failure for all tested temperatures. The fatigue strength in each direction depends clearly on temperature. Figure 8.8a shows the strength degradation of CFRP laminates versus the number of cycles to failure. The strength degradation of CFRP laminates f_C can be found by subtracting the strength variation by the viscoelastic compliance of the matrix resin from the fatigue strength in Figure 8.7. Figure 8.4c presents the Weibull distribution of strength degradation of CFRP laminates for all tested temperatures. From these results, the scale and shape parameters σ_{0f} and α_f for fatigue strengths can be determined as shown in Table 8.2. The parameter α_f of fatigue strength for each direction is much smaller than α_r of static strength for all tested temperatures. It is assumed that the fatigue failure mechanism for each direction at high temperature (150 °C) differs from those at 25 °C

Table 8.2 Parameters for the master curve of static and fatigue strength for unidirectional CFRP laminates.

	X	X′	Y	Y′
n_R	0.165	0.605	1.54	2.11
n_F	0.0903	0.0455	0.0785	0.0335
n_F for fatigue strength except 150 °C	0.0728	0.0343	0.122	0.0417
σ_{0s} (MPa) for static strength at room temperature (RT)	2923	2305	115	225
σ_{0r} (MPa) for static strength at all temperatures	2919	2349	116	225
σ_{0f} (MPa) for fatigue strength	3098	2394	116	229
σ_{0f} (MPa) for fatigue strength except 150 °C	3016	2367	113	223
α_s for static strength at RT	19.1	23.1	6.81	15.0
α_r for static strength at all temperatures	22.3	18.9	8.39	17.3
α_f for fatigue strength	7.26	7.55	4.39	7.61
α_f for fatigue strength except 150 °C	10.6	13.6	9.31	15.7

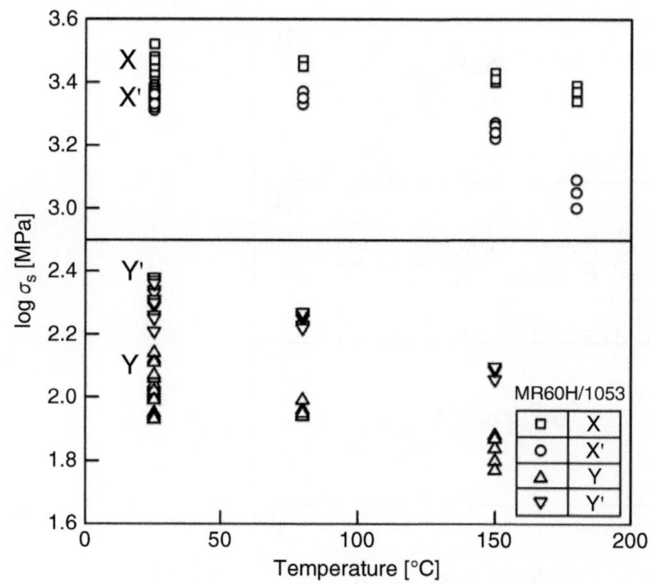

Figure 8.5 Temperature dependence for the static strength of CFRP laminates.

and 80 °C. Therefore, we determined the strength degradation of CFRP laminates except for the data measured at 150 °C. Figures 8.8b and 8.4d, respectively, show the strength degradation of CFRP laminates versus the number of cycles to failure and the Weibull distribution of strength degradation at 25 and 80 °C. From these results, the scale and shape parameters σ_{0f} and α_f for fatigue strength can be found, which are presented in Table 8.2. The value of α_f for each direction is similar to that of

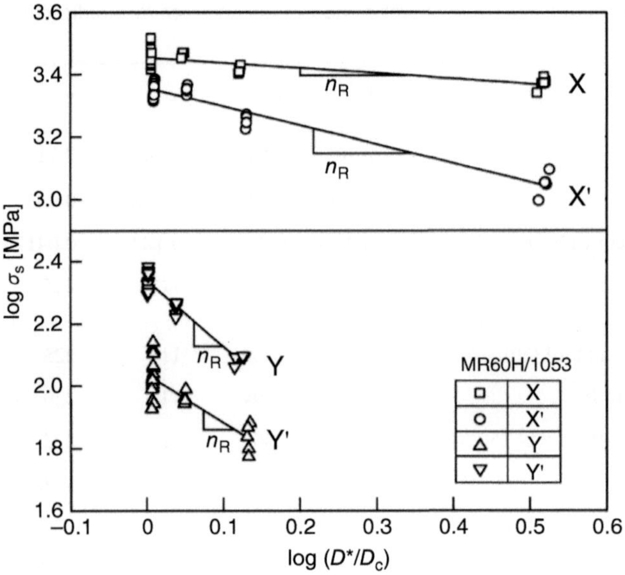

Figure 8.6 Static strength of CFRP laminates versus viscoelastic compliance of matrix resin.

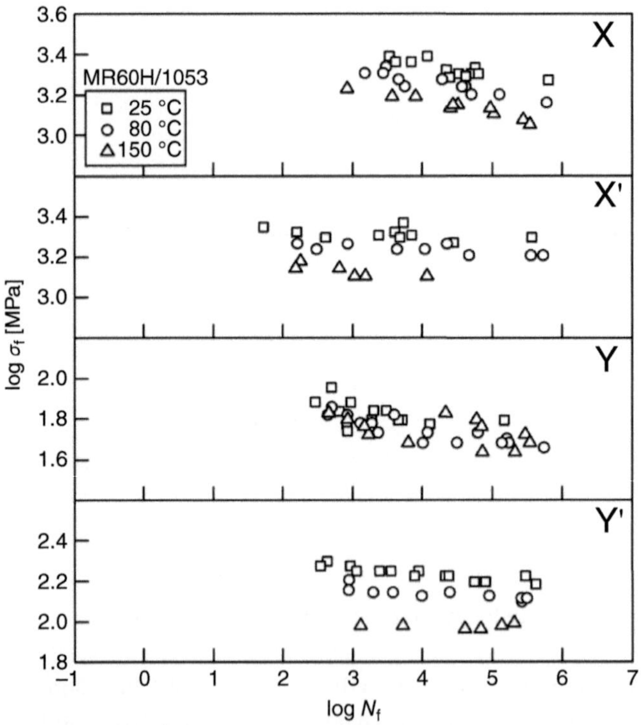

Figure 8.7 Fatigue strength of CFRP laminates versus number of cycles to failure at various temperatures.

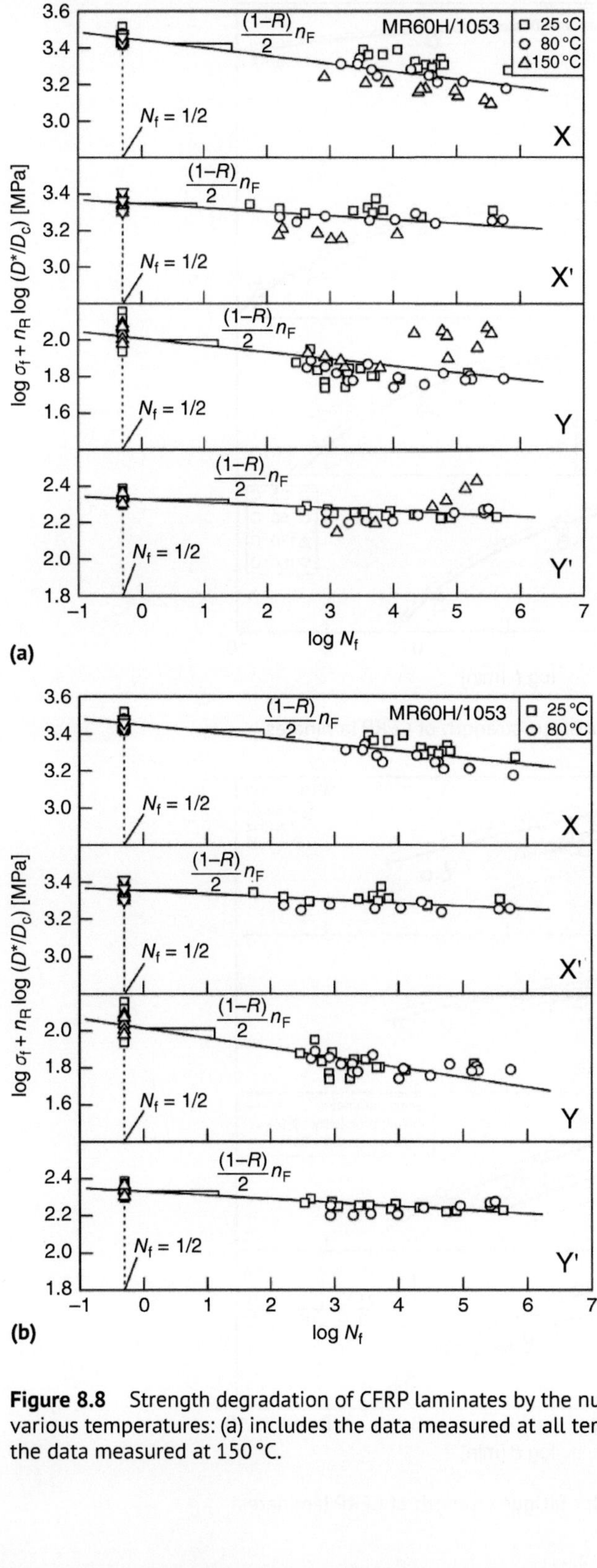

Figure 8.8 Strength degradation of CFRP laminates by the number of cycles to failure at various temperatures: (a) includes the data measured at all temperatures tested; (b) excepts the data measured at 150 °C.

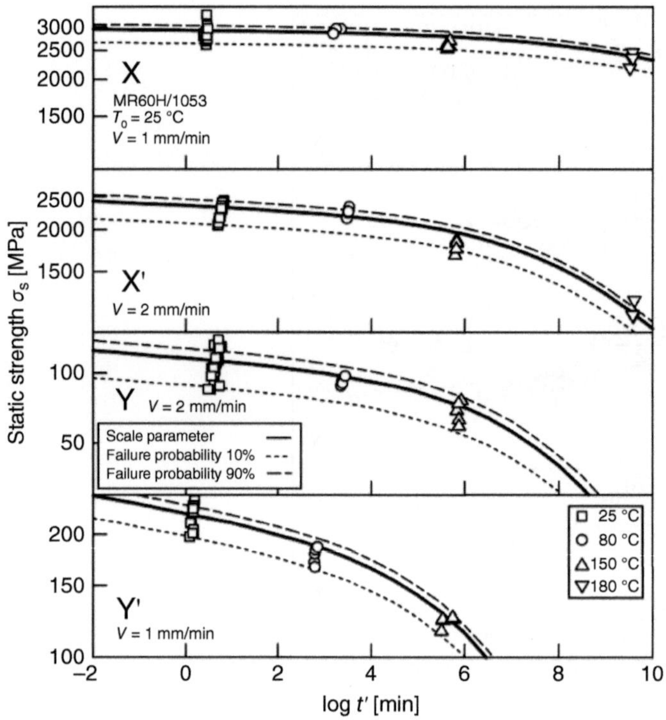

Figure 8.9 Master curves for static strength of CFRP laminates.

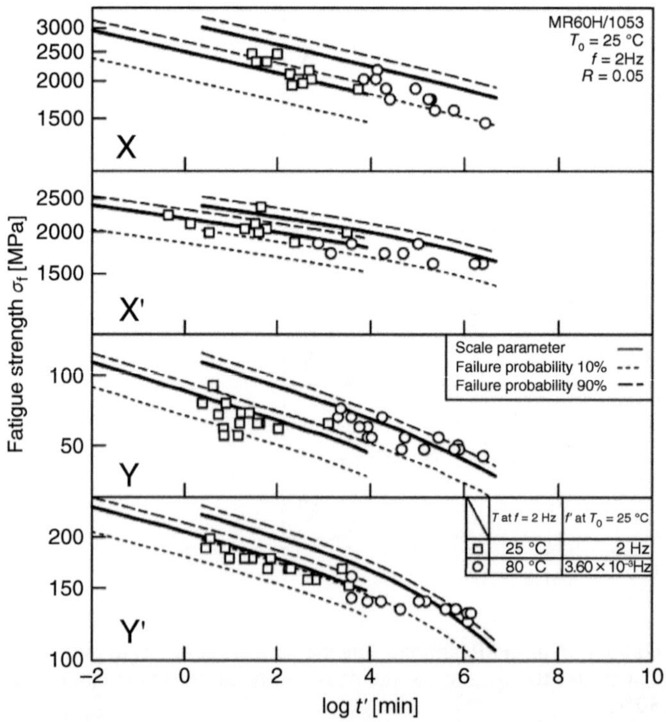

Figure 8.10 Master curves for fatigue strength of CFRP laminates.

the static strength α_r, except in the X-direction. It can be presumed that the tensile fatigue strengths in the X-direction are affected by stress concentration in the grip of the specimen. Therefore, the scatter of fatigue strength is large.

Figures 8.9 and 8.10, respectively, present master curves of static and fatigue strengths of CFRP laminates for each direction. The solid, dashed, and dotted lines, respectively, show the scale parameter and failure probability levels for 90 and 10%. The master curves of static strength for each direction agree well with the experimental results obtained for all tested temperatures. The master curves of fatigue strength for each direction agree well with the experimentally obtained data except those measured at 150 °C. The failure mechanism of CFRP laminates for cyclic loading at high temperature near the glass transition temperature should be examined in future studies.

8.4 Conclusion

Formulation of the long-term fatigue strength of FRP was performed based on the formulation of the static strength of the FRP. A feature of this evaluation method is that the fatigue strength of FRP can be determined statistically by considering the effect of repeated loading in addition to the effect of the viscoelasticity of the matrix resin on static strength.

The evaluation method was applied to the tensile and compressive fatigue strengths in the longitudinal and transverse directions of unidirectional CFRP laminates, proving the usefulness of the proposed evaluation method while also finding that fatigue characteristics differ significantly depending on the loading direction.

References

1 Nakada, M. and Miyano, Y. (2015) Advanced accelerated testing methodology for long-term life prediction of CFRP laminates. *Journal of Composite Materials*, **49**, 163–175.

2 Miyano, Y., Nakada, M. and Cai, H. (2008) Formulation of long-term creep and fatigue strengths of polymer composites based on accelerated testing methodology. *Journal of Composite Materials*, **42**, 1897–1919.

3 Christensen, R.M. (1982) *Theory of Viscoelasticity*, 2nd edn, Dover Publications, Inc., Mineola, NY.

4 Tsai, S.W. (1992) *Theory of Composites Design*, Think Composites, Dayton, OH.

9

Formulation of Creep Strength of Fiber-reinforced Polymer

9.1 Introduction

Formulation of the long-term creep strength of the fiber-reinforced polymer (FRP) is performed based on the formulation of the static strength of the FRP presented in Chapter 7. This formulation is demonstrated by statistical prediction of the long-term creep failure time under longitudinal tension loading for unidirectional carbon-fiber-reinforced polymer (CFRP) from the statistical static strength measured for the same loading direction at various temperatures and creep compliance of the matrix resin [1].

9.2 Formulation

The formulation of static strength σ_s established in Chapter 7 is shown again below.

$$\sigma_s = \sigma_0 \cdot f_A \cdot f_B \quad \log \sigma_s = \log \sigma_0 + \log f_A + \log f_B$$

The formulation of creep strength σ_c can be represented as the following equation and Figure 9.1 based on the conditions of A, B, and the additional condition of B′.

Condition A: The failure probability is independent of temperature and load histories.

Condition B: Time and temperature dependence of strength is controlled by the viscoelasticity of the matrix resin.

Condition B′: Strength degradation under creep loading occurs according to Christensen's viscoelastic crack kinetics [2].

$$\sigma_c = \sigma_0 \cdot f_A \cdot f'_B \quad \log \sigma_c = \log \sigma_0 + \log f_A + \log f'_B \tag{9.1}$$

In the above expressions, σ_0 represents the static strength at reference time and temperature determined by the type of fiber, type of weave, the volume fraction, the load direction, and other factors. Also, f_A denotes the scatter of strength as a function of the failure probability P_f determined by the type of fiber, type of weave,

Properties and Testing of Fiber-Reinforced Polymers, First Edition. Yasushi Miyano and Masayuki Nakada. © 2026 Wiley-VCH GmbH. All rights reserved, including rights for text and data mining and training of artificial intelligence technologies or similar technologies. Published 2026 by Wiley-VCH GmbH.

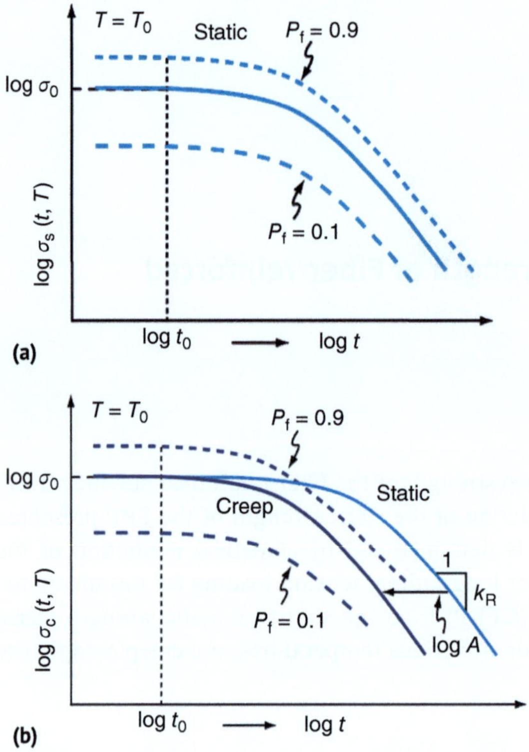

Figure 9.1 Master curves of static and creep strengths.

the volume fraction, the load direction, and other factors; and f'_B is the time- and temperature-dependent strength degradation under creep loading determined by the viscoelastic compliance $D^*(t,T)$ of the matrix resin. Details of Eq. (9.1) are shown by the following equation:

$$\log \sigma_c(P_f, t, T) = \log \sigma_0(t_0, T_0) + \frac{1}{\alpha}\log[-\ln(1 - P_f)] - n_R \log\left[\frac{D^*(At,T)}{D_c(t_0,T_0)}\right] \tag{9.2}$$

Here, D^* and A are expressed by the following equations for creep strength:

$$D^*(t) = D_c\left(\frac{t}{2}\right) \tag{9.3}$$

$$\log A = \log(1 + 1/k_R) \tag{9.4}$$

Figure 9.1a displays the static strength against time to failure at the reference temperature T_0. Figure 9.1b shows the creep strength against time to failure at T_0, which is obtainable by horizontal shifting along the logarithmic time scale with $\log A$, shown in Eq. (9.4). Parameter k_R is the slope of the static strength master curve.

The relation of failure probability P_f against time to failure t for creep stress σ_{c0} obtained from Eq. (9.2) is presented as the following equation and Figure 9.2.

$$P_f = 1 - \exp\left\{-\left[\frac{\sigma_{c0}}{\sigma_0}\right]^{\alpha}\left[\frac{D^*(At,T)}{D_c(t_0,T_0)}\right]^{n_R\alpha}\right\} \tag{9.5}$$

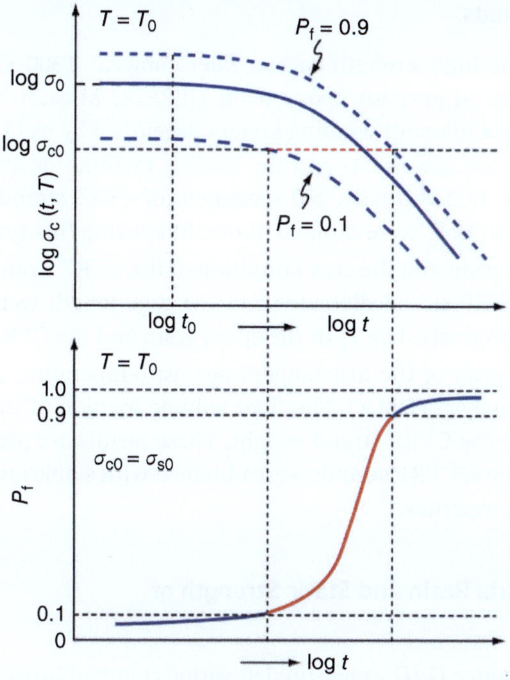

Figure 9.2 Failure probability against failure time under a creep loading.

This equation indicates that the statistical creep failure time for FRP under an arbitrary creep load and temperature can be predicted from the statistical static strengths of FRP at various temperatures and the creep compliance of the matrix resin.

9.3 Application of Formulation

As described here, the proposed method of predicting the statistical creep failure time under the tension loading along the longitudinal direction of the unidirectional CFRP from the statistical static strengths of unidirectional CFRP measured at various temperatures is clearly valid quantitatively. First, a method of predicting the statistical creep failure time of the CFRP from the statistical static strengths of the CFRP measured at various temperatures is proposed based on Christensen's model of viscoelastic crack kinetics [2]. Second, many resin-impregnated carbon fiber strands (CFRP strands) combined with T300-3000 and epoxy resin as specimens for the unidirectional CFRP are prepared using simultaneous molding to elicit stable and uniform mechanical and thermal properties. Third, the static strengths of the unidirectional CFRP are measured experimentally and statistically at various temperatures using these CFRP strands. Then, the creep failure time of the unidirectional CFRP is predicted statistically using the statistical static strengths at various temperatures. Finally, the creep failure times of the unidirectional CFRP at a constant load and a temperature are measured experimentally and probabilistically using these CFRP strands for comparison with predicted ones.

9.3.1 Specimens and Test Methods

A CFRP strand that consists of the high-strength carbon fiber Torayca® T300-3K (Toray Industries, Inc.) and a general-purpose epoxy resin (jER828; Mitsubishi Chemical Corp.) was molded using a filament winding system developed by us [3]. The details of molding method of test specimens and the loading method for test specimens are described in Section 12.2. Actually, 200 specimens of CFRP strands for stable glass transition temperatures T_g were molded at one time using this system. The composition of the epoxy resin and the cure condition of the CFRP strand are presented in Table 9.1. The CFRP strand diameter and the gage length were approximately 1 and 200 mm, respectively. The T_g of the epoxy resin and the CFRP strand were determined from the peak of the loss tangent against temperature at 1 Hz using dynamic mechanical analysis (DMA). The fiber volume fraction of the CFRP strand was ascertained from the CFRP strand weight. These results are presented in Table 9.2. As the table shows, CFRP strands were obtained with stable and uniform mechanical and thermal properties.

9.3.2 Creep Compliance of Matrix Resin and Static Strength of CFRP Strand

The non-dimensional creep compliance D_c/D_{c0} measured at various temperatures is shown on the left of Figure 9.3. Long-term D_c/D_{c0} at $T = 120\,°\text{C}$ is obtained by horizontally shifting those at various temperatures, as shown at the right of Figure 9.3. The reference temperature and time were selected for this study as $T_0 = 25\,°\text{C}$ and

Table 9.1 Composition and cure schedule of CFRP strand.

Carbon fiber strand	Composition of resin (weight ratio)	Cure schedule
T300-3000	Epoxy: jER828 (100)	100 °C × 5 h
	Hardener: MHAC-P (103.6)	+150 °C × 4 h
	Cure accelerator: 2E4MZ (1)	+190 °C × 2 h

Table 9.2 Mechanical and thermal properties of CFRP strand and matrix resin.

Materials		CFRP strand	Matrix resin
Volume fraction of fibers	Average (%)	58.5	–
	Standard deviation (%)	2.4	–
	Coefficient of variation (%)	4.1	–
	Number of specimens	190	–
Peak temperature of tan δ at 1 Hz	Average (°C)	165	166
	Standard deviation (°C)	1.22	–
	Coefficient of variation (%)	0.74	–
	Number of specimens	12	1

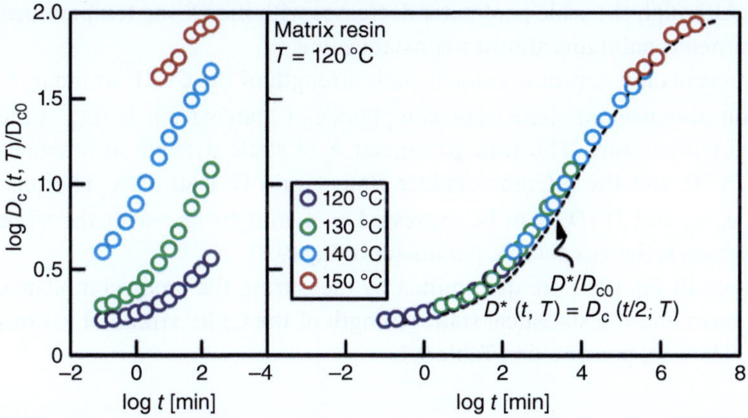

Figure 9.3 Non-dimensional creep compliance of matrix resin at $T = 120\,°C$.

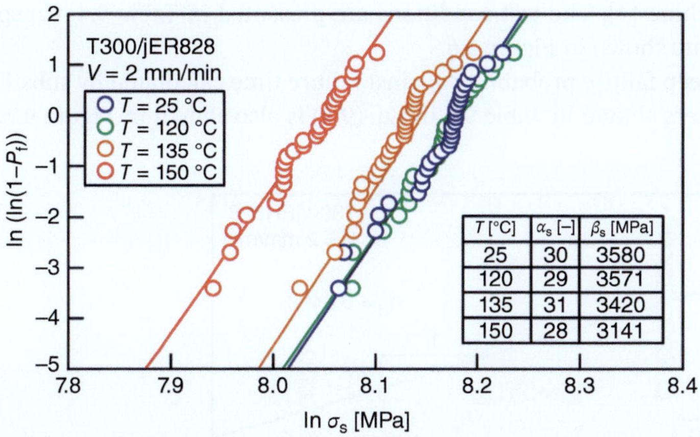

Figure 9.4 Weibull distributions of static tensile strength of CFRP strand at four temperatures.

$t_0 = 1\,\text{min}$. The creep compliance at reference temperature and reference time, D_{c0}, is 0.33/GPa. The dashed curve is the non-dimensional viscoelastic compliance D^*/D_{c0} of the matrix resin under the constant strain rate at $T = 120\,°C$.

The static tension tests for CFRP strand were conducted at four temperatures 25, 120, 135, and 150 °C with the cross-head speed 2 mm/min. The tensile strength of the CFRP strand σ_s was obtained using the following equation:

$$\sigma_s = \frac{P_{max}}{t_e}\rho \tag{9.6}$$

Here, P_{max} is the maximum load (N). ρ and t_e, respectively, represent the density of the carbon fiber (kg/m^3) and the tex of the carbon fiber strand (g/1000 m). The Weibull distributions of the static strength of CFRP strand at four temperatures are shown in Figure 9.4, where α_s is the shape parameter and β_s is the scale parameter of

CFRP strand. Although the scale parameter decreases with increasing temperature, the shape parameter maintains almost a constant value.

Figure 9.5 presents the non-dimensional static strength of the CFRP strand σ_s/σ_0 against the non-dimensional viscoelastic compliance of matrix resin D^*/D_{c0} at the same time and temperature. The scale parameter σ_0 of static strength at reference temperature 25 °C and the reference failure time 1 min is 3580 MPa. The relation between σ_s/σ_0 and D^*/D_{c0} can be expressed as a straight line with the slope $n_R = 0.0488$, which is the viscoelastic parameter in Eq. (9.5).

All parameters in Eq. (9.5) are determined by measuring the creep compliance of the matrix resin and the statistical static strength of the CFRP strand at various temperatures. They are presented in Table 9.3.

9.3.3 Creep Failure Time of CFRP Strand

Creep failure tests of CFRP stand were conducted using a specially designed creep failure testing machine [4]. The test conditions are presented in Table 9.4. Creep failure test results are shown in Figure 9.6.

The predicted creep failure probability against failure time calculated by substituting the parameters shown in Table 9.3 in Eq. (9.5) is also shown in Figure 9.6.

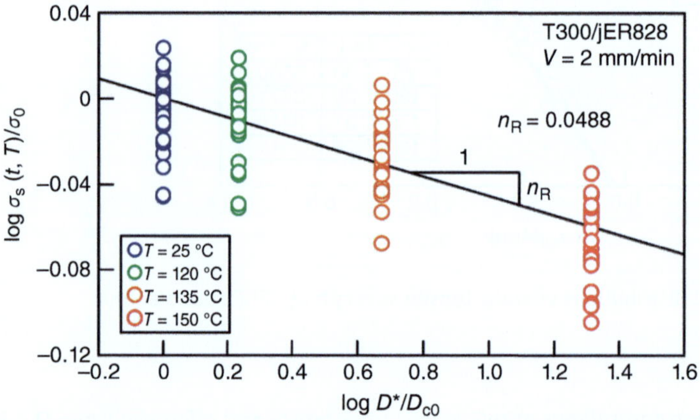

Figure 9.5 Statistical static strength of CFRP strand against the viscoelastic compliance of resin.

Table 9.3 Parameters for statistical creep failure time prediction.

Scale parameter of static strength of CFRP strand at 25 °C: σ_0	3580 MPa
Shape parameter of static strength of CFRP strand: α	29
Viscoelastic parameter of matrix resin: n_R	0.0488
Slope of viscoelastic compliance of matrix resin: m_R	0.28
Slope of static strength of CFRP strand against failure time: k_R	0.0137
Logarithmic time shifting factor: $\log A$	1.87

Table 9.4 Conditions of creep failure tests of T300 CFRP strands.

Temperature (°C)	Creep stress σ_{c0} (MPa)	Number of specimens
120	3007[a]	30

a. 84% of scale parameter of static strength at 25 °C.

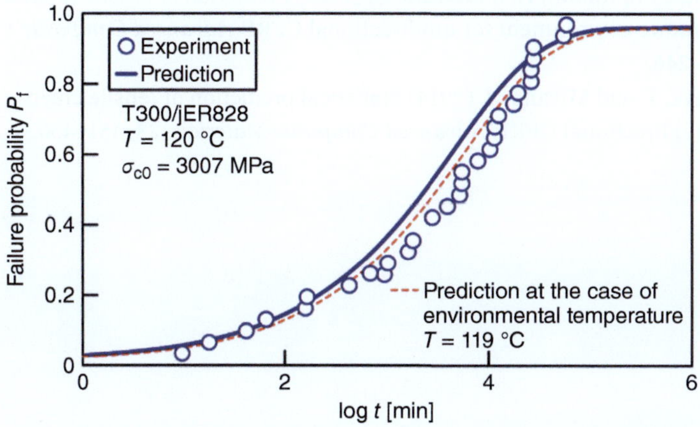

Figure 9.6 Failure probability against the creep failure time of T300 CFRP strand.

The predicted statistical creep failure time agrees with the experimentally obtained data. The dashed curve is the predicted curve when the environmental temperature $T = 119$ °C $= 120 - 1.0$ °C is assumed. Actually, $\Delta T = 1.0$ °C is the difference of the glass transition temperatures of the matrix resin and the CFRP strand shown in Table 9.2. The dashed curve agrees well with the experimentally obtained data. This fact clarified that the statistical creep failure time of the unidirectional CFRP can be predicted quantitatively from the statistical static strengths of the unidirectional CFRP and the creep compliances of the matrix resin at various temperatures.

9.4 Conclusion

The formulation of the long-term creep strength of FRP was conducted based on the formulation of the static strength of FRP presented in Chapter 7. This formulation was applied to the statistical prediction of the long-term creep failure time under longitudinal tension loading for the unidirectional CFRP from the statistical static strength measured for the same loading direction at various temperatures and the creep compliance of the matrix resin.

References

1 Nakada, M. and Miyano, Y. (2016) Statistical creep failure time of unidirectional CFRP. *Experimental Mechanics*, **56**, 653–658.
2 Christensen, R. and Miyano, Y. (2006) Stress intensity controlled kinetic crack growth and stress history dependent life prediction with statistical variability. *International Journal of Fracture*, **137**, 77–87.
3 Miyano, Y., Nakada, M., Kudoh, H. and Muki, R. (1999) Prediction of tensile fatigue life under temperature environment for unidirectional CFRP. *Advanced Composite Materials*, **8**, 235–246.
4 Nakada, M., Okuya, T. and Miyano, Y. (2014) Statistical prediction of tensile creep failure time for unidirectional CFRP. *Advanced Composite Materials*, **23**, 451–460.

10

Application 1 of Advanced Accelerated Testing Methodology: Static Strengths in Various Load Directions of Unidirectional Carbon-fiber-reinforced Polymer Laminates Under Water Absorption Condition

10.1 Introduction

This chapter presents the advanced accelerated testing methodology (Advanced ATM) evaluation of the tensile and compressive static strengths in the longitudinal and transverse directions of two types of unidirectional carbon-fiber-reinforced polymer (CFRP) laminates under wet conditions. The applicability of Advanced ATM and the effects of water absorption on the time and temperature dependence of these static strengths are discussed here [1].

10.2 Experimental Procedures

Two types of unidirectional CFRP laminates were used in this study. One is T300/EP, which consists of carbon fiber T300 and epoxy resin 2500 (Toray Industries, Inc.). The laminates were cured using an autoclave technique at 135 °C for 2 hours. Then they were post-cured at 160 °C for 2 hours. The aging treatment for post-cured specimens was carried out at 110 °C for 50 hours. This specimen was designated as the "dry" specimen. The "wet" specimens were produced by soaking a dry specimen in hot water of 95 °C for 121 hours for a 1-mm-thick specimen in the longitudinal direction, 95 °C for 144 hours for a 2-mm-thick specimen in the longitudinal direction, and 95 °C for 121 hours for a 2-mm-thick specimen in the transverse direction. The water content of all wet specimens was 1.9 wt%.

The other type of laminate was T700/VE, which consisted of the carbon fiber T700 unidirectional non-crimp fabric (Toray Industries, Inc.) and the vinylester resin Neopol 8250L (Japan U-Pica Co. Ltd.). The laminates were molded using vacuum-assisted resin transfer molding and were then cured at room temperature for 24 hours. The post-curing operation was conducted at 150 °C for 2 hours. This specimen is designated as a "dry" specimen. The "wet" specimens were prepared by

Properties and Testing of Fiber-Reinforced Polymers, First Edition. Yasushi Miyano and Masayuki Nakada. © 2026 Wiley-VCH GmbH. All rights reserved, including rights for text and data mining and training of artificial intelligence technologies or similar technologies. Published 2026 by Wiley-VCH GmbH.

soaking a dry 1-mm-thick specimen in hot water of 95 °C for 25 hours for measurements in the longitudinal direction and 2-mm-thick specimens at 95 °C for 50 hours for measurements in the longitudinal and transverse directions. The water content of the wet specimens was 0.5 wt%.

Dynamic mechanical analysis (DMA) tests in the transverse direction of unidirectional CFRP laminates were conducted at various frequencies and temperatures to construct the master curve of creep compliance for the matrix resin. Static tests for the four typical directions of unidirectional CFRP laminates were conducted at various temperatures to construct the master curves of static strength for such materials. Longitudinal tension tests were conducted according to SACMA 4R-94. Longitudinal bending tests were conducted according to ISO 14125 to ascertain the longitudinal compressive static strength. Transverse bending tests were conducted according to ISO 14125 to obtain the transverse tensile static strength. Transverse compression tests were conducted according to SACMA 1R-94.

10.3 Viscoelastic Behavior of Matrix Resin

The left side of Figure 10.1 shows the loss tangent tan δ for the transverse direction of two kinds of unidirectional CFRP laminates (dry specimen) versus time t, where t is the inverse of frequency. The right side shows the master curve of tan δ, which is constructed by shifting tan δ at various constant temperatures along the logarithmic scale of t until they overlapped, for the reduced time t' at the reference temperature $T_0 = 25\,°\mathrm{C}$. Because tan δ at various constant temperatures can be superimposed to construct a smooth curve, the modified time–temperature superposition principle (modified TTSP) is applicable for tan δ for the transverse direction of unidirectional CFRP of two kinds. Figure 10.1 shows that the master curve of tan δ for the wet specimens can also be constructed. The modified TTSP is also applicable for tan δ under wet conditions. Figure 10.1 shows that the master curve of tan δ is shifted to the left side by water absorption.

The left side of Figure 10.2 shows the storage modulus E' for the transverse direction of two kinds of unidirectional CFRP laminates (dry specimen) versus time t. The right side shows the master curve of E', which is constructed by shifting E' at various constant temperatures along the logarithmic scale of t using the same shift amount for tan δ and logarithmic scale of E' until they mutually overlap, for the reduced time t' at the reference temperature $T_0 = 25\,°\mathrm{C}$. Because E' at various constant temperatures can be superimposed so that a smooth curve is constructed, the modified TTSP is applicable for E' for the transverse direction of the two types of unidirectional CFRP laminates. Figure 10.2 shows that the master curve of E' for

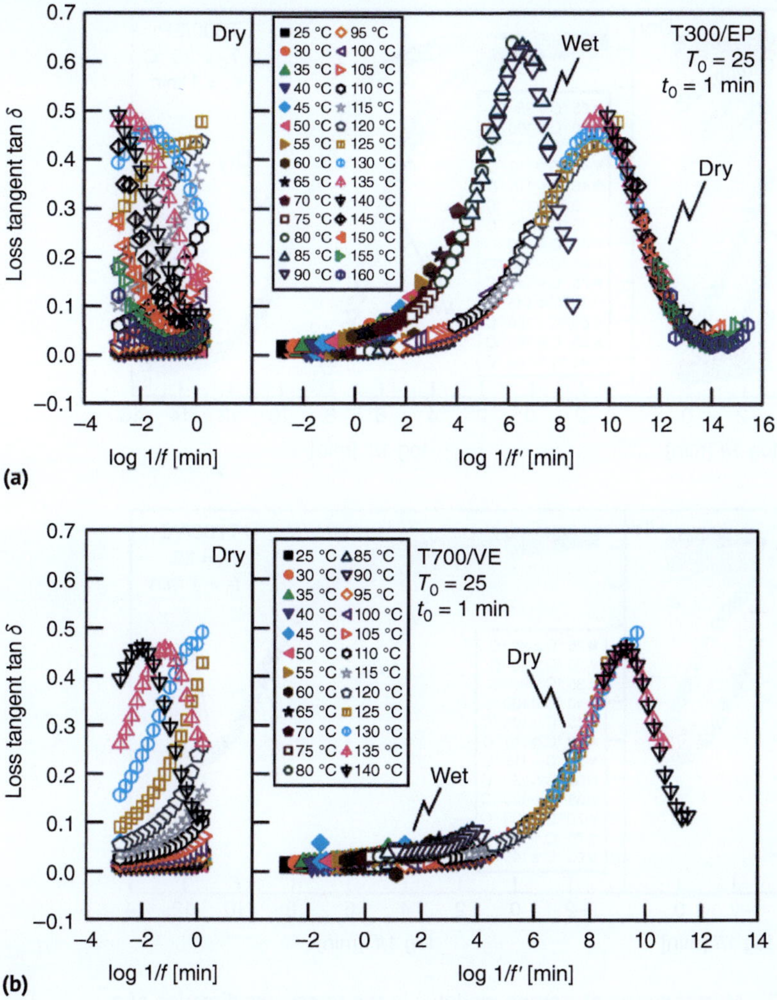

Figure 10.1 Master curves of loss tangent for the transverse direction of a unidirectional CFRP: (a) T300/EP and (b) T700/VE.

wet specimens can also be constructed. Modified TTSP is also applicable for E' under wet conditions.

The time–temperature shift factors $a_{T_0}(T)$, the horizontal shift amount shown in the upper portion of Figure 10.3, are formulated using Eq. (2.6) in Chapter 2 and the temperature shift factor $b_{T_0}(T)$, which is the amount of vertical shift shown in the lower portion of Figure 10.3. They are fitted with Eq. (2.7) in Chapter 2.

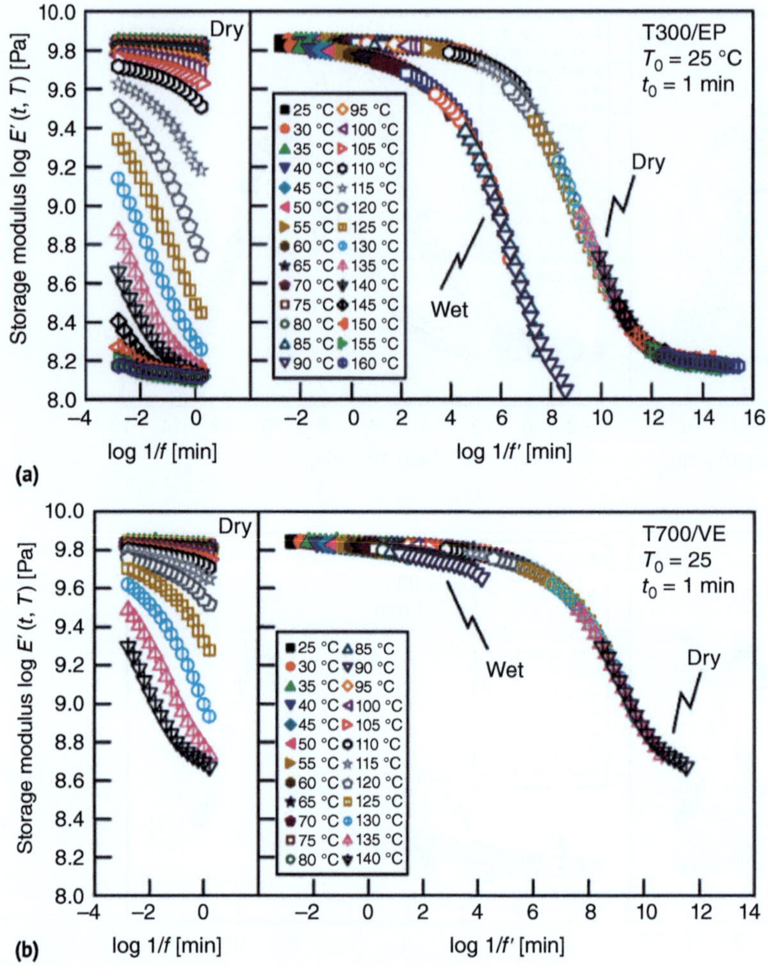

Figure 10.2 Master curves of storage modulus in the transverse direction of a unidirectional CFRP: (a) T300/EP and (b) T700/VE.

The creep compliance D_c of the matrix resin is back-calculated from the storage modulus E' for the transverse direction of two kinds of unidirectional CFRP laminates using Eq. (2.4) in Chapter 2 and the approximate averaging method reported by Uemura and Yamada [2]. The master curves of back-calculated D_c of two types of the matrix resin are shown in Figure 10.4. The master curve of D_c can be formulated using Eq. (2.5) in Chapter 2. Parameters obtained from the formulations of $a_{T_0}(T)$, $b_{T_0}(T)$, and D_c are presented in Table 10.1.

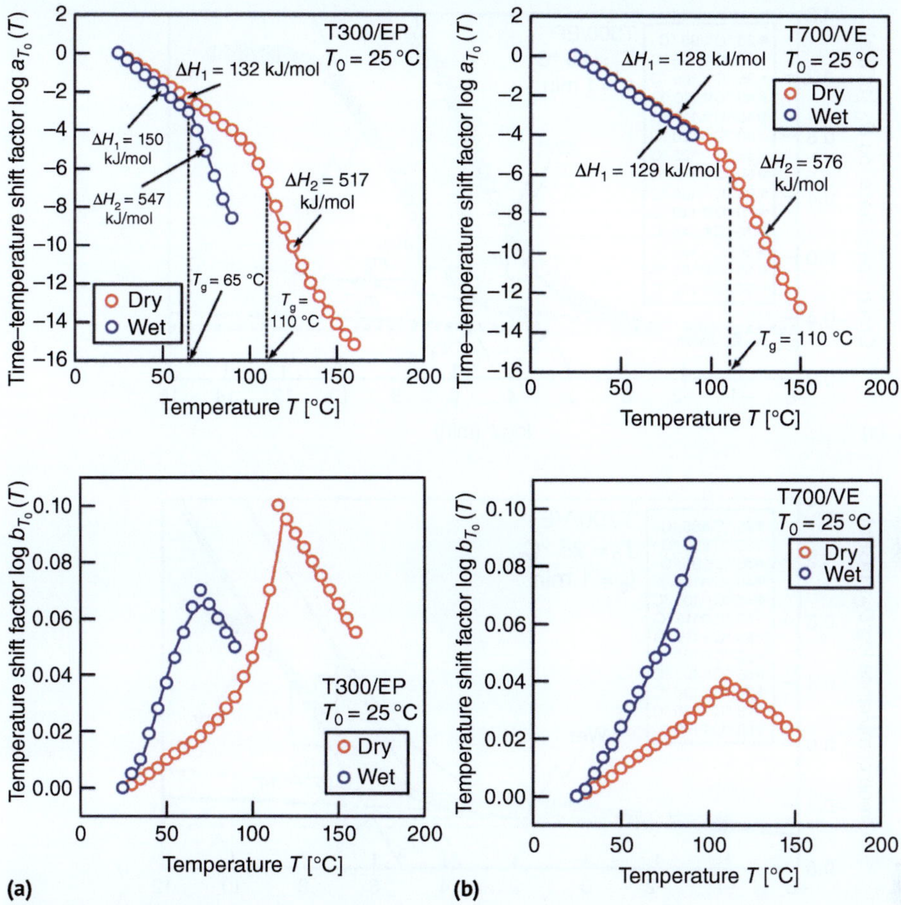

Figure 10.3 Shift factors of storage modulus in the transverse direction of a unidirectional CFRP: (a) T300/EP and (b) T700/VE.

10.4 Master Curves of Static Strengths for Unidirectional CFRP Laminates

Figures 10.5 and 10.6 depict the master curves of static strengths for longitudinal tension X, longitudinal compression X′, transverse tension Y, and transverse compression Y′ for dry and wet specimens of unidirectional CFRP laminates of two kinds obtained from strength data at various temperatures using the time–temperature shift factors a_{T_0} shown in Figure 10.3. The solid and dotted curves in

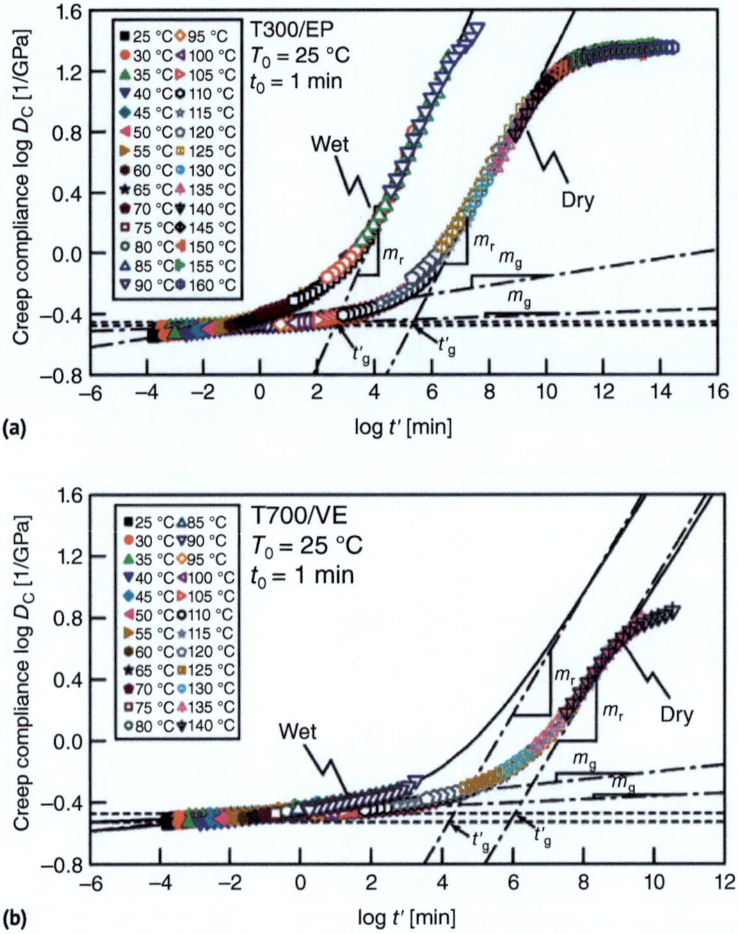

Figure 10.4 Master curves of creep compliance for matrix resin calculated from the storage modulus in the transverse direction of a unidirectional CFRP: (a) T300/EP and (b) T700/VE.

these figures show the fitting curves by Eq. (7.4) using the master curves of creep compliance of the matrix resin in Figure 10.4. The parameters obtained using this formulation are shown in Table 10.2. Figures 10.5 and 10.6 show that the static strengths of unidirectional CFRP laminates of two kinds decrease concomitantly

Table 10.1 Parameters for the master curve and shift factors of creep compliance for matrix resin.

	T300/EP		T700/VE	
	Dry	Wet	Dry	Wet
T_0 (°C)	25	25	25	25
T_g (°C)	110	65	110	(90)
D_{c0} (per GPa)	0.337	0.351	0.337	0.339
t'_0 (min)	1	1	1	1
t'_g (min)	1.54×10^6	2.34×10^3	1.36×10^6	(1.80×10^4)
m_g	0.0101	0.0348	0.00893	0.0195
m_r	0.405	0.466	0.373	(0.373)
ΔH_1 (kJ/mol)	132	150	128	129
ΔH_2 (kJ/mol)	517	547	576	–
b_0	1.65×10^{-2}	0.15	3.24×10^{-4}	-8.77×10^{-3}
b_1	-1.86×10^{-3}	-1.39×10^{-2}	-1.99×10^{-4}	1.45×10^{-4}
b_2	6.64×10^{-5}	4.26×10^{-4}	8.64×10^{-6}	9.63×10^{-6}
b_3	-8.29×10^{-7}	-4.71×10^{-6}	-1.75×10^{-8}	-
b_4	3.81×10^{-9}	1.73×10^{-8}	-1.71×10^{-10}	-

Note that ( ) means that the value was determined by extrapolation.

with increasing time, temperature, and water absorption. The time, temperature, and water absorption dependences of static strength of unidirectional CFRP laminates differ clearly according to the loading direction. The measured data of static strength in the transverse compression direction Y′ do not agree well with the fitted curves.

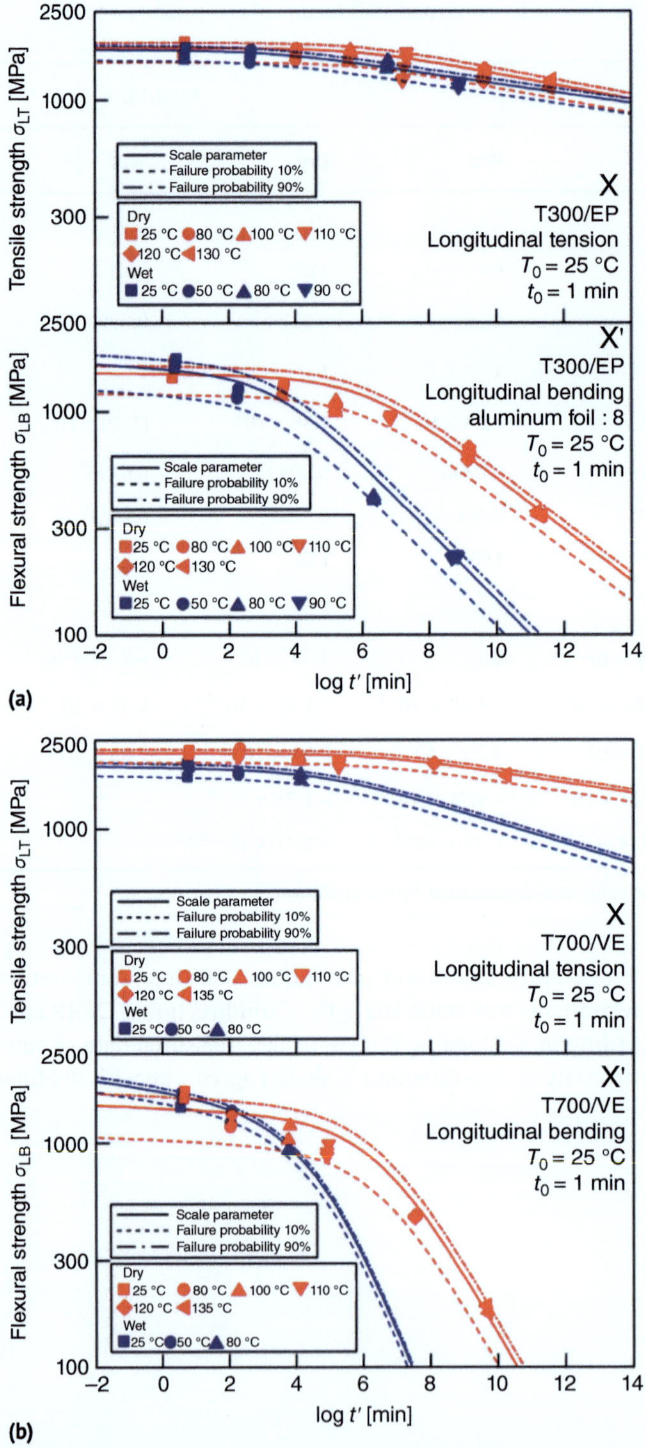

Figure 10.5 Master curves of tensile and compressive strengths in the longitudinal direction of a unidirectional CFRP: (a) T300/EP and (b) T700/VE.

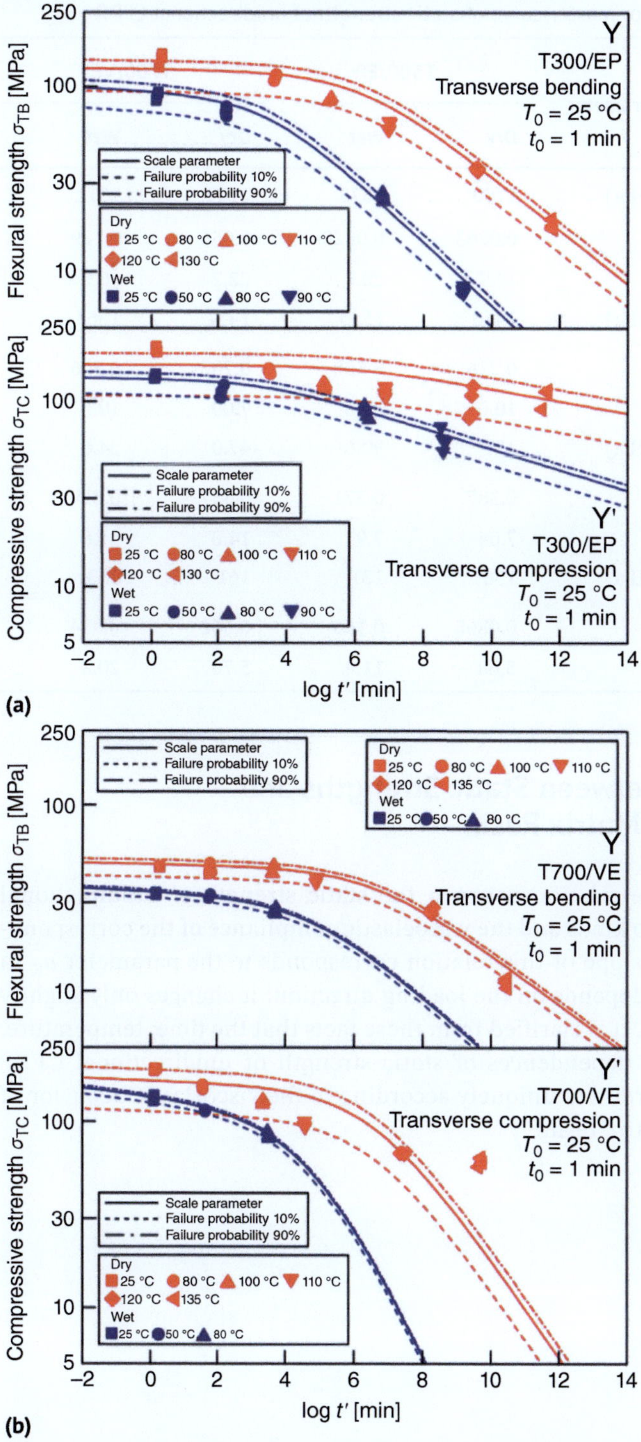

Figure 10.6 Master curves of tensile and compressive strengths in the transverse direction of a unidirectional CFRP: (a) T300/EP and (b) T700/VE.

Table 10.2 Parameters for master curve of static strength of unidirectional CFRP.

Load direction	Parameter	T300/EP		T700/VE	
		Dry	Wet	Dry	Wet
X	σ_0 (MPa)	1700	1675	2169	1911
	n_R	0.0762	0.0528	0.056	0.129
	α	14.7	20.7	22.2	20.6
X′	σ_0 (MPa)	1446	1535	1431	1363
	n_R	0.316	0.356	0.782	0.956
	α	10.0	7.18	7.00	19.6
Y	σ_0 (MPa)	121	90.6	47.0	34.0
	n_R	0.387	0.371	0.337	0.319
	α	7.04	7.97	14.0	21.8
Y′	σ_0 (MPa)	156	131	161	133
	n_R	0.0868	0.130	0.662	0.934
	α	5.68	11.4	5.70	20.8

10.5 Relation Between Static Strengths and Viscoelasticity of Matrix Resin

Figure 10.7 shows the relation between the static strength of unidirectional CFRP laminates of two kinds and the viscoelastic compliance of the corresponding matrix resin. The slope of this relation corresponds to the parameter n_R in Table 10.2. The slope depends on the loading direction; it changes only slightly with water absorption. It is clarified from these facts that the time, temperature, and water absorption dependences of static strength of unidirectional CFRP laminates can be determined uniquely according to the viscoelastic behavior of the corresponding matrix resin.

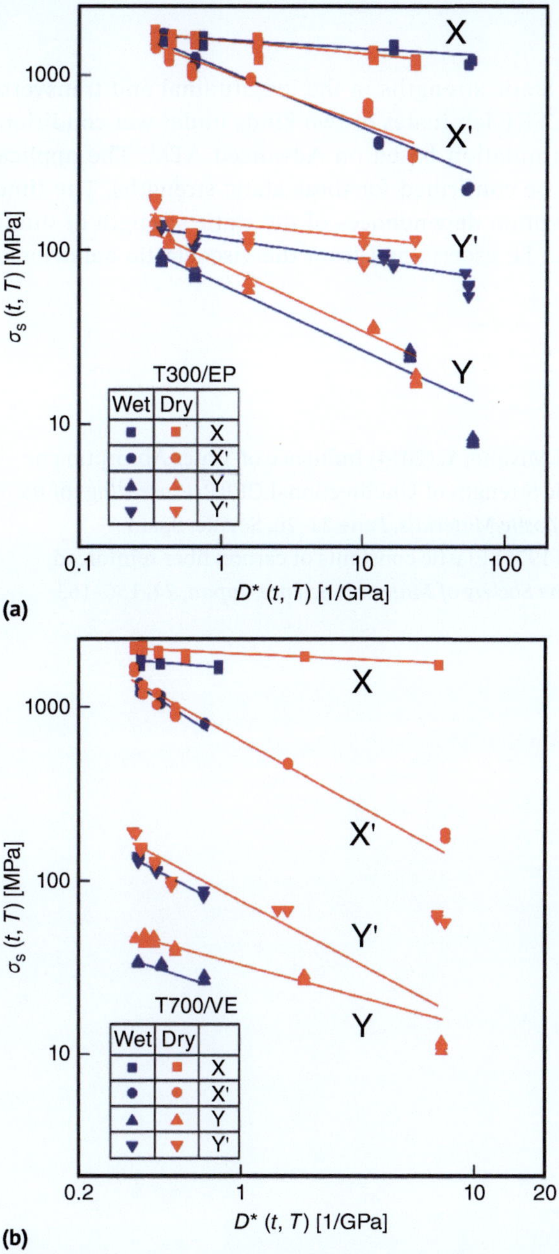

Figure 10.7 Static strengths for four directions of unidirectional CFRP versus viscoelastic compliance of matrix resin: (a) T300/EP and (b) T700/VE.

10.6 Conclusion

The tensile and compressive static strengths in the longitudinal and transverse directions of unidirectional CFRP laminates of two kinds under wet conditions were evaluated using the formulation based on Advanced ATM. The applicability of Advanced ATM can be confirmed for these static strengths. The time, temperature, and water absorption dependences of the static strength of unidirectional CFRP laminates can be ascertained from the viscoelastic behavior of the matrix resin.

References

1 Yamakita, Y., Nakada, M. and Miyano, Y. (2014) Influence of Water Absorption on Temperature Dependent Static Strength of Unidirectional CFRP. *Proceedings of the European Conference on Composite Materials*, June 22–26, Seville, Spain.

2 Uemura, M. and Yamada, N. (1975) Elastic constants of carbon fiber reinforced plastic materials. *Journal of the Society of Materials Science, Japan*, **24**, 156–163 (in Japanese).

11

Application 2 of Advanced Accelerated Testing Methodology: Life Prediction of Carbon-fiber-reinforced Polymer Structures

11.1 Introduction

The advanced accelerated testing methodology (Advanced ATM) was proposed for the prediction of long-term fatigue strength of carbon-fiber-reinforced polymer (CFRP) laminates based on the modified time–temperature superposition principle (modified TTSP). Based on Advanced ATM, the long-term fatigue strength for CFRP laminates can be predicted by measuring the short-term fatigue strengths at elevated temperatures [1, 2].

The failure criteria of separated fiber and matrix in polymer composites have been developed. The failure of composite structures has been predicted based on analyses of micromechanics, laminates, and structural levels. Recently, the stress-based micromechanics of failure (MMF) has been proposed by Ha et al. for fiber-reinforced polymers (FRPs) [3].

In this chapter, we describe the procedure of MMF/ATM combined with Advanced ATM and MMF for fatigue life prediction of structures made of CFRP laminates. The validity of the MMF/ATM is confirmed through the following two steps. As the first step, the master curves of MMF/ATM critical parameters of CFRP laminates are determined by measuring the static and fatigue strengths at elevated temperatures in the longitudinal and transverse, tension, and compression directions of unidirectional CFRP laminates. As the second step, the open-hole compression (OHC) fatigue strengths of quasi-isotropic CFRP laminates (QILs), as one example of CFRP structures, are measured at elevated temperatures. Then, these experimentally obtained data are compared with the predicted results based on the MMF/ATM using the master curves of MMF/ATM critical parameters of CFRP laminates.

11.2 Procedure of MMF/ATM

The procedure of the proposed MMF/ATM is presented schematically in Figures 11.1 and 11.2. Figure 11.1 shows the first step for the prediction procedure by the MMF/ATM, that is, the process of determining the MMF/ATM critical parameters. First, the

Properties and Testing of Fiber-Reinforced Polymers, First Edition. Yasushi Miyano and Masayuki Nakada. © 2026 Wiley-VCH GmbH. All rights reserved, including rights for text and data mining and training of artificial intelligence technologies or similar technologies.
Published 2026 by Wiley-VCH GmbH.

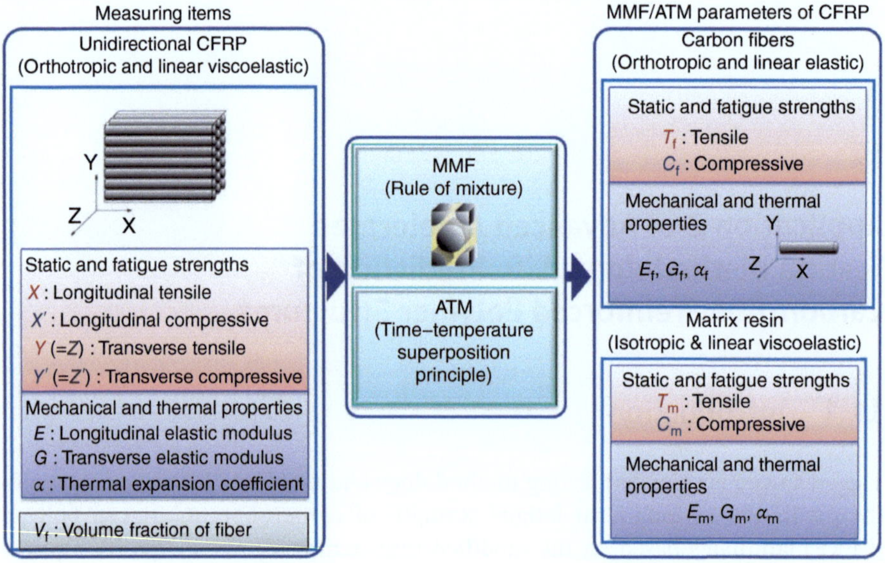

Figure 11.1 First step in the prediction procedure by the MMF/ATM: determination of MMF/ATM critical parameters.

viscoelastic modulus in the transverse direction of unidirectional CFRP laminates is measured at various temperatures. The master curve and the time–temperature shift factor are ascertained using these test data based on the modified TTSP. Second, the static and fatigue strengths in the four typical directions of the unidirectional CFRP laminates are measured, respectively, at various temperatures at a single loading rate and a single loading frequency. The strengths are assessed in the following four directions: longitudinal tension X, longitudinal compression X′, transverse tension Y, and transverse compression Y′. Third, the master curves of these strengths are determined using the measured data and the time–temperature shift factor for the viscoelastic modulus. Fourth, the master curves of four MMF/ATM critical parameters, namely the fiber tensile strength T_f, the fiber compressive strength C_f, the matrix tensile strength T_m, and the matrix compressive strength C_m, are determined using the method described in the literature [2].

Figure 11.2 presents the second step in the prediction using the MMF/ATM, which is the lifetime determination of CFRP structures. With the master curves of the MMF/ATM critical parameters, long-term strength prediction of CFRP structures becomes possible. Three-step stress analyses are necessary to obtain the test results, including stress analysis for "homogeneous" CFRP structures and CFRP laminates at the macro level and stress analysis for the constituents at the micro level by stress amplification. From the master curves of the MMF/ATM critical parameters and failure criteria for the fiber and matrix, the CFRP structural strength can be determined under any arbitrary temperature and time to failure.

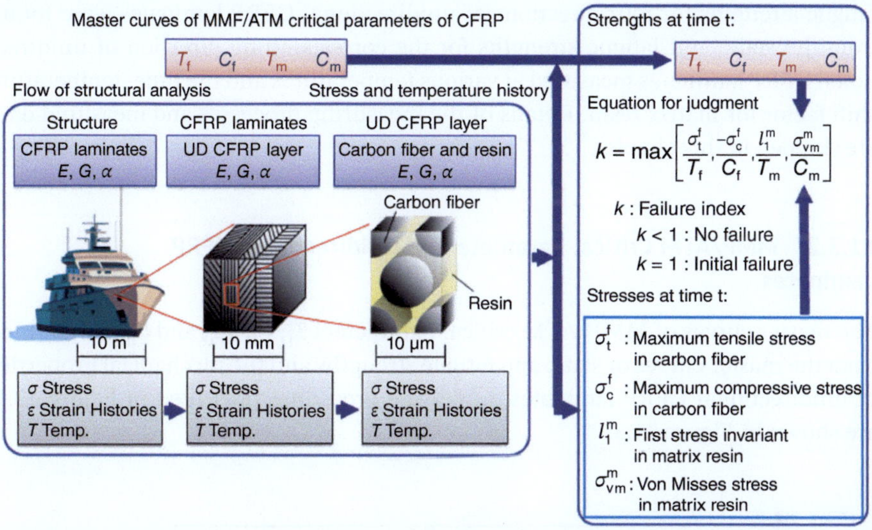

Figure 11.2 Second step in the prediction procedure by the MMF/ATM: life determination of CFRP structures.

11.3 Determination of MMF/ATM Critical Parameters

11.3.1 Long-term Static and Fatigue Strengths of Unidirectional CFRP Laminates

The test specimens were fabricated from unidirectional CFRP laminates of T800S/3900-2B, which were employed in Chapter 3. All the laminates were cured at 180 °C for 2 hours using autoclave technique. The volume fraction of the fiber is approximately 58%. Dynamic mechanical analysis (DMA) tests were applied for various frequencies and temperatures for the transverse direction of unidirectional CFRP laminates to obtain the viscoelastic compliance of the matrix resin. The static and fatigue tests for the following four directions of unidirectional CFRP laminates were conducted at various temperatures: Longitudinal tension tests under static and fatigue loadings were conducted according to SACMA 4R-94; longitudinal bending tests under static and fatigue loadings were conducted according to ISO 14125 to obtain the longitudinal compressive static and fatigue strengths; transverse bending tests under static and fatigue loadings were conducted according to ISO 14125 to ascertain the transverse tensile static and fatigue strengths; transverse compression tests under static and fatigue loadings were conducted according to SACMA 1R-94.

The master curve of creep compliance and the time–temperature shift factor for matrix resin was inferred from the storage modulus measured at various temperatures and frequencies based on modified TTSP. The master curves of static and fatigue strengths for four directions of unidirectional CFRP laminates were found from the static and fatigue strengths for the corresponding direction of unidirectional CFRP laminates measured at various temperatures and the time–temperature shift factor for matrix resin. Details of the measuring processes and measured data are skipped in this chapter.

11.3.2 MMF/ATM Critical Parameters of Unidirectional CFRP Laminates

The master curves of MMF/ATM critical parameters T_f, C_f, T_m, and C_m determined from the master curves of static and fatigue strengths and the mechanical properties of unidirectional CFRP laminates, under the procedure presented in Figure 11.1, are shown in Figure 11.3.

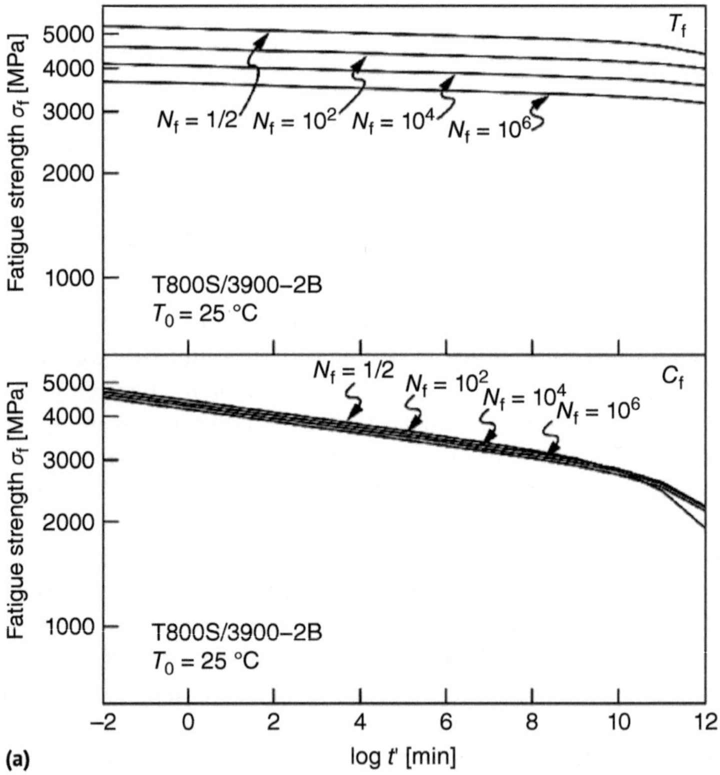

(a)

Figure 11.3 Master curves of MMF/ATM critical parameters: (a) T_f and C_f and (b) T_m and C_m.

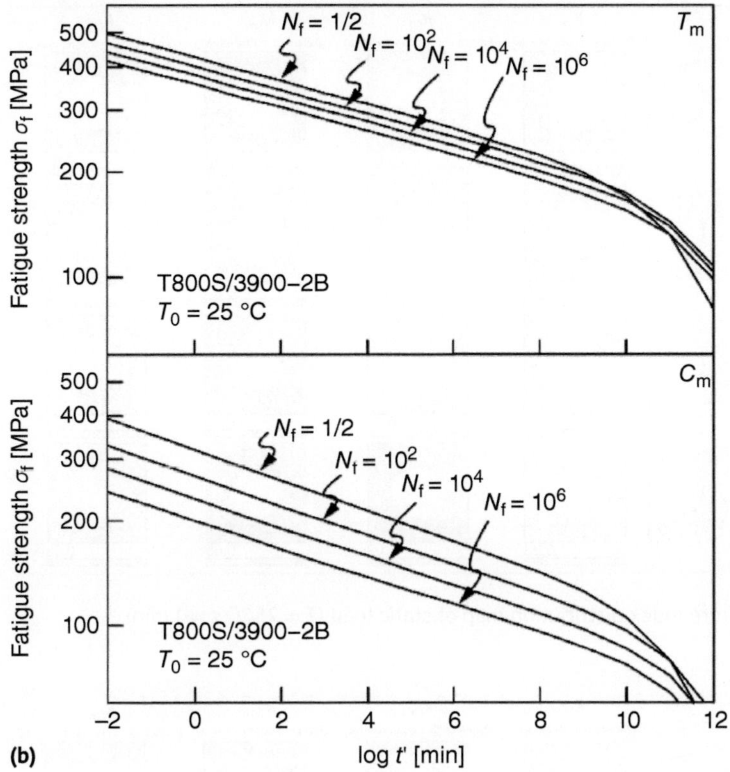

Figure 11.3 (Cont'd)

11.4 Life Determination of CFRP Structure Based on MMF/ATM

As an example of life determination of CFRP structures based on MMF/ATM, the long-term OHC strength for QILs was predicted using the master curves of MMF/ATM critical parameters. The stacking sequence of QILs is $[45/0/-45/90]_{2S}$. The open-hole specimen for prediction has configurations with 3 mm thickness, 40 mm width, and 6.35 mm diameter.

Figures 11.4 and 11.5 show failure index distribution maps for each layer and the failure mode for static loading and cyclic loading at room temperature, where k_{Tf} and k_{Cf}, respectively, represent the failure indexes of carbon fibers under tension and compression and where k_{Tm} and k_{Cm}, respectively, represent the failure indexes of the matrix resin under tension and compression. When one of these four failure indexes at one of the four layers of 45°, 0°, −45°, and 90° reaches unity, the initial failure of laminates occurs. Numbers in the maps show the maximum of the

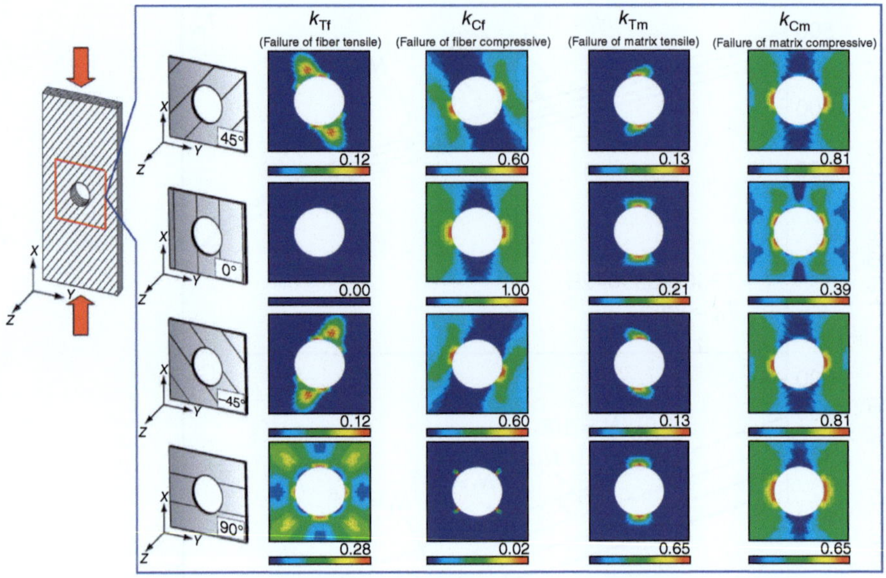

Figure 11.4 Failure index distribution map of static load ($T = 25\,°C$, $t = 1\,min$).

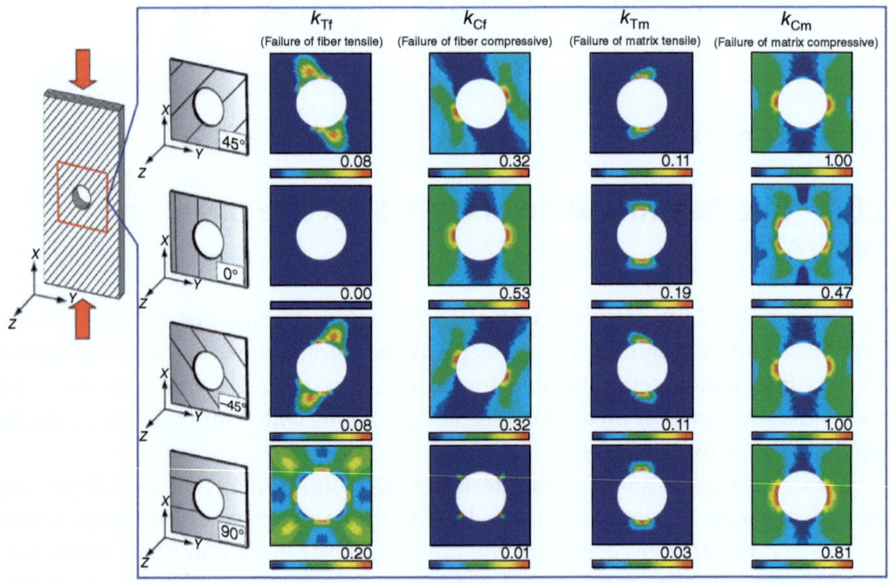

Figure 11.5 Failure index distribution map of cyclic load ($T = 25\,°C$, $f = 2\,Hz$, $N = 1 \times 10^5$).

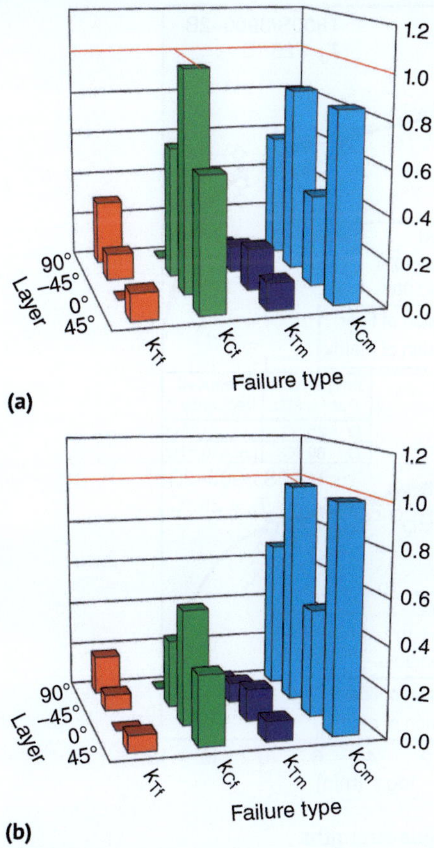

Figure 11.6 Failure indexes at the corner of the open hole under static and cyclic loadings: (a) static loading ($T = 25\,°C$, $t' = 1\,min$) and (b) cyclic loading ($T = 25\,°C$, $f = 2\,Hz$, $N = 1 \times 10^5$).

failure index in each layer and the failure mode. These maps show that the failure index of 0° layer and k_{Cf} reaches first unity at the corner of the open hole for static loading, and that the failure index of 45° layer and k_{Cm} reaches first unity at the corner of the open hole for cyclic loading. Figure 11.6 shows the failure indexes for four failure modes and four layers at the corner of the open hole under static and cyclic loadings. Results clarified that the OHC static failure of QILs was triggered by the fiber's compressive failure in the 0° layer, as shown in Figure 11.6a, and that the OHC fatigue failure of QILs was triggered by the matrix's compressive failure in the 45° layer, as shown in Figure 11.6b.

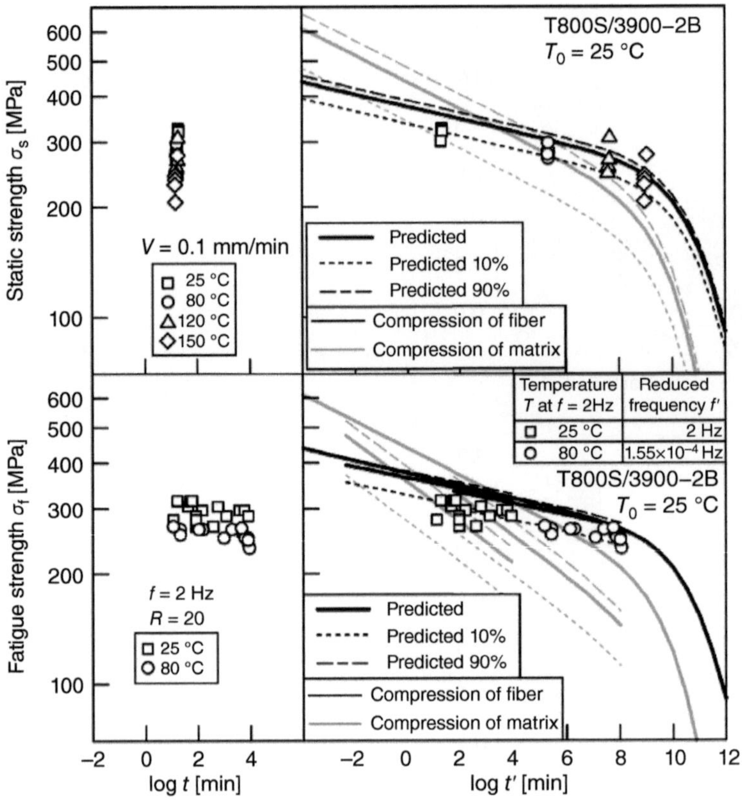

Figure 11.7 Prediction of OHC static and fatigue strengths.

The predicted master curves of static and fatigue OHC strengths for QILs were calculated based on the results obtained using the MMF/ATM with master curves of MMF/ATM critical parameters, where the initiation of failure is defined as both compressive failure of fibers of 0° layers and the compressive failure of the matrix at the corner of an open hole. These predicted results are presented in Figure 11.7.

11.5 Experimental Confirmation for OHC Static and Fatigue Strengths of CFRP QILs

Figure 11.8 shows that static and fatigue compression tests were performed for CFRP QILs with a central hole. Figure 11.9 presents fractographs of the corner of the open hole under static loading at $T = 25\,°C$, in which the compressive static load was stopped at 95% of the failure load level. Microbuckling of fibers in the 0° layer (second and sixth) and in the transverse crack in the 45° layer (first) of the surface layer was observed near the hole edge. In the 45° (third, fifth, and seventh) and 90° layers (fourth and eighth), no failure was observed. Figure 11.10 presents fractographs of the corner of the open hole under cyclic loading at $T = 25\,°C$, in which the maximum load of 80% level of the static failure load was stopped at $N = 1 \times 10^5$. The transverse cracks in the 45° layers (first and fifth) were observed near the hole edge. In the 0°,

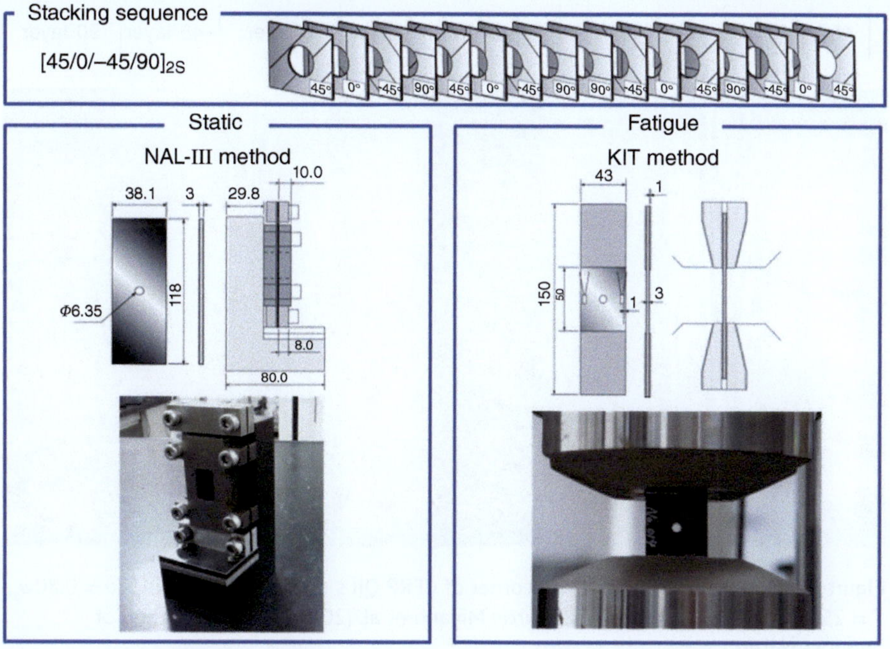

Figure 11.8 Static and cyclic compression tests for CFRP QILs with a central hole.

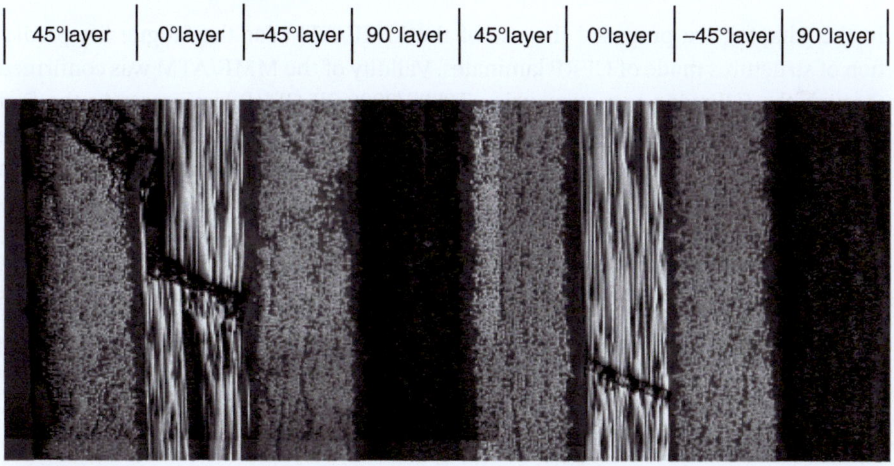

Figure 11.9 Initial failure at the corner of the hole of CFRP QILs under static loading ($\sigma = 0.95\,\sigma_s$, $T = 25\,°C$, $V = 0.1$ mm/min). *Source*: Miyano et al. (2020)/with permission of Springer Nature.

−45°, and 90° layers, no failure was observed. It was clear from these results that the predicted initial failure position, layer, and failure mode agreed well with the experimentally obtained ones for static and cyclic loading.

The measured static and fatigue strengths are shown in Figure 11.7. All these experimentally obtained results are triggered by the compression of fibers in the 0° layers at the hole corner. These results agree well with predicted values for all time to failure t'.

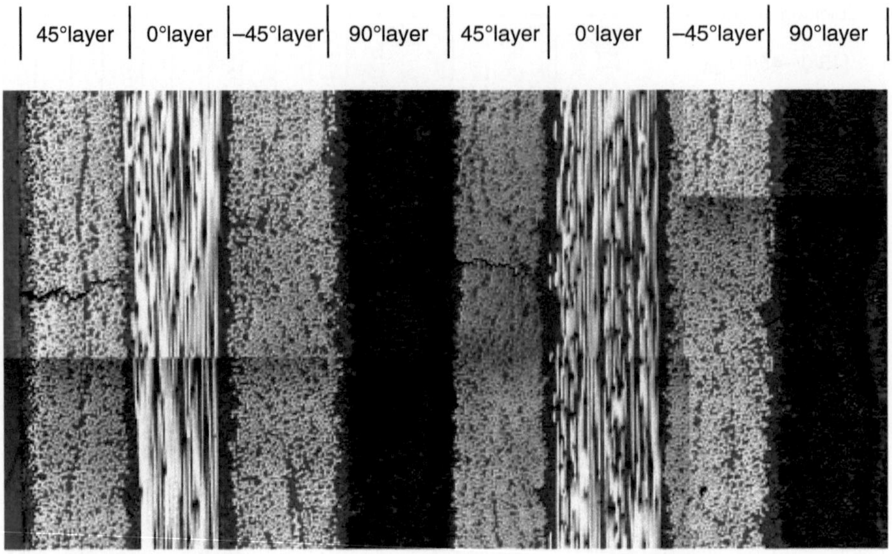

Figure 11.10 Initial failure at the corner of CFRP QILs under cyclic loading ($\sigma = 0.80\,\sigma_S$, $T = 25\,°C$, $f = 2\,Hz$, $N = 1 \times 10^5$). *Source*: Miyano et al. (2020)/with permission of Springer Nature.

11.6 Conclusion

In this chapter, we proposed the use of the MMF/ATM for the fatigue life prediction of structures made of CFRP laminates. Validity of the MMF/ATM was confirmed through the following two steps using T800S/3900-2B CFRP laminates: As the first step, the master curves of MMF/ATM critical parameters of CFRP laminates were determined by measuring the static and fatigue strengths at elevated temperatures in the longitudinal and transverse as well as tension and compression directions of unidirectional CFRP laminates. As the second step, the OHC static and fatigue strengths of CFRP QILs with failure modes were predicted based on MMF/ATM using the master curves of MMF/ATM critical parameters of CFRP laminates. These OHC strengths were measured at elevated temperatures for comparison with the predicted results. The predicted initial failure position, layer, and failure mode, in addition to the predicted strengths, agreed well with the experimental ones for static and cyclic loadings.

References

1 Cai, H., Miyano, Y., Nakada, M. and Ha, S.K. (2008) Long-term fatigue strength prediction of CFRP structure based on micromechanics of failure. *Journal of Composite Materials*, **42**, 825–844.

2 Nakada, M. and Miyano, Y. (2018) Multiscale modeling for long-term life prediction of CFRP structures under cyclic loading. *Multiscale and Multidisciplinary Modeling Experiments and Design*, **1**, 19–32.

3 Ha, S.K., Jin, K.K. and Huang, Y. (2008) Micro-mechanics of failure (MMF) for continuous fiber reinforced composites. *Journal of Composite Materials*, **42**, 1873–1895.

12

Application 3 of Advanced Accelerated Testing Methodology: Effect of Molding Condition on Statistical Static and Creep Strengths of Carbon-fiber-reinforced Polymer Strand

12.1 Introduction

In Chapter 9, it was clear that the proposed method of predicting the statistical creep failure time under the tension loading along the longitudinal direction of unidirectional carbon-fiber-reinforced polymer (CFRP) from the statistical static strengths of resin-impregnated carbon fiber strands (CFRP strands) measured at various temperatures based on Christensen's model of viscoelastic crack kinetics is clearly valid quantitatively. The approach was as follows. First, many CFRP strands were prepared using simultaneous molding to elicit stable and uniform mechanical and thermal properties. Second, the static strengths were experimentally and statistically measured at various temperatures using these CFRP strands. Then, the creep failure time of CFRP strands was predicted statistically using the statistical static strengths at various temperatures based on the predicting method. Finally, the creep failure times of CFRP strands at a constant load and a temperature are measured experimentally and probabilistically using these CFRP strands for comparison with the predicted ones. And it was cleared that the time- and temperature-dependent tensile static and creep strengths of CFRP strands are controlled by the viscoelastic behavior of matrix resin based on the Rosen's shear lag model.

In this chapter, two kinds of CFRP strands molded by different cure conditions as molding conditions are prepared. The statistical static tensile strengths for both CFRP strands are measured at various temperatures, and the statistical creep failure times for both CFRP strands are predicted by the same procedure as Chapter 9. The effect of molding condition on the statistical static and creep strengths of CFRP strands is discussed theoretically and experimentally [1].

12.2 Experiments

A CFRP strand, which consists of high-strength type carbon fiber T300-3000 (Toray Industries, Inc.) and a general-purpose epoxy resin jER828 (Mitsubishi Chemical Corp.) was molded using a filament winding system developed by the authors, as

Properties and Testing of Fiber-Reinforced Polymers, First Edition. Yasushi Miyano and Masayuki Nakada. © 2026 Wiley-VCH GmbH. All rights reserved, including rights for text and data mining and training of artificial intelligence technologies or similar technologies.
Published 2026 by Wiley-VCH GmbH.

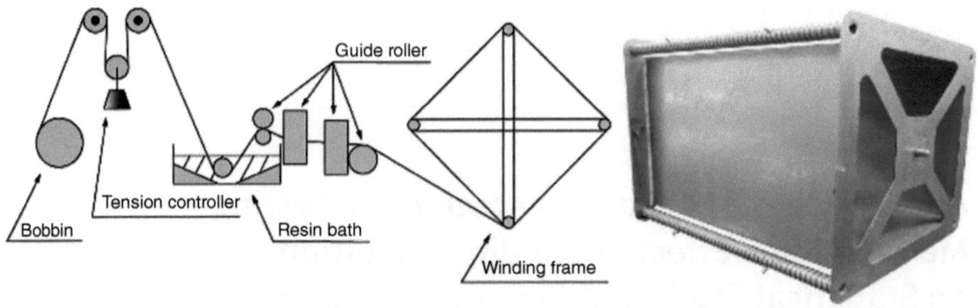

Figure 12.1 Molding system for CFRP strands. *Source*: Miyano et al. (2020)/with permission of Materials System Research Laboratory, Kanazawa Institute of Technology.

Table 12.1 Composition and cure schedule of CFRP strands.

Carbon fiber strand	Composition of resin (weight ratio)	Cure schedule A	Cure schedule B
T300-3000	Epoxy: jER828 (100)	100 °C × 5 h	70 °C × 12 h
	Hardener: MHAC-P (103.6)	+150 °C × 4 h	+150 °C × 4 h
	Cure accelerator: 2E4MZ (1)	+190 °C × 2 h	+190 °C × 2 h

shown in Figure 12.1. Actually, 200 specimens of CFRP strands for a stable glass transition temperature T_g were molded at one time using this system. The composition of epoxy resin and the cure condition of CFRP strand are presented in Table 12.1. As shown in this table, two kinds of cure schedules having different cure time and temperature at the first curing step are employed for studying the effect of molding conditions on the statistical static and creep strengths of CFRP strands predicted by our proposed procedures. The cure schedule B has the first cure step of a relatively long time and low temperature compared with that of the cure schedule A. This difference in curing temperature causes the difference of thermal expansion of winding frame shown in Figure 12.1. Therefore, the tensile strength and its scatter of CFRP strand depend on the curing condition as mentioned later. The diameter and the gage length of CFRP strands are approximately 1 and 200 mm, respectively.

The static tension tests for two kinds of CFRP strands were conducted at four temperature levels, 25, 120, 135, and 150 °C, with cross-head speed 2 mm/min as shown in Figure 12.2. The tensile strength of CFRP strand σ_s is obtained using Eq. (9.6).

The creep tension tests for two kinds of CFRP strands were conducted at a constant temperature 120 °C with a constant stress selected as 84% of the scale parameter of static strength σ_0 at the reference failure time and temperature. The static and creep tension testing machines are shown in Figure 12.3.

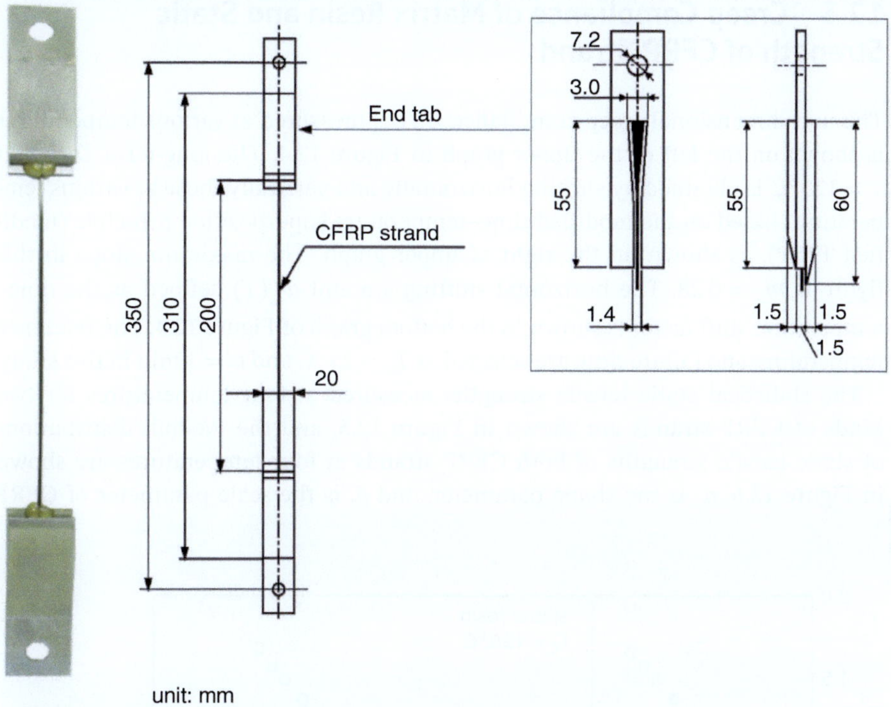

Figure 12.2 Specimen configuration of CFRP strands. *Source*: Miyano et al. (2020)/with permission of Materials System Research Laboratory, Kanazawa Institute of Technology.

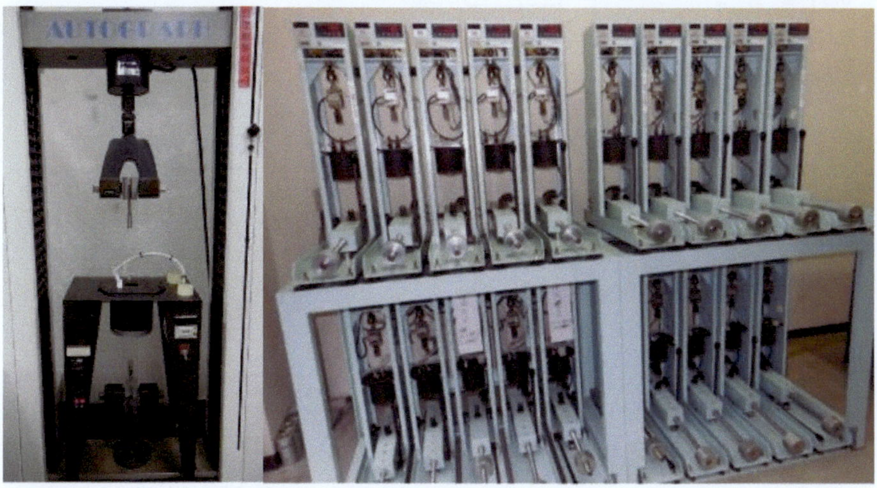

Figure 12.3 Static and creep tension testing machines for CFRP strands. *Source*: Miyano et al. (2020)/with permission of Materials System Research Laboratory, Kanazawa Institute of Technology.

12.3 Creep Compliance of Matrix Resin and Static Strength of CFRP Strand

The non-dimensional creep compliance D_c/D_{c0} measured at various temperatures is shown on the left of the upper graph in Figure 12.4. The long-term D_c/D_{c0} at $T = 120\,°C$ is obtained by shifting horizontally and vertically those at various temperatures based on the modified time–temperature superposition principle (modified TTSP), as shown on the right of upper graph. The maximum slope in this figure is $m_R = 0.28$. The horizontal shifting amount $a_{T_0}(T)$ defined as the time–temperature shift factor is shown in the bottom graph of Figure 12.4. The reference temperature and failure time are selected as $T_0 = 25\,°C$ and $t_0 = 1$ min in this study.

The statistical static tensile strengths measured at four temperatures for two kinds of CFRP strands are shown in Figure 12.5, and the Weibull distributions of static tensile strengths of both CFRP strands at four temperatures are shown in Figure 12.6. α_s is the shape parameter, and β_s is the scale parameter of CFRP

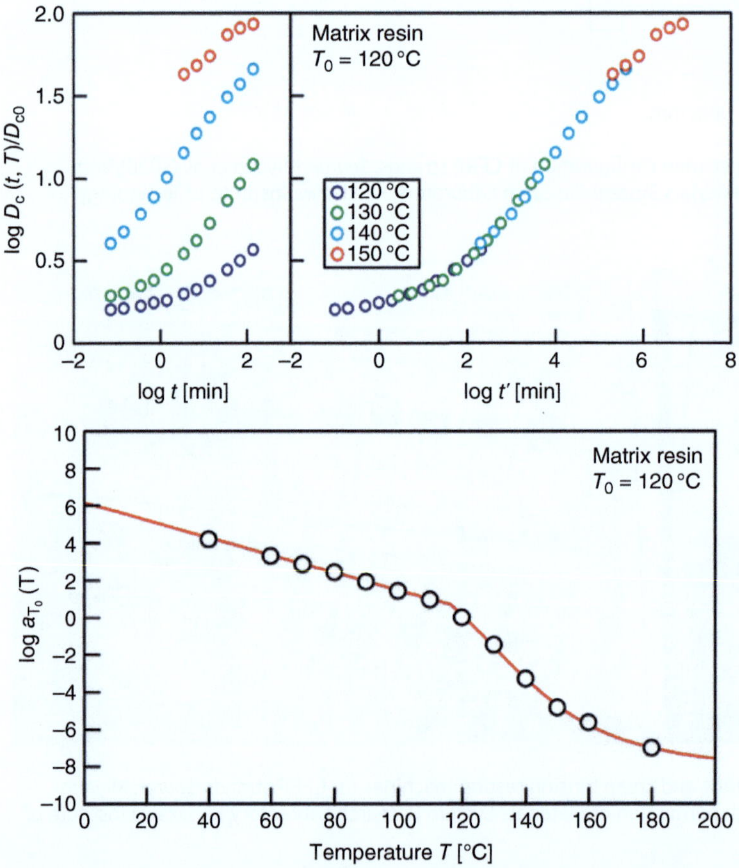

Figure 12.4 Non-dimensional creep compliance and time–temperature shift factor of matrix resin at $T_0 = 120\,°C$.

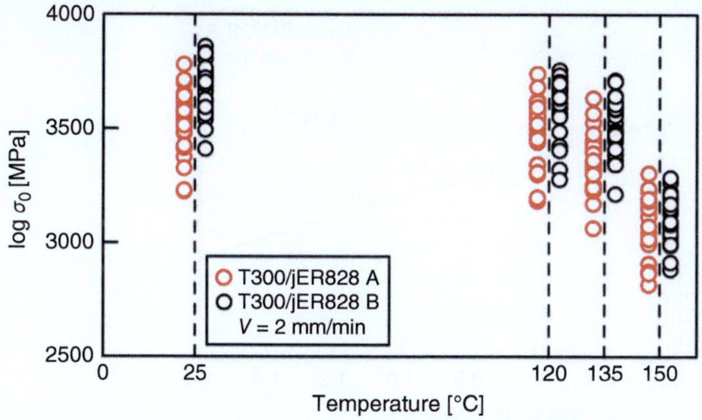

Figure 12.5 Statistical static tensile strength versus temperature for CFRP strands.

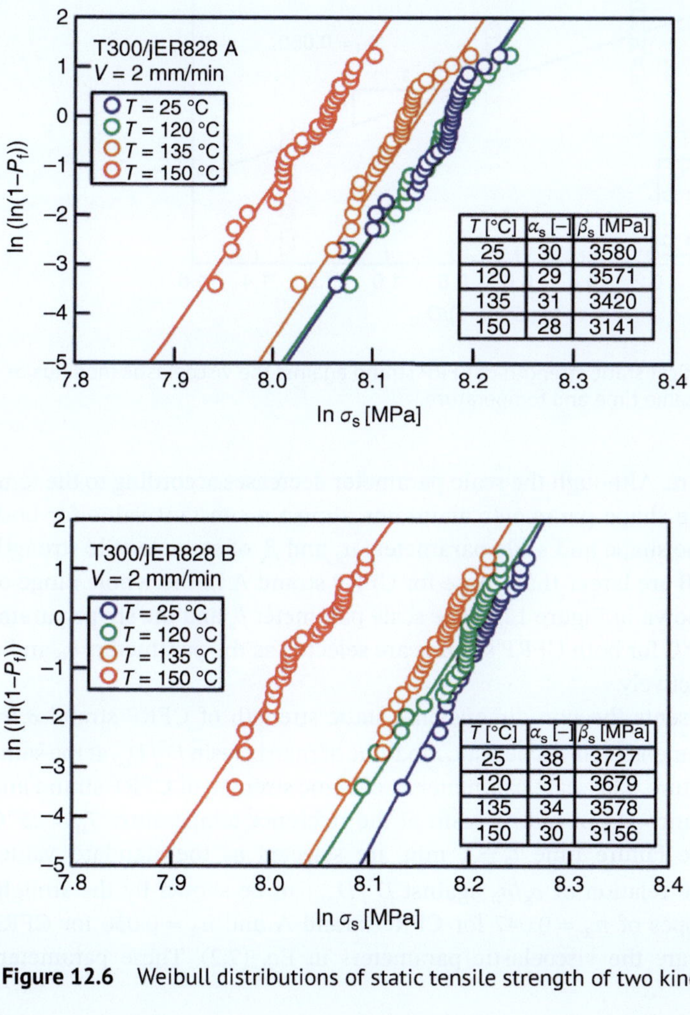

Figure 12.6 Weibull distributions of static tensile strength of two kinds of CFRP strands.

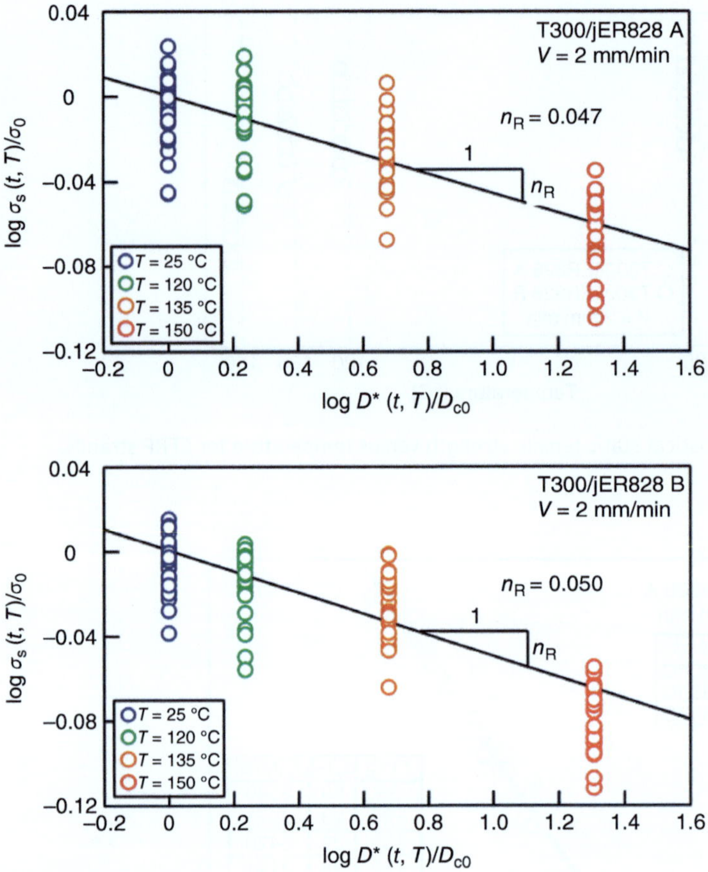

Figure 12.7 Statistical static strength of CFRP strand against the viscoelastic modulus of matrix resin at the same time and temperature.

strand in this figure. Although the scale parameter decreases according to the temperature raise, the shape parameter maintains almost a constant value for both CFRP strands. The shape and scale parameters α_s and β_s of static tensile strength for CFRP strand B are larger than those for CFRP strand A in the whole range of temperature as shown in Figure 12.6. The scale parameter β_s and the shape parameter α_s at $T_0 = 25\,°C$ for both CFRP strands are selected as the parameters σ_0 and α in Eq. (7.2), respectively.

Figure 12.7 presents the non-dimensional static strength of CFRP strand σ_s/σ_0 against the non-dimensional viscoelastic modulus of matrix resin D^*/D_{c0} at the same time and temperature. The scale parameter σ_0 of static strength of CFRP strand and the creep compliance D_{c0} of matrix resin at the reference temperature $T_0 = 25\,°C$ and the reference failure time $t_0 = 1$ min are selected as the standard values in this figure. The relation of σ_s/σ_0 against D^*/D_{c0} can be shown by the straight lines with the slopes of $n_R = 0.047$ for CFRP strand A and $n_R = 0.050$ for CFRP strand B, which are the viscoelastic parameters in Eq. (7.2). These parameters

Table 12.2 Parameters for statistical static and creep strengths prediction.

Parameter	A	B
Scale parameter of static strength of CFRP strand at 25 °C: σ_0(MPa)	3580	3727
Shape parameter of static strength of CFRP strand at 25 °C: α	30	38
Viscoelastic parameter: n_R	0.047	0.050
Slope of viscoelastic compliance of matrix resin: m_R	0.28	0.28
Slope of static strength of CFRP strand against failure time: k_R	0.0132	0.0140
Logarithmic time shifting factor: log A	1.88	1.86

are very close to the viscoelastic parameter $n_R = 0.0625$ predicted by substituting the shape parameter $\alpha_c = 8.0$ of tensile strength for T300 single carbon fiber into Eq. (12.1) based Rosen model [2].

$$n_R = 1/(2\alpha_c) \tag{12.1}$$

All parameters in Eqs. (7.2) and (9.2) have been determined for two kinds of CFRP strands, as shown in Table 12.2.

12.4 Master Curves of Statistical Static and Creep Strengths of CFRP Strands

The master curve of statistical static strength of two kinds of CFRP strands at $T = 120\,°\text{C}$ constructed by substituting all parameters of both CFRP strands in Table 12.2 into Eq. (7.2) is shown in Figure 12.8. The statistical static tensile strengths of both CFRP strands measured at various temperatures are converted to $T = 120\,°\text{C}$ by using the time–temperature shift factor shown in the bottom graph in Figure 12.4 and also indicated in Figure 12.8. The predicted master curve of the static tensile strengths for both CFRP strands agrees well with the experimental results. The shape and scale parameters of static tensile strength for CFRP strand B are clearly higher than those for CFRP strand A because of the slight difference of cure schedule for both CFRP strands.

The master curve of statistical creep tensile strength of both CFRP strands at $T = 120\,°\text{C}$ predicted by substituting all parameters of both CFRP strands in Table 12.2 into Eq. (9.2) is shown in the upper graph of Figure 12.9. And the failure probability versus failure time under the load condition of constant stress $\sigma_{c0}(= 0.84\sigma_0)$ predicted by substituting all parameters in Table 12.2 into Eq. (9.5) is shown at the bottom of this figure. The creep failure times under the same load condition of constant stress $\sigma_{c0}(= 0.84\sigma_0)$ were statistically measured at $T = 120\,°\text{C}$, and these measured creep failure times are shown in both graphs of this figure. As a result, the predicted master curve of the creep tensile strengths for both CFRP strands also agrees well with the experimental results, respectively. The shape and scale parameters of creep tensile strength for CFRP strand B are clearly also higher than those for CFRP strand A

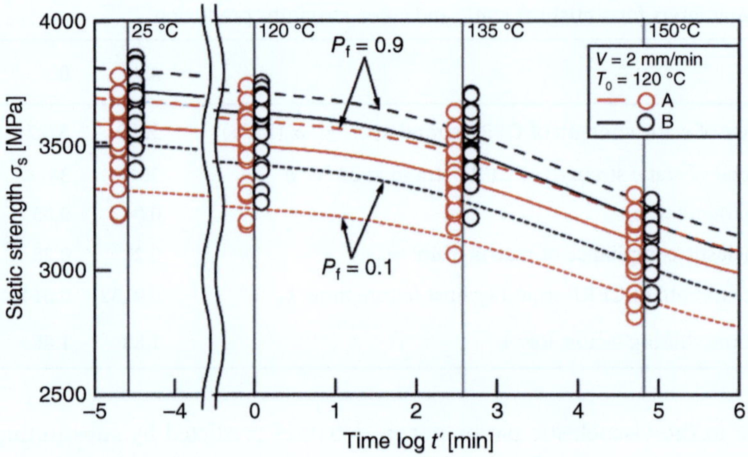

Figure 12.8 Comparison of experimental and predicted results for statistical static strength of two kinds of CFRP strands A and B at $T_0 = 120\,°C$.

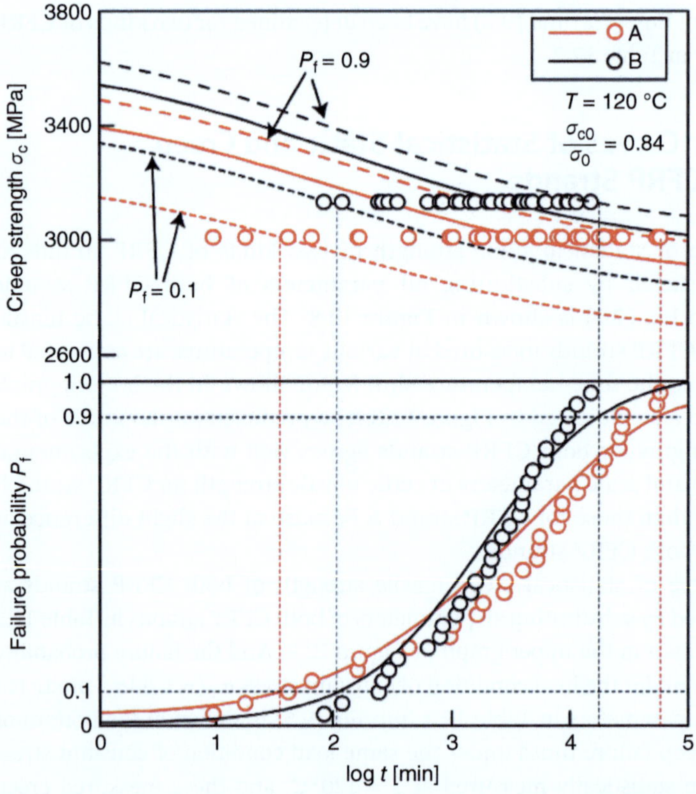

Figure 12.9 Comparison of experimental and predicted results for statistical creep failure time of two kinds of CFRP strands at $T = 120\,°C$.

because of the slight difference of cure schedule for both CFRP strands. It was clear that our proposed prediction method is separable for the slight effect of cure schedule as molding conditions on the statistical static and creep strengths of CFRP strands, and as a result, has high accuracy and reliability.

12.5 Conclusion

Two kinds of CFRP strands molded by different cure conditions as molding conditions were prepared. The effect of molding condition on the statistical static and creep strengths of CFRP strands was theoretically and experimentally discussed. It was clear that our proposed prediction method is separable for the slight effect of cure schedule as molding conditions on the statistical static and creep strengths of CFRP strands, and as a result, has high accuracy and reliability.

References

1 Nakada, M. and Miyano, Y. (2020) Effect of molding condition on statistical static and creep strengths of resin impregnated CFRP strand. *Materials System*, **37**, 53–58.

2 Nakada, M., Miyano, Y., Kinoshita, M., Koga, R., Okuya, T. and Muki, R. (2002) Time–temperature dependence of tensile strength of unidirectional CFRP. *Journal of Composite Materials*, **36**, 2567–2581.

13

Application 4 of Advanced Accelerated Testing Methodology: Effect of Carbon Fiber on Statistical Static and Creep Strengths of Carbon-fiber-reinforced Polymer Strand

13.1 Introduction

In this chapter, the statistical static strengths under tension loading along the longitudinal direction of unidirectional carbon-fiber-reinforced polymer (CFRP) with carbon fibers of various kinds are measured at various temperatures and are discussed in terms of the role of the matrix resin viscoelasticity. First, the statistical static strengths of carbon fibers of four kinds are measured using many specimens of high-strength and high-modulus PAN-based carbon fibers and low- and high-modulus pitch-based carbon fibers. Second, the temperature-dependent tensile static strengths of unidirectional CFRP with carbon fibers of four kinds are measured statistically at various temperatures using resin-impregnated carbon fiber strands (CFRP strands) as unidirectional CFRP specimens. The fractographs of CFRP strands with carbon fibers of four kinds are shown at various temperatures. Finally, the effect of carbon fiber on statistical static and creep strengths is discussed based on the Rosen's shear lag model [1].

13.2 Molding of CFRP Strands and Testing Methods

Carbon fibers of four kinds were used in this study: high-strength PAN-based carbon fiber T300-3000 and high-modulus PAN-based carbon fiber M40J-6000 (Toray Industries Inc.); and low-modulus pitch-based carbon fiber XN05-3000 and high-modulus pitch-based carbon fiber XN50-3000 (Nippon Graphite Fibers Corp.). The carbon fibers used in this study and their mechanical properties, as referred from catalogs are shown in Table 13.1.

We molded the CFRP strands of four kinds combined with the respective fibers and a generally used epoxy resin (jER828; Mitsubishi Chemical Corp.) using a filament winding system developed by the authors, as shown in Figure 12.1. The epoxy resin composition and the CFRP strand curing conditions are presented in Table 13.2. The gage lengths of CFRP strands are approximately 200 mm, as shown in Figure 12.2.

Properties and Testing of Fiber-Reinforced Polymers, First Edition. Yasushi Miyano and Masayuki Nakada. © 2026 Wiley-VCH GmbH. All rights reserved, including rights for text and data mining and training of artificial intelligence technologies or similar technologies.
Published 2026 by Wiley-VCH GmbH.

Table 13.1 Carbon fiber of four kinds and their mechanical properties.

Carbon fiber	PAN or pitch	Density (g/cm^3)	Tex (g/km)	Elastic modulus (GPa)	Tensile strength (MPa)
T300-3000	PAN	1.76	198	230	3530
M40J-6000	PAN	1.75	225	377	4400
XN05-3000	Pitch	1.65	410	54	1100
XN50-3000	Pitch	2.14	450	520	3830

Table 13.2 Composition and cure schedule of CFRP strand of four kinds.

CFRP strand	Carbon fiber	Composition of resin (weight ratio)	Cure schedule
T300/EP	T300-3000	Epoxy: jER828 (100)	70 °C × 12 h
M40J/EP	M40J-6000	Hardener: MHAC-P (103.6)	+150 °C × 4 h
XN05/EP	XN05-3000	Cure accelerator: 2E4MZ (1)	+190 °C × 2 h
XN50/EP	XN50-3000		

The glass transition temperatures $T_g = 160\,°C$ of the epoxy resin were ascertained from the peak of loss tangent against temperature at 1 Hz using a dynamic mechanical analysis (DMA) testing machine. The fiber volume fraction $V_f = 55\%$ of CFRP strand is ascertained from the CFRP strand weight.

The original static testing machine with the temperature chamber shown in the left of Figure 12.3 can achieve a constant elongation rate in a wide range and a constant temperature in the range of 25–170 °C. The original creep testing machine with temperature chamber shown in the right of Figure 12.3 can realize a constant load in the maximum load 1 kN and a constant temperature in the range of 25–150 °C.

13.3 Results and Discussion

13.3.1 Creep Compliance of Matrix Resin and Static Strength of Carbon Fibers

The non-dimensional creep compliance D_c/D_{c0} measured at various temperatures is shown on the left of Figure 13.1. Long-term D_c/D_{c0} at $T = 120°C$ is obtained by horizontally shifting those at various temperatures, as shown at the right of Figure 13.1. The reference temperature and time are selected for this study as $T_0 = 25°C$ and $t_0 = 1$ min. The creep compliance at reference temperature and reference time D_{c0} is 0.33 GPa^{-1}. The dashed curve is the non-dimensional viscoelastic compliance D^*/D_{c0} of matrix resin under the constant strain rate at $T = 120°C$.

The carbon fiber tensile strength σ_t measured statistically at 25 °C is obtained. The carbon fiber gage length is 25 mm. The number of specimens for each fiber is 50. Figure 13.2 shows the Weibull distributions for the tensile strengths of the carbon fibers. The table in this figure presents shape parameters α_c and scale parameters β_c for these carbon fibers.

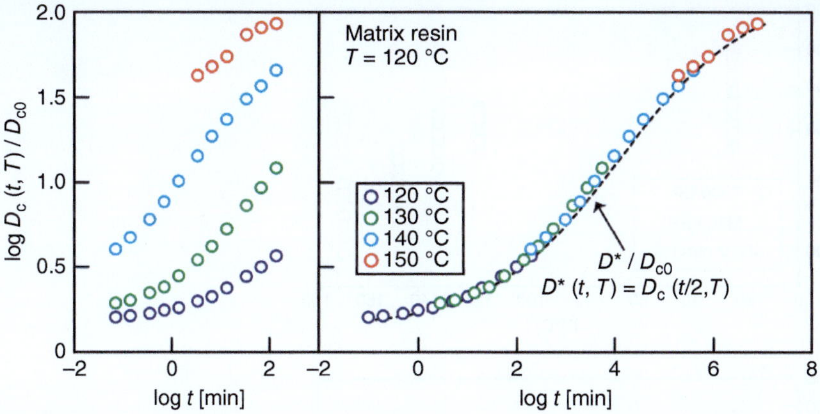

Figure 13.1 Non-dimensional creep compliance of matrix resin at $T = 120\,°C$.

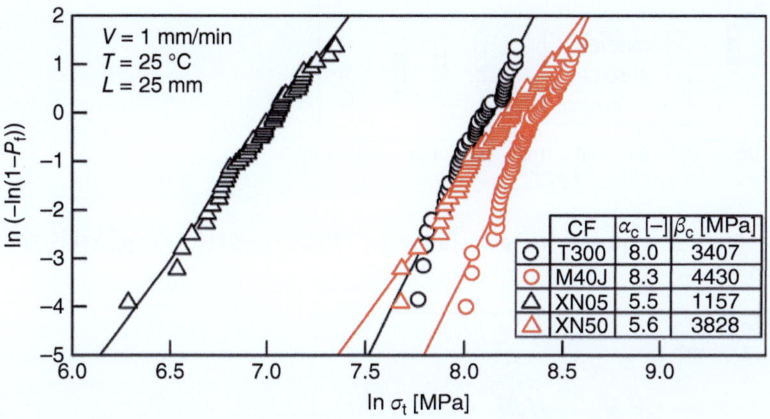

Figure 13.2 Weibull distributions of tensile strengths for carbon fiber of four kinds.

13.3.2 Static Tensile Strengths of CFRP Strands at Various Temperatures

Static tension tests for CFRP strands of four kinds were conducted at four or five temperatures of 25, 120, 135, 150, and 170 °C with cross-head speed of 2 mm/min. The tensile strength σ_s of CFRP strand was obtained using Eq. (7.7).

Figure 13.3 shows static strengths versus temperature for the CFRP strands of four kinds. The static strengths for T300/EP and XN05/EP decrease markedly with increasing temperature. This strength for M40J/EP decreases only slightly with increasing temperature. That for XN50/EP remains constant with increasing temperature.

The Weibull distributions for static strength at various temperatures are depicted in Figure 13.4 for the respective CFRP strands. In this figure, α_s is the shape parameter and β_s is the scale parameter of CFRP strand. Although the scale parameter decreases or keeps a constant value according to the temperature rise, the shape parameter maintains almost a constant value for all CFRP strands of four kinds to

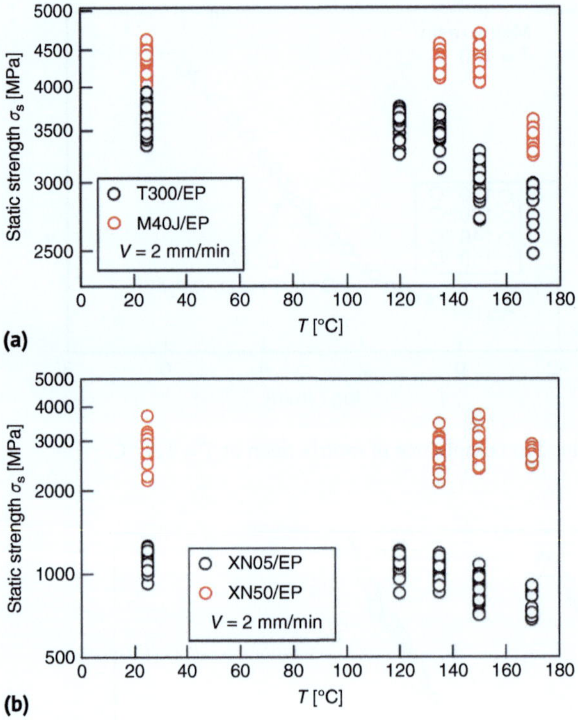

Figure 13.3 Static strengths versus temperature: (a) T300/EP and M40J/EP; (b) XN05/EP and XN50/EP.

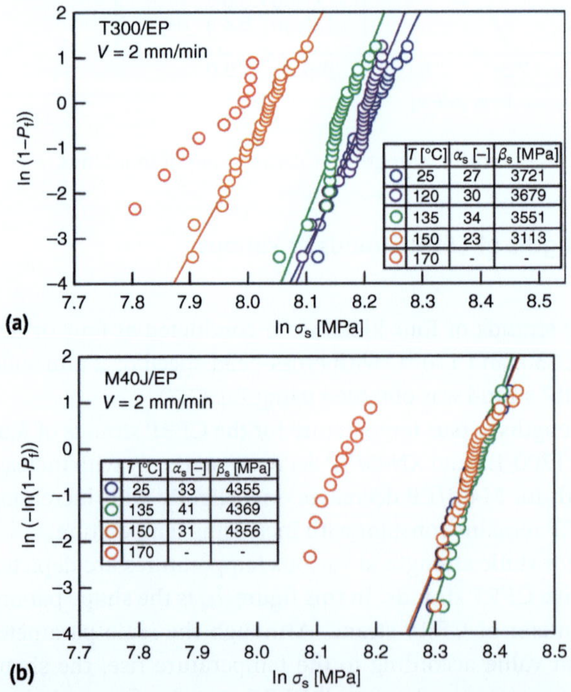

Figure 13.4 Weibull distributions of static tensile strength at various temperatures: (a) T300/EP, (b) M40J/EP, (c) XN05/EP, and (d) XN50/EP.

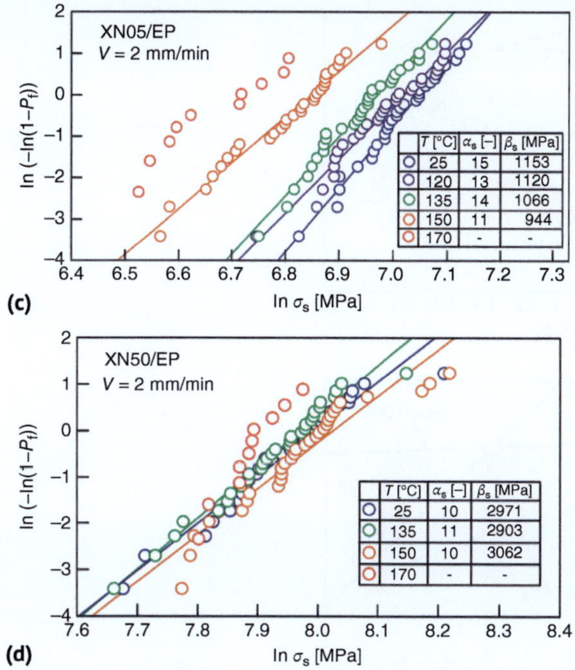

Figure 13.4 (Cont'd)

the temperature raise. Shape parameter α_s and the scale parameter β_s at the temperature $T = 25\,°C$ in these figures can be inferred as shape parameter α_s and scale parameter σ_0 of static strength at the reference temperature $T_0 = 25\,°C$ and the reference failure time $t_0 = 1\,\text{min}$ used in Eq. (7.2).

13.3.3 Static Tensile Strength of CFRP Strand Against Viscoelastic Compliance of Matrix Resin

Figure 13.5 shows the non-dimensional static strength σ_s/σ_0 of a CFRP strand against the non-dimensional viscoelastic compliance of matrix resin D^*/D_{c0} at the same time and temperature. The relation of σ_s/σ_0 against D^*/D_{c0} can be represented as one or two solid straight lines with the slope of n_R, which is the viscoelastic parameter in Eq. (7.2), as obtained from Christensen's model of viscoelastic crack kinetics. The dotted lines in these figures are straight lines with the slope of the viscoelastic parameters n_R obtained by substituting the shape parameter α_c of single carbon fiber strength into the following equation based on Rosen's model [2].

$$n_R = 1/(2\alpha_c) \tag{13.1}$$

The slopes of CFRP strands of four kinds in Figure 13.5 show characteristic behaviors. The slopes of T300/EP and XN05/EP coincide well with those obtained from the shape parameter α_c of single carbon fiber strength based on Rosen's model. The slopes of M40J/EP and XN50/EP are approximately zero in the wide range of compliance except for high compliance of the matrix resin.

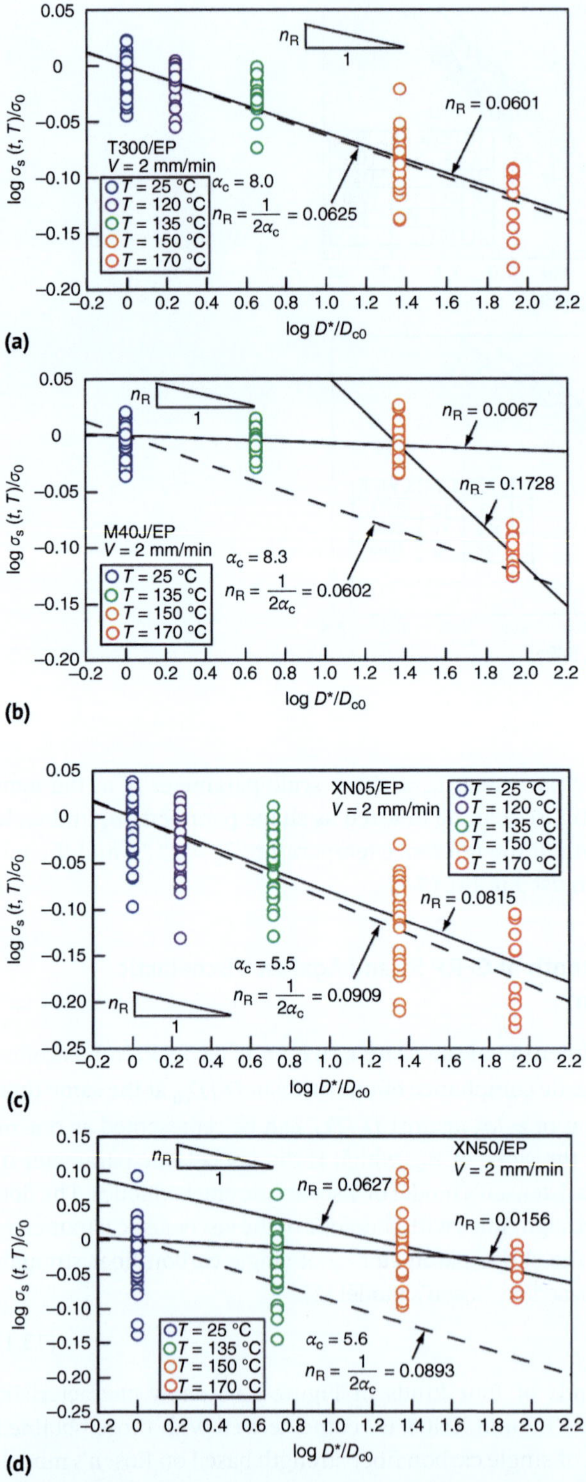

Figure 13.5 Static strength against viscoelastic compliance of the matrix resin: (a) T300/EP, (b) M40J/EP, (c) XN05/EP, and (d) XN50/EP.

13.3.4 Master Curves of Static Tensile Strength for Various CFRP Strands

All parameters in Eq. (7.2) for the master curve of static strength are determined for CFRP strands of four kinds as shown in Table 13.3: shape parameter α_s and scale parameter σ_0 of static strength at the reference temperature $T_0 = 25\,°\mathrm{C}$; reference failure time $t_0 = 1\,\mathrm{min}$; viscoelastic parameter n_R.

The master curves of statistical static strengths for CFRP strands are shown in Figure 13.6 with experimental data measured at various temperatures. Each master curve in this figure shows characteristic behavior. The master curves of T300/EP and XN05/EP show a smooth decreasing single curve with increasing failure time.

Table 13.3 Parameters of statistical static strengths for various CFRP strands.

CFRP strand	σ_0 (MPa)	α_s (–)	n_R (–)
T300/EP	3721	27	0.0601
M40J/EP	4355	33	0.0067 (low temperature) 0.1728 (high temperature)
XN05/EP	1153	15	0.0815
XN50/EP	2971	10	0.0156 (low temperature) 0.0627 (high temperature)

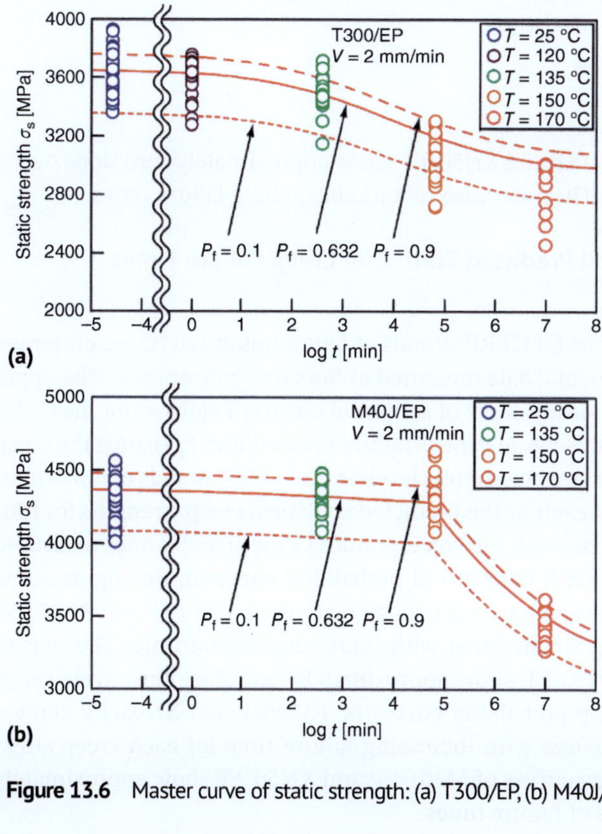

Figure 13.6 Master curve of static strength: (a) T300/EP, (b) M40J/EP, (c) XN05/EP, and (d) XN50/EP.

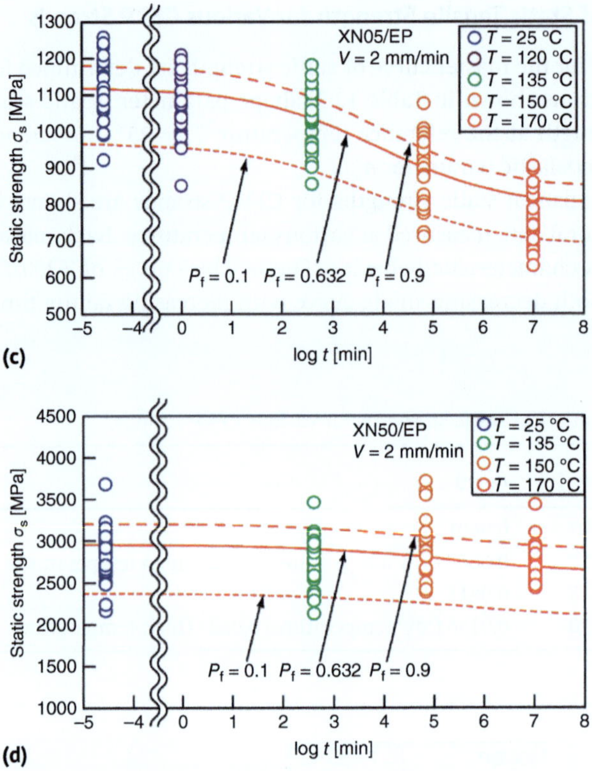

Figure 13.6 (Cont'd)

The master curves of M40 J/EP and XN50/EP show approximately zero slope over a wide range of failure time. They decrease remarkably at long failure times.

13.3.5 Experimental and Predicted Statistical Creep Failure Times for Various CFRP Strands

The statistical creep strengths for CFRP strands of four kinds at 120 °C are presented in Figure 13.7 with experimental data measured at the same temperature. The upper side of this figure portrays master curve of statistical creep strengths σ_c for the CFRP strands. The lower side of this figure depicts failure probabilities P_f against the creep failure time t at several constant creep stress levels σ_{c0} for CFRP strands of four kinds.

These figures clarify that each of the predicted statistical creep strengths for four kinds of CFRP strands agrees with the corresponding experimental ones measured directly using creep tests. Each master and probability curves in this figure show characteristic behavior. The creep strength master curves of T300/EP and XN05/EP show a smooth decreasing single curve with increasing failure time. The master curves of M40J/EP and XN50/EP show approximately zero slope for a wide range of failure times. The failure probability curves of T300/EP and XN05/EP show a smooth increasing single curve with increasing failure time for each creep stress level. The failure probability curves of M40J/EP and XN50/EP show approximately zero slope for a wide range of failure times.

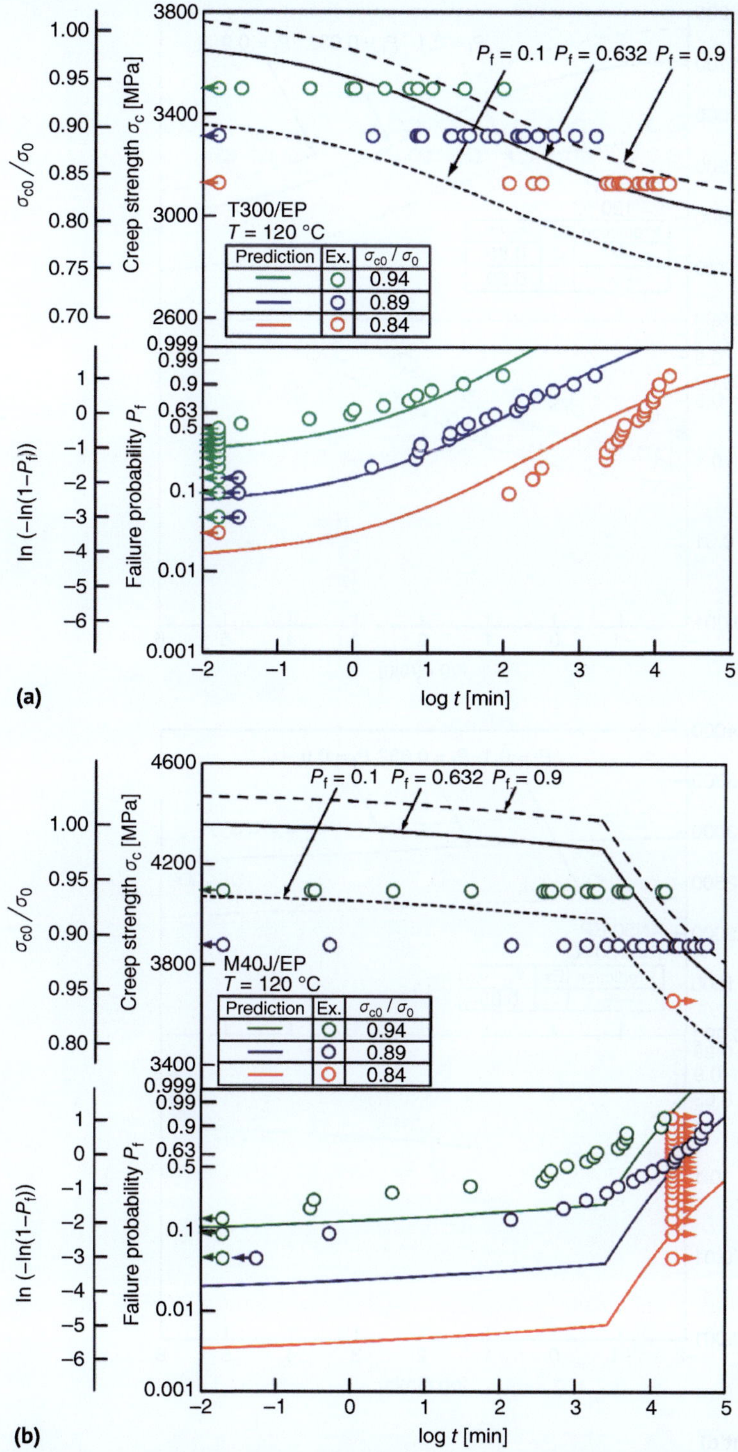

Figure 13.7 Experimental and predicted statistical creep failure time: (a) T300/EP, (b) M40J/EP, (c) XN05/EP, and (d) XN50/EP.

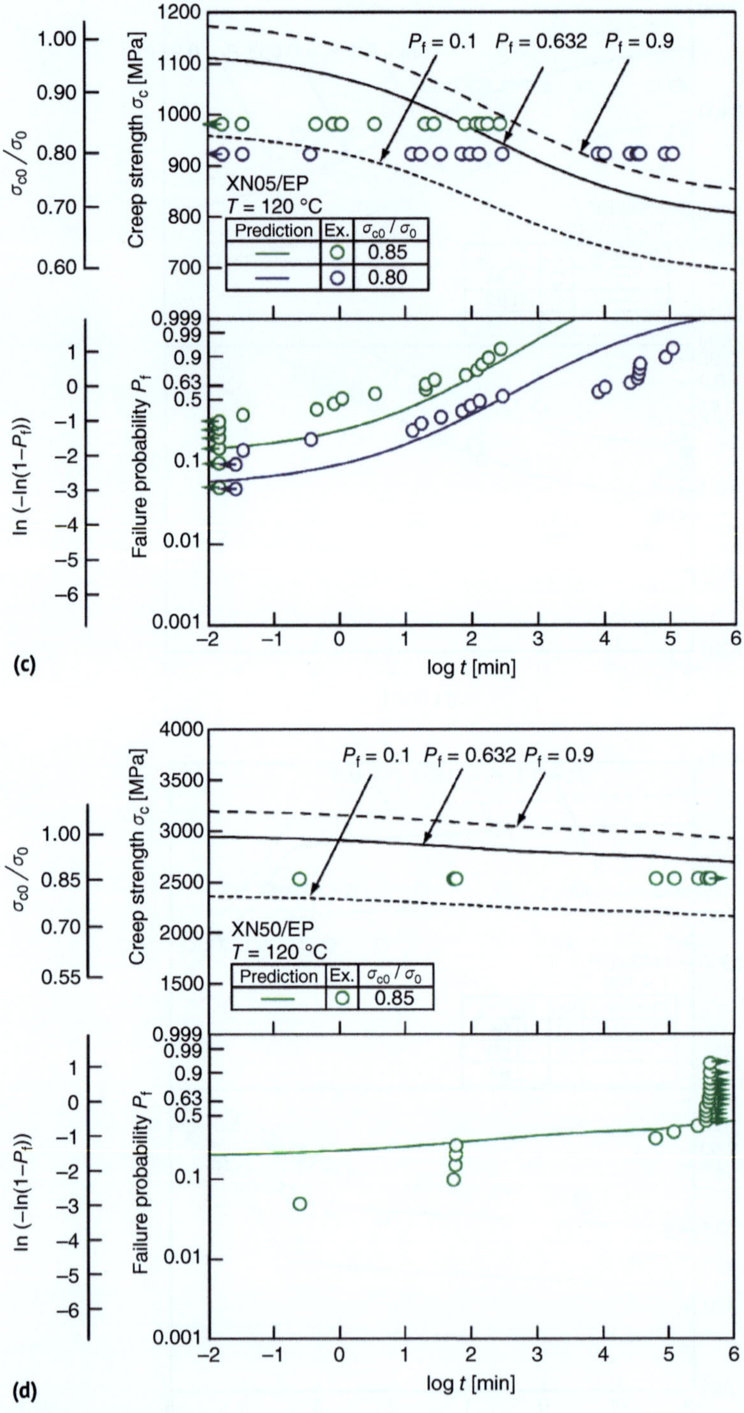

Figure 13.7 (Cont'd)

13.3.6 Fractographs Obtained After Static and Creep Tests

The viscoelastic parameter n_R of T300/EP and XN05/EP with low-modulus fibers coincide well with the values predicted from the shape parameter α_c of carbon fiber strength as shown in Figure 13.5a,c. The n_R of M40J/EP and XN50/EP with high-modulus fibers are approximately zero in a wide range of compliance of the matrix resin, and then n_R changes drastically at high temperature, as shown in Figure 13.5b,d. Figure 13.8 shows the fractographs after static tension tests for the four kinds of CFRP strands. The fracture mode for T300/EP and XN05/EP is the cumulative fracture at 25 and 170 °C. Two fracture modes exist for M40J/EP and XN50/EP, that is, the brittle fracture at 25 °C and the cumulative fracture at 170 °C.

Figure 13.9 shows the definition for depth of tensile fracture surface of CFRP strand. Figure 13.10 shows the depth of tensile fracture surface for CFRP strands.

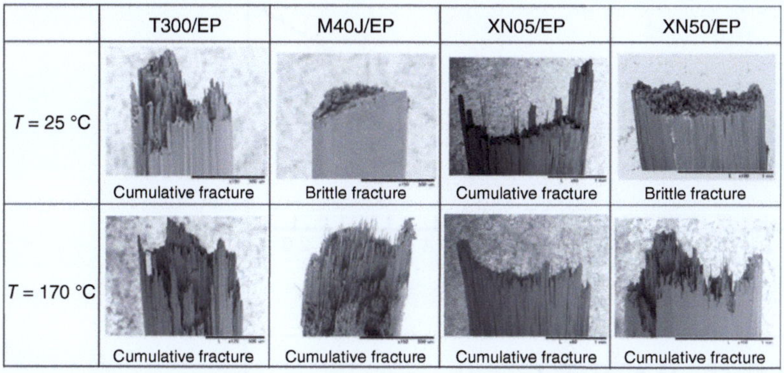

Figure 13.8 Fractographs after static tension tests for various CFRP strands. *Source*: Miyano et al. (2020)/with permission of Springer Nature.

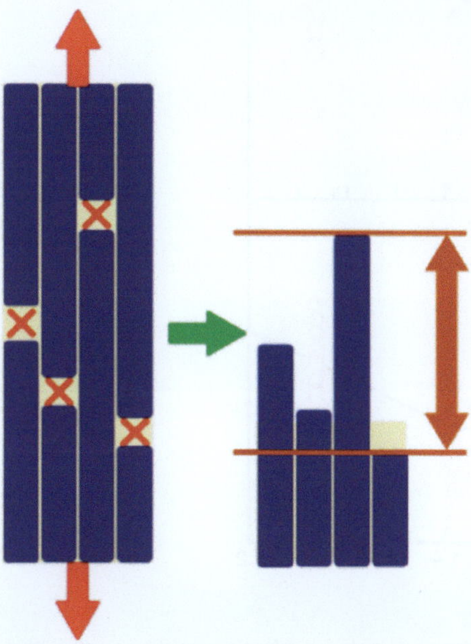

Figure 13.9 Definition of depth for tensile fracture surface for CFRP strands.

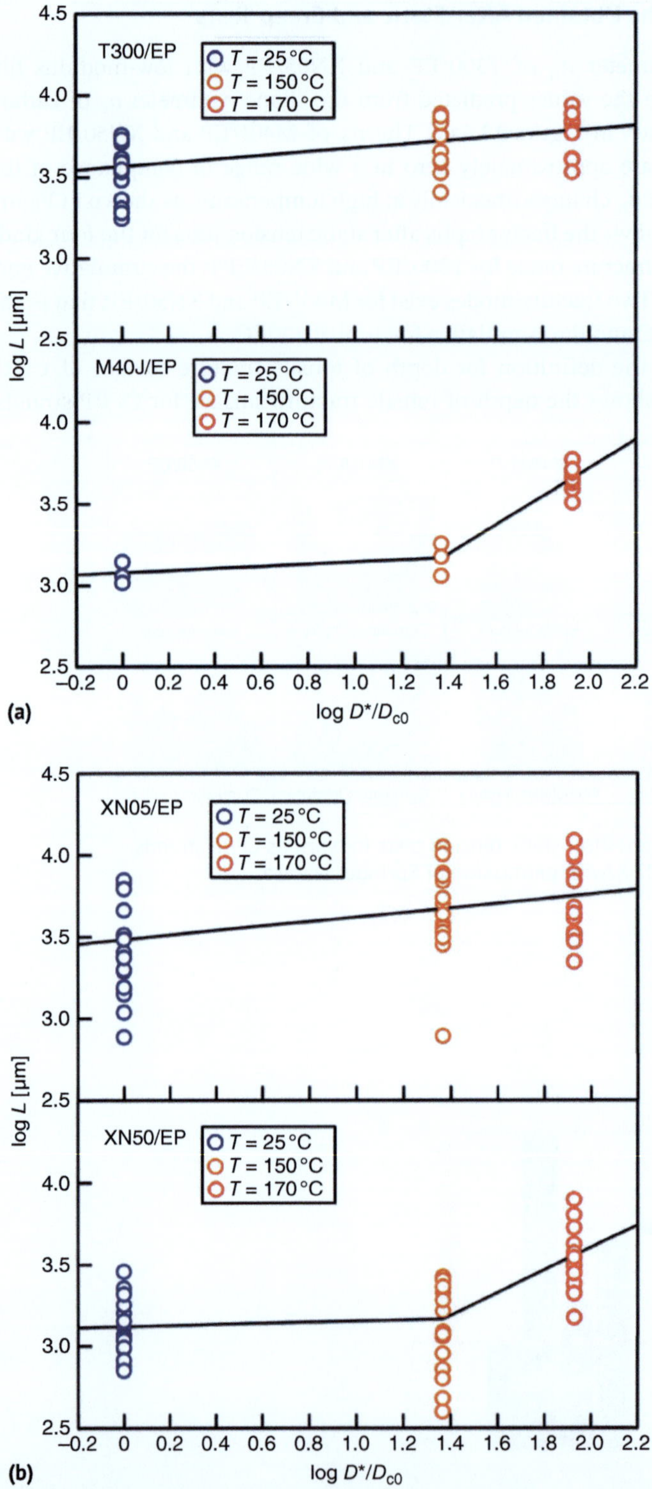

Figure 13.10 Depth of tensile fracture surface against viscoelastic compliance of matrix resin: (a) T300/EP and M40J/EP; (b) XN05/EP and XN50/EP.

These behaviors correspond respectively to the behaviors of the static strength against the viscoelastic compliance of matrix resin shown in Figure 13.5.

13.4 Conclusion

Formulation for the statistical time- and temperature-dependent static and creep strengths for CFRP based on the matrix resin viscoelasticity was proposed in Chapters 7 and 9. The validity of the formulation was clarified experimentally for tension loading along the longitudinal direction of unidirectional CFRP using our developed CFRP strand system.

Results clarified that the long-term creep failure time of unidirectional CFRP under tension loading can be predicted statistically and easily using static tests for CFRP strands at various temperatures. Furthermore, results clarified that two fracture modes exist in CFRP strands under static and creep loadings. Viscoelastic parameter n_R showing sensitivity to the matrix resin viscoelasticity changes drastically with the fracture mode.

References

1 Miyano, Y. and Nakada, M. (2020) Statistical long-term creep failure time of unidirectional CFRP. 529–549. In *Advances in Thick Section Composite and Sandwich Structures*, Edited by Sung W. Lee, Springer Nature Switzerland AG.

2 Nakada, M., Miyano, Y., Kinoshita, M., Koga, R., Okuya, T. and Muki, R. (2002) Time–temperature dependence of tensile strength of unidirectional CFRP. *Journal of Composite Materials*, **36**, 2567–2581.

Part 3

Integrated Accelerated Testing Methodology

Introduction

This part aims to propose an integrated accelerated testing methodology (Integrated ATM) that comprehensively evaluates the long-term life of carbon-fiber-reinforced polymers (CFRP) under static, fatigue, and creep loads and to introduce examples of its application.

In Chapter 14, we present an Integrated ATM formulation that uniformly expresses static, creep, and fatigue strengths of CFRP, and apply this method to the longitudinal tensile strength of unidirectional CFRP. Various applications for the Integrated ATM are introduced in Chapters 15–20.

In Chapter 15, the creep and fatigue lives of unidirectional CFRP laminates subjected to bending load in the longitudinal direction are evaluated. In Chapter 16, the influence of the mechanical properties of carbon fiber on statistical creep and fatigue lives of resin-impregnated carbon fiber strands (CFRP strands) as the tensile test specimens for unidirectional CFRP with various carbon fibers is studied. In Chapter 17, the influence of the mechanical properties of matrix resin on statistical creep and fatigue lives of CFRP strands with thermoplastic epoxy resin as matrix is studied. In Chapter 18, the statistical tensile and flexural creep and fatigue lives of unidirectional CFRP laminates with polypropylene as matrix are evaluated. In Chapter 19, the creep failure life of CFRP strands with heat-resistant epoxy resin as matrix exposed to high temperature under tension load is evaluated. In Chapter 20, the effect of annealing on statistical creep life of CFRP strands with thermoplastic epoxy resin as matrix is evaluated.

14

Integrated Accelerated Testing Methodology

14.1 Introduction

In this chapter, we propose a new integrated accelerated testing methodology (Integrated ATM) for evaluating the durability of carbon-fiber-reinforced polymers (CFRPs) and their structures. The unique feature of Integrated ATM is that it expresses the failure life under any temperature environment in a range of strain ratio R, from a constant load at $R = 1$ to a cyclic load at $R = 0$, in a single equation. First, we introduce the new formulation of Integrated ATM. Next, as an application example of Integrated ATM, we experimentally determine the important parameters that control the life of unidirectional CFRP under the tension load in the fiber direction. Finally, we consider the effect of these parameters on the life of unidirectional CFRP [1].

14.2 Formulation

There are many different failure modes in CFRP structures. The most important types of failure in CFRP structures are tensile failure and compressive failure of carbon fibers, as they become unable to maintain their structural functions. These failure modes can be realized through tension and compression tests of unidirectional CFRP. The stress–strain relationship, tensile strength, and compressive strength of carbon fiber alone are not time- or temperature-dependent. On the other hand, the tensile and compressive strengths of unidirectional CFRP, in which carbon fiber bundles are aligned in one direction and impregnated with resin, exhibit time, and temperature dependence due to the influence of the time- and temperature-dependent stress–strain relationship of the matrix resin, which is responsible for stress transmission between fibers. The shear lag model, typified by the Rosen model, is the failure mode for tensile strength of unidirectional CFRP, while the microbuckling model is the failure mode for compressive strength.

As shown in Figure 14.1, when a load is applied to a CFRP structure, it can be assumed that the deformation of the matrix resin filling the gaps between the carbon

Properties and Testing of Fiber-Reinforced Polymers, First Edition. Yasushi Miyano and Masayuki Nakada. © 2026 Wiley-VCH GmbH. All rights reserved, including rights for text and data mining and training of artificial intelligence technologies or similar technologies. Published 2026 by Wiley-VCH GmbH.

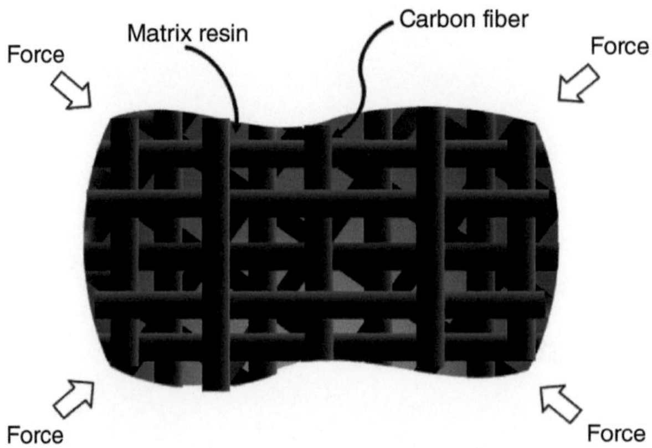

Figure 14.1 Mechanical model of CFRP.

fibers is completely restrained by the highly rigid carbon fibers. Therefore, the time- and temperature-dependent strength of a CFRP structure is thought to be controlled by the relaxation modulus of the matrix resin, not by its creep compliance.

The statistical time- and temperature-dependent strength of CFRP and its structures is formulated using the viscoelasticity of the matrix resin. The most important condition in this formulation is the modified time–temperature superposition principle (modified TTSP), which applies to the viscoelasticity of the matrix resin. The following two conditions are necessary for the formulation to be valid:

A. The failure probability is independent of temperature and loading history.
B. The time- and temperature-dependent strength is controlled by the viscoelasticity of the matrix resin.

14.2.1 Viscoelasticity of Matrix Resin

The time- and temperature-dependent stress–strain relationship of the matrix resin plays an important role in the static strength as well as creep and fatigue strengths of CFRP.

The evaluation method and formulation of the viscoelastic properties of matrix resins are described in detail in Section 2.4 of Part 1. Since its contents form the core of Integrated ATM, we will explain its details again.

An accurate relaxation modulus of the matrix resin over a wide time range at a reference temperature and the accurate time–temperature shift factor are the most important factors in predicting the life of CFRP. In this study, we propose a process to determine these properties using the following two types of tests.

The short-term relaxation modulus E_r of the matrix resin can be obtained by substituting the storage modulus E' at various frequencies f and temperatures T shown on the left side of Figure 14.2, measured using DMA, into the following equation [2]:

$$E_r(t,T) = E'(f,T),\ t = 1/(\pi^2 f) \tag{14.1}$$

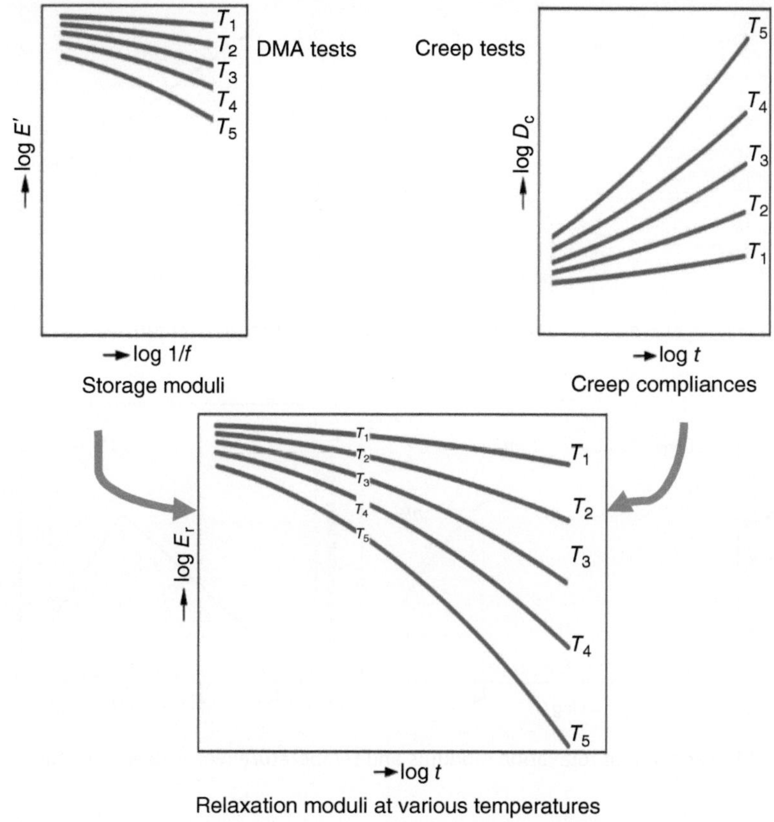

Figure 14.2 Relaxation moduli for a wide range of times.

The long-term relaxation modulus E_r of the matrix resin can be obtained by substituting the creep compliance D_c into the following equation at various elapsed times t and temperatures T shown on the right side of Figure 14.2, measured using a creep testing machine.

$$E_r(t,T) = \frac{a}{D_c(t,T)}, \quad a = \frac{1}{\Gamma(1+m)\Gamma(1-m)} \tag{14.2}$$

where Γ is the gamma function and parameter m is the slope of the curve that shows the relationship between log D_c and log t. Eq. (14.2) can be easily obtained from linear viscoelastic theory.

Combining the relaxation modulus E_r obtained using DMA over short term at various temperatures with E_r obtained over the long-term at various temperatures using a creep testing machine yields E_r values over a wide range of times at various temperatures, as shown in the bottom part of Figure 14.2.

The time–temperature shift factor a_{T_0} of the modified TTSP [3] can be determined by horizontally shifting the loss tangent tan δ at various temperatures T measured by DMA, as shown in Figure 14.3. The temperature shift factor b_{T_0} can be determined

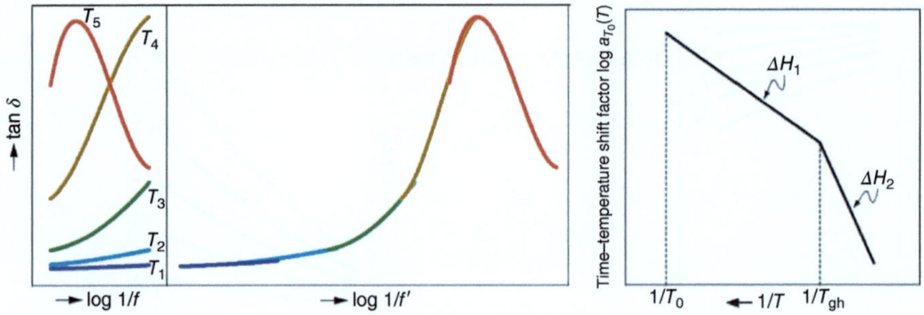

Figure 14.3 Master curve of loss tangent and time–temperature (horizontal) shift factor.

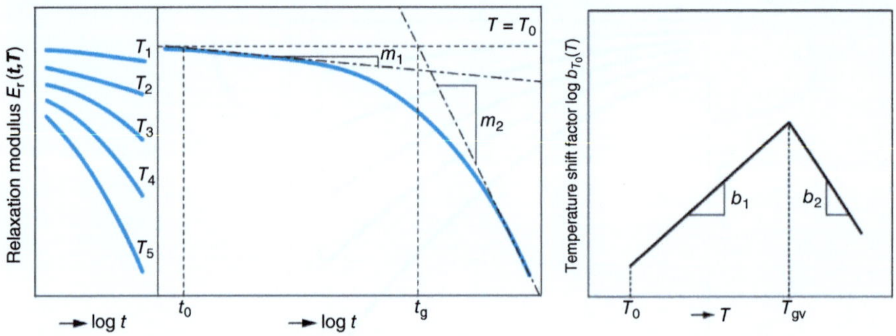

Figure 14.4 Master curve of relaxation modulus and temperature (vertical) shift factor.

by vertically shifting the relaxation modulus E_r at various temperatures T, as shown in Figure 14.4. These shift factors can be expressed by Eqs. (14.3) and (14.4).

$$\log a_{T_0}(T) = \frac{\Delta H_1}{2.303G}\left(\frac{1}{T}-\frac{1}{T_0}\right)H\left(T_{gh}-T\right) + \left[\frac{\Delta H_1}{2.303G}\left(\frac{1}{T_{gh}}-\frac{1}{T_0}\right) + \frac{\Delta H_2}{2.303G}\left(\frac{1}{T}-\frac{1}{T_{gh}}\right)\right]\left(1-H\left(T_{gh}-T\right)\right) \tag{14.3}$$

$$\log b_{T_0}(T) = b_1(T-T_0)H\left(T_{gv}-T\right) + \left[b_1\left(T_{gv}-T_0\right) + b_2\left(T-T_{gv}\right)\right]\left(1-H\left(T_{gv}-T\right)\right) \tag{14.4}$$

where ΔH_1 and ΔH_2 are the activation energies in the low and high temperature regions. Furthermore, H is the Heaviside step function. G is the gas constant. T_0 is the reference temperature. T_{gh} and T_{gv} are the glass transition temperatures at a_{T_0} and b_{T_0}, respectively. b_1 and b_2 are parameters in the vertical shift factor.

The master curve of the relaxation modulus E_r at the reference temperature T_0 can be formulated as Eq. (14.5), as shown on the left side of Figure 14.4.

$$\log E_r(t,T_0) = \log E_r(t_0,T_0) - \log\left[\left(\frac{t}{t_0}\right)^{m_1} + \left(\frac{t}{t_g}\right)^{m_2}\right] \tag{14.5}$$

where t_0 and t_g are the reference time and the glass transition time, respectively, and m_1 and m_2 are the slopes in the short and long-term domains, respectively.

The viscoelasticity of a matrix resin can be expressed by the aforementioned three equations and the 10 parameters contained in these equations.

14.2.2 General Formulation of CFRP Strength

The statistical strength σ^* of CFRP for any loading direction and type is generally defined as follows based on Christensen's viscoelastic crack kinetics theory [4, 5].

$$\log \sigma^*(t, f, T, R) = \log \sigma_0(t_0, T_0) + \frac{1}{\alpha}\log[-\ln(1 - P_f)] + n_R \log\left[\frac{E^*(t, f, T, R)}{E_r(t_0, T_0)}\right] \\ -F_f(t, f, T, R), \quad N_f = \frac{t}{f}, \quad N_f \geq \frac{1}{2} \tag{14.6}$$

Here, the symbols in Eq. (14.6) are defined as shown in the following table.

t:	Failure time
t_0:	Reference time
T:	Temperature
T_0:	Reference temperature
P_f:	Failure probability
f:	Frequency
N_f:	Number of cycles to failure
R:	Strain ratio
σ_0 and α:	Scale and shape parameters of static strength at t_0 and T_0
n_R:	Viscoelastic parameter determined by failure mode
E^*:	Viscoelastic modulus of matrix resin
E_r:	Relaxation modulus of matrix resin
F_f:	Fatigue degradation parameter

Figure 14.5 shows a graph of Eq. (14.6). The red letters in the figure represent the first term on the right-hand side of Eq. (14.6). The blue, green, and purple arrows represent the second, third, and fourth terms, respectively.

14.2.3 Formulation of Static and Creep Strengths

The statistical static strength σ_s at any constant strain rate and temperature is easily derived from the general formulation (14.6) as follows.

$$\log \sigma_s(t, T) = \log \sigma_0(t_0, T_0) + \frac{1}{\alpha}\log[-\ln(1 - P_f)] + n_R \log\left[\frac{E_s(t, T)}{E_r(t_0, T_0)}\right], \\ E_s(t, T) = E_r(t/2, T) \tag{14.7}$$

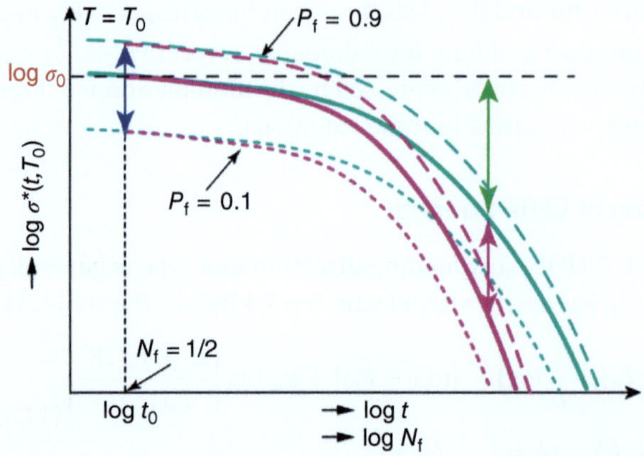

Figure 14.5 CFRP strength versus failure time and number of cycles to failure.

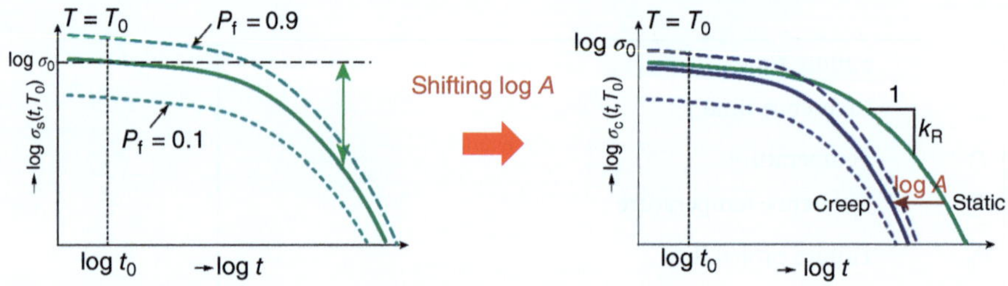

Figure 14.6 Determination of the master curve of statistical creep strength of CFRP.

All the parameters in Eq. (14.7) can be determined through static tests of CFRP at a constant strain rate and various temperatures, as well as viscoelastic tests of the matrix resin.

The statistical creep strength σ_c can be calculated by shifting the static strength by $\log A$ based on Christensen's viscoelastic crack kinetics theory [4, 5]. Therefore, the creep strength σ_c can be expressed by replacing the subscript "s" with "c" in Eq. (14.7). In this equation, E_c represents the viscoelastic modulus at a constant load. The master curve for creep strength can be obtained by shifting the master curve for static strength by $\log A$, as shown in Figure 14.6.

$$\log \sigma_c(t,T) = \log \sigma_0(t_0,T_0) + \frac{1}{\alpha}\log[-\ln(1-P_f)] + n_R \log\left[\frac{E_c(t,T)}{E_c(t_0,T_0)}\right], \qquad (14.8)$$
$$E_c(t,T) = E_s(At,T) \doteq E_r(At/2,T)$$

The master curve of viscoelastic modulus E_c at the reference temperature T_0 can be expressed as follows using Eqs. (14.5) and (14.8).

$$\log E_c(t,T_0) = \log E_r\left(\frac{A_1 t_0}{2}, T_0\right) - \log\left[\left(\frac{A_1 t}{2t_0}\right)^{m_1} + \left(\frac{A_2 t}{2t_g}\right)^{m_2}\right] \qquad (14.9)$$

When a constant tension load is applied to unidirectional CFRP, the shifting amount $\log A$ can be calculated from the slope k_R of the logarithmic static strength against the logarithmic time-to-failure using the following equation.

$$\log A_i = \log(1 + 1/k_{Ri}),\ k_{Ri} = m_i \times n_R,\ i = 1,2 \tag{14.10}$$

where m_i represents the slope of the logarithmic relaxation modulus of matrix resin versus logarithmic time curve. In the case of compressive loading of unidirectional CFRP, the failure mode is microbuckling rather than crack propagation, so the parameter A is 2 [6].

From the explanation of the formulation above, it can be seen that the long-term creep failure time of CFRP can be statistically predicted by substituting the relaxation modulus E_r of the matrix resin and the three parameters σ_0, α, and n_R of the static strength of CFRP into Eq. (14.8). These three parameters can be easily determined from the static tests of CFRP.

14.2.4 Formulation of Fatigue Strength

We proposed the following formulation of the statistical fatigue strength σ_f of CFRP using fatigue degradation parameter F_f based on the viscoelasticity of the matrix resin.

$$\begin{aligned}\log \sigma_f(t, f, T, R) = \log \sigma_0(t_0, T_0) + \frac{1}{\alpha}\log[-\ln(1 - P_f)] + n_R \log\left[\frac{E_f(t, f, T, R)}{E_r(t_0, T_0)}\right] \\ - F_f(t, f, T, R),\ t = N_f/f\end{aligned} \tag{14.11}$$

Here, R represents the strain ratio for the cyclic load shown in Figure 14.7, and E_f is the viscoelastic modulus of the matrix resin when a cyclic load is applied to the CFRP.

As shown in Figure 14.8, the fatigue strength can be expressed in two parts. One is the reduction in strength due to the viscoelasticity of the matrix resin. The other is the reduction in nondimensional fatigue strength due to the energy loss of the matrix resin. The amount of strength reduction σ_f^* due to the viscoelasticity of the matrix resin can be calculated as follows.

$$\begin{aligned}\log \sigma_f^*(t, f, T, R) = \log \sigma_0(t_0, T_0) + n_R \log\left[\frac{E_f(t, f, T, R)}{E_f(t_0, T_0)}\right], \\ E_f(t, f, T, R) = \frac{1}{2}\left[(1 - R)E_r\left(1/(\pi^2 f), T\right) + (1 + R)E_r(At/2, T)\right], t = N_f/f\end{aligned} \tag{14.12}$$

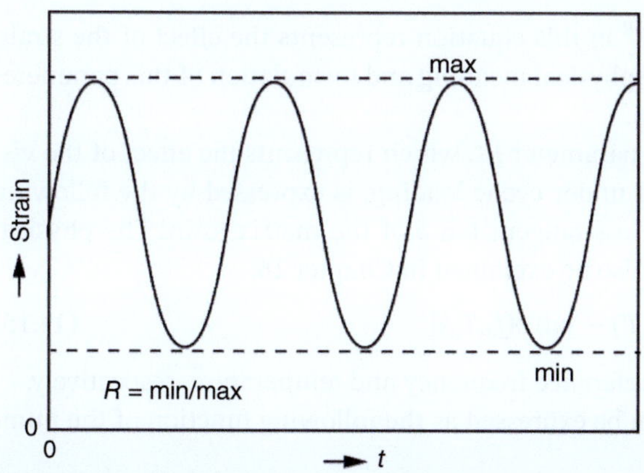

Figure 14.7 Explanation of strain ratio.

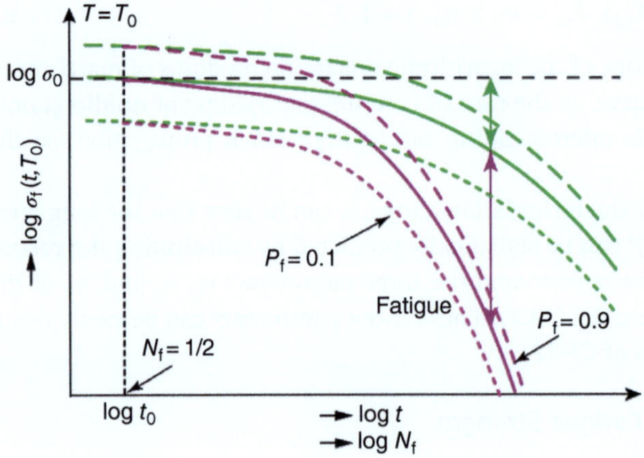

Figure 14.8 Strength reduction of CFRP caused by the viscoelasticity of the matrix resin and cyclic load.

All parameters of σ_f^* are determined by static tests of CFRP at different temperatures and the viscoelasticity of the matrix resin.

The nondimensional fatigue strength S_f, which includes variation, is expressed by the following equation.

$$\log S_f\ (f,T,R,N_f) = \frac{1}{\alpha}\log[-\ln(1-P_f)] - F_f\ (f,T,R,N_f)$$
$$= \log\frac{\sigma_f(f,T,R,N_f)}{\sigma_0(t_0,T_0)} - n_R \log\left[\frac{E_f\ (f,T,R,N_f)}{E_r(t_0,T_0)}\right],\quad N_f = tf \tag{14.13}$$

The nondimensional fatigue strength S_f can be obtained by substituting the fatigue strength σ_f and the number of cycles to failure N_f measured at each frequency f and temperature T into the right-hand side of Eq. (14.13).

The fatigue degradation parameter F_f at the center of Eq. (14.13) is expressed as follows.

$$F_f\ (f,T,R,N_f) = F_f^R F_f^*(f,T)F_f'(N_f) \tag{14.14}$$

The strain ratio parameter F_f^R in this equation represents the effect of the strain ratio due to cyclic loading. The physical meaning and formulation of this parameter will be explained in Chapter 16.

The frequency–temperature parameter F_f^*, which represents the effect of the viscoelasticity of the matrix resin under cyclic loading, is expressed by the following equation, which includes the loss tangent tan δ of the matrix resin. The physical meaning of this equation will also be explained in Chapter 16.

$$F_f^*(f,T) = 1 + a_f[\tan\delta(f,T) - \tan\delta(f_0,T_0)] \tag{14.15}$$

where f_0 and T_0 represent the reference frequency and temperature, respectively.

The fatigue parameter F_f' can be expressed as the following function of the number of cycles to failure N_f.

$$F_f'(N_f) = \sum_{i=1}^{n} n_{fi} \log(2N_f)^i, \ N_f \geq 1/2 \tag{14.16}$$

Here, the fatigue strength at $N_f = 1/2$ is equal to the static strength when the time to failure t is equal to $1/(2f)$.

Furthermore, S_{f0} defined as the following equation is the master curve of non-dimensional fatigue strength versus number of cycles to failure ($S_{f0} - N_f$ master curve) at the reference conditions of $f = f_0$, $T = T_0$, and $R = 0$, and S_{f0} statistically generates one curve for different frequencies and temperatures.

$$\log S_{f0}(N_f) = \log S_f + F_f'(N_f)\left[F_f^R F_f^*(f,T) - 1\right] \tag{14.17}$$

The modified TTSP for the viscoelasticity of the matrix resin is expected to hold for the frequency–temperature fatigue parameter $F_f^*(f,T)$.

The coefficient n_{fi} in Eq. (14.16) is determined by the fatigue strength-number of cycles to failure (S–N) curve measured at a reference frequency f_0 and a reference temperature T_0. The coefficient a_f in Eq. (14.15) is also determined by fatigue tests at various frequencies and temperatures.

Finally, we emphasize that the long-term statistical strength of CFRP at any frequency f, temperature T, and strain ratio R can be estimated by experimentally determining the viscoelastic parameters of the matrix resin and the static and fatigue strength parameters of the CFRP.

14.3 Application of Integrated ATM

Integrated ATM described in the previous section is applied to the statistical long-term prediction of creep and fatigue strengths of resin-impregnated carbon fiber strands (CFRP strands), which are unidirectional CFRPs.

14.3.1 CFRP Strand and Testing Method

The carbon fiber used in this study was a high-strength PAN-based carbon fiber (T300-3000; manufactured by Toray Industries, Inc.). The mechanical properties listed in the catalog are shown in Table 14.1.

CFRP strands, which were composites of carbon fiber and general-purpose epoxy resin (jER828; manufactured by Mitsubishi Chemical Corp.), were molded using a filament winding system [7] developed by the authors. The composition of the epoxy resin and the curing conditions for the CFRP strands are shown in Table 14.2. The gauge length of the CFRP strands was approximately 200 mm, as shown in

Table 14.1 Carbon fiber and the mechanical properties.

Carbon fiber	Density (g/cm^3)	Tex (g/km)	Elastic modulus (GPa)	Tensile strength (MPa)
T300-3000	1.76	198	230	3530

Table 14.2 Composition and cure schedule of CFRP strand.

CFRP strand	Carbon fiber	Composition of resin (weight ratio)	Cure schedule
T300/EP	T300-3000	Epoxy: jER828 (100) Hardener: MHAC-P (103.6) Cure accelerator: 2E4MZ (1)	70 °C × 12 h + 150 °C × 4 h + 190 °C × 2 h

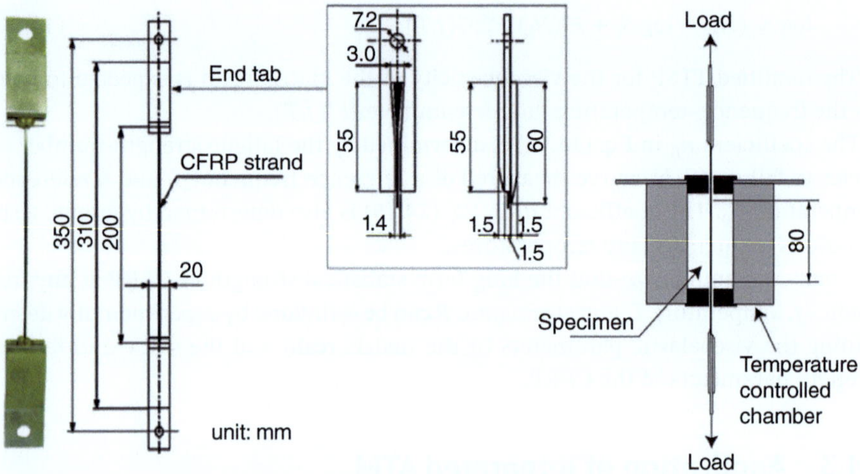

Figure 14.9 Specimen configuration and temperature-controlled chamber.

Figure 14.9. A mechanism for gripping the CFRP strands was developed to perform static, creep, and fatigue tests at various temperatures.

The glass transition temperature of the epoxy resin, $T_g = 160\,°C$, was determined from the peak value of the loss tangent at 1 Hz using a DMA testing machine. The fiber volume fraction of the CFRP strand, $V_f = 55\%$, was determined from the weight of the CFRP strand. These tests were performed using the original static, creep, and fatigue testing machines equipped with a maximum load of 1 kN with constant temperature chambers ranging from 25 °C to 150 °C.

The CFRP strand tensile strength σ_s was obtained using the following equation.

$$\sigma_s = \frac{W_{max}}{t_e}\rho \tag{14.18}$$

where W_{max} is the maximum load [N], ρ and t_e are the density of the carbon fiber (kg/m^3) and the tex of the carbon fiber strand (g/1000 m), respectively.

14.3.2 Master Curve of Relaxation Modulus of Matrix Resin

Based on the above modified TTSP, the storage modulus and loss tangent versus frequency measured at different temperatures using DMA testing machine, and creep compliance versus time measured at different temperatures using a creep testing

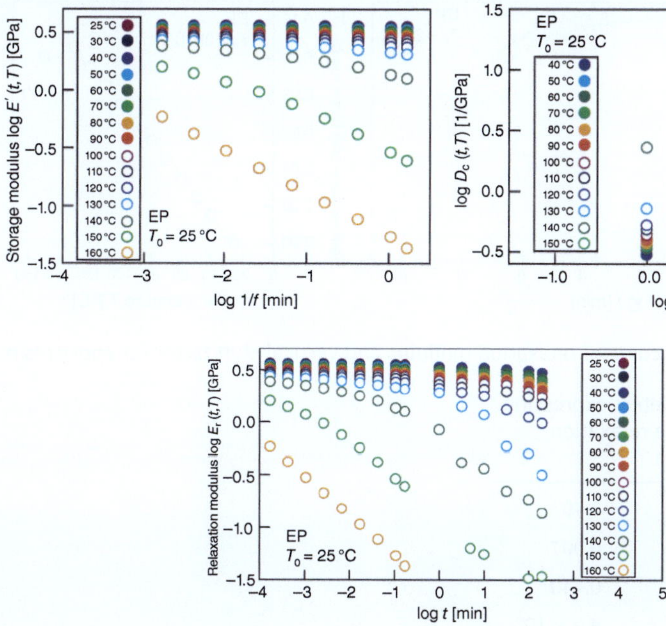

Figure 14.10 Relaxation moduli for a wide range of times for epoxy resin.

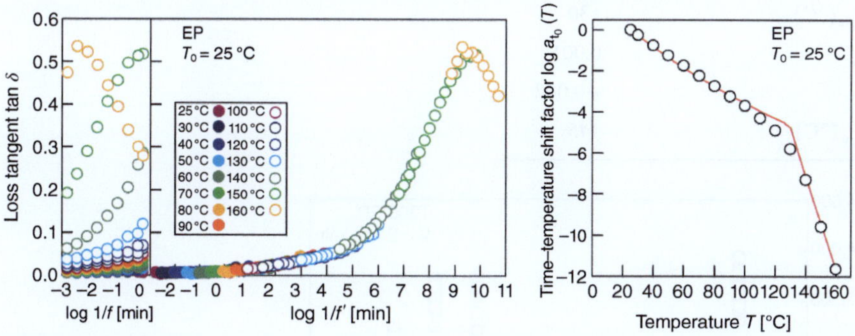

Figure 14.11 Master curve of loss tangent and horizontal shift factor for epoxy resin.

machine, are used to determine the master curves of relaxation modulus and time–temperature (horizontal) and temperature (vertical) shift factors.

Figure 14.10 shows the relaxation modulus of epoxy resin over a wide time range at various temperatures, obtained from the results of DMA and creep tests using the procedure in Figure 14.2. Figure 14.11 shows the master curves for the loss tangent and horizontal shift factor of epoxy resin measured from DMA tests. Figure 14.12 shows the master curves for the relaxation modulus and vertical shift factor of epoxy resin.

The master curve for the relaxation modulus of epoxy resin at a reference time of 1 minute and a reference temperature of 25 °C is expressed by Eqs. (14.3–14.5). All parameters in the equations are shown in Table 14.3.

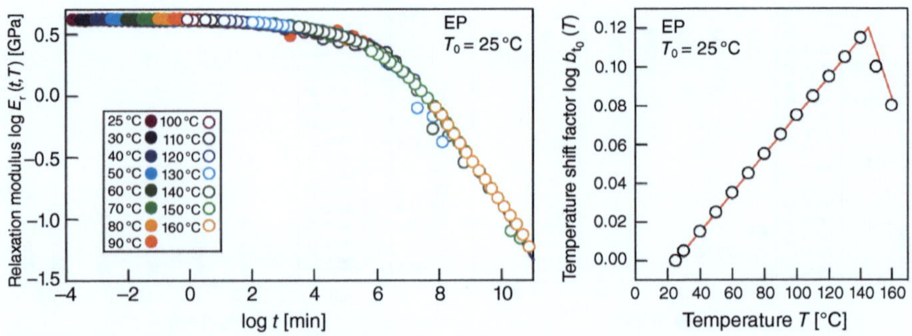

Figure 14.12 Master curve of relaxation modulus and vertical shift factor for epoxy resin.

Table 14.3 All parameters expressing the master curve of the relaxation modulus of epoxy resin.

$E_r(t_0, T_0)$ [GPa]	3.40
m_1	0.007
m_2	0.440
t_g(min)	4.0×10^6
ΔH_1 (kJ/mol)	102
ΔH_2 (kJ/mol)	770
T_{gh} (°C)	130
b_1	0.001
b_2	−0.002
T_{gv} (°C)	145

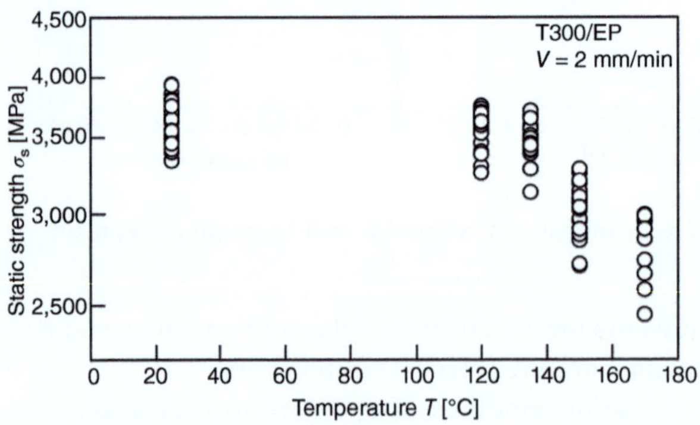

Figure 14.13 Static strengths of CFRP strand against temperature.

14.3.3 Static Strength of CFRP Strand

Static tensile tests on CFRP strands were conducted at five temperatures 25, 120, 135, 150, and 170 °C, with a crosshead speed of 2 mm/min. Twenty specimens were tested at each temperature. Figure 14.13 shows the relationship between the static strength of CFRP strands and temperature. The static strength of CFRP strands decreases significantly with increasing temperature.

Figure 14.14 shows the Weibull distribution for the static strength of CFRP strand at each temperature. In this figure, α_s and β_s show the shape parameter and scale parameter, respectively, of the static strength of CFRP strand. The scale parameter decreases with increasing temperature, but the shape parameter maintains a nearly constant value for the static strength of CFRP strand with increasing temperature. The shape parameter α_s and scale parameter β_s in this figure at temperature $T = 25\,°\text{C}$ can be used as the shape parameter α and scale parameter σ_0 of the static strength at the reference temperature $T_0 = 25\,°\text{C}$ and the reference time to failure $t_0 = 1$ min used in Eq. (14.7).

Figure 14.15 shows the relationship between the non-dimensional static strength σ_s/σ_0 of a CFRP strand and the non-dimensional viscoelastic coefficient E_s^*/E_{r0} of the matrix resin under the same temperature and time conditions. The relationship between σ_s/σ_0 and E_s^*/E_{r0} can be expressed by a single solid line with a slope of the viscoelastic parameter n_R in Eq. (14.2). The dotted line in the figure is a straight line with the slope of the viscoelastic parameter n_R obtained by substituting the shape parameter α_c of the single carbon fiber strength into the equation ($n_R = 1/(2\alpha_i)$) based on the Rosen model [8]. The two straight lines match well with each other.

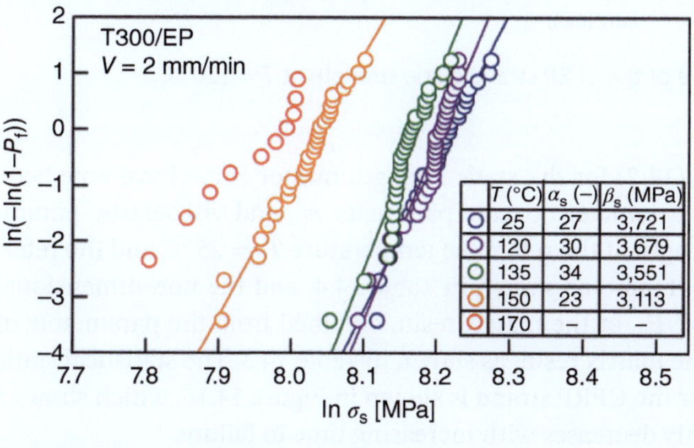

Figure 14.14 Weibull distributions of the static strength of a CFRP strand at various temperatures.

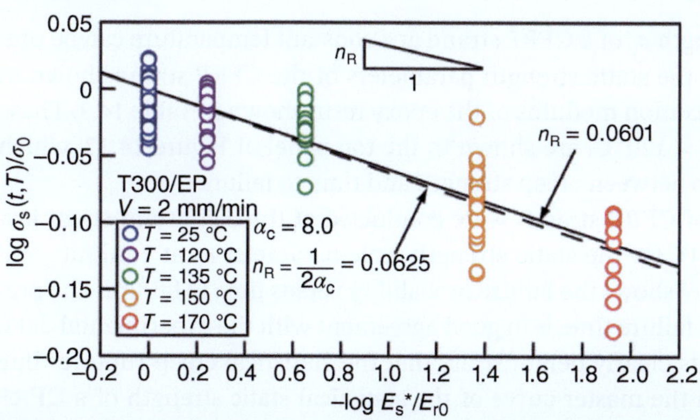

Figure 14.15 Static strength of CFRP strands against viscoelastic modulus of the matrix resin.

Table 14.4 Parameters of statistically determined tensile static strength for CFRP strand.

σ_0(MPa)	α	n_R
3721	27	0.0601

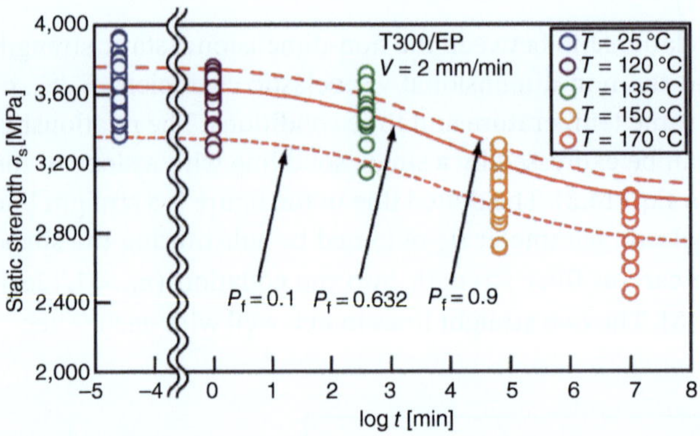

Figure 14.16 Master curve of the CFRP strand static strength at $T = 120$ °C.

All parameters of Eq. (14.7) for the static strength master curve have now been determined. The shape parameter α, scale parameter σ_0, and viscoelastic parameter n_R for the static strength at the reference temperature $T_0 = 25$ °C and the reference time to failure $t_0 = 1$ min, as shown in Table 14.4, and the non-dimensional viscoelastic coefficient E_s^*/E_{r0} of the matrix resin, obtained from the parameters of relaxation modulus of the matrix resin, as shown in Table 14.3. The statistical static strength master curve for the CFRP strand is shown in Figure 14.16, which shows a single curve that smoothly decreases with increasing time to failure.

14.3.4 Creep Strength of CFRP Strand

The tensile creep strength σ_c of a CFRP strand at a constant temperature can be predicted by substituting the static strength parameters of the CFRP strand shown in Table 14.4 and the relaxation modulus of the epoxy resin shown in Table 14.3. These results, predicted at $T = 120$ °C, are shown in the top panel of Figure 14.17, which shows the relationship between creep strength and time to failure.

Creep failure tests of CFRP strands were conducted at three constant stress levels σ_c of 94, 89, and 84% for the static strength scale parameter σ_0 at 1 minute and 25 °C. The bottom panel shows the failure probability versus time to failure. The predicted statistical creep failure time is in good agreement with the experimental data. Therefore, these results clearly demonstrate that the statistical creep rupture time can be predicted from the master curve of the statistical static strength of a CFRP strand and the relaxation modulus of the matrix resin at various temperatures.

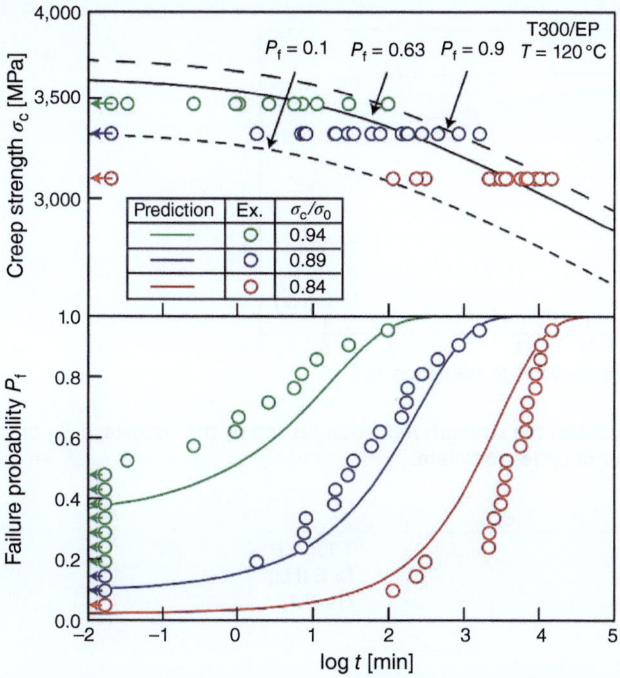

Figure 14.17 Creep strength and failure probability against failure time at $T = 120$ °C.

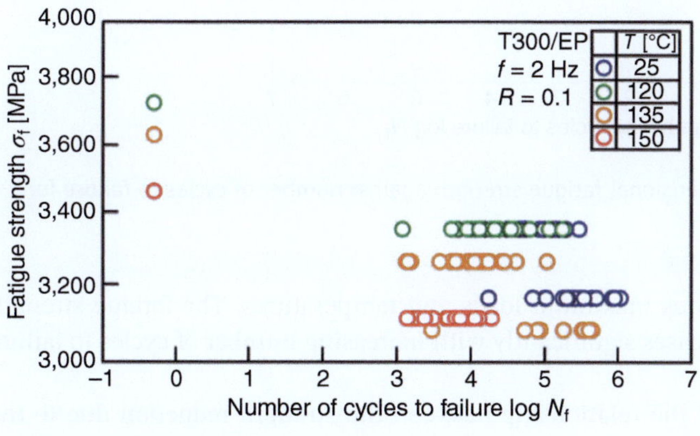

Figure 14.18 Fatigue strengths of CFRP strands against the number of cycles to failure.

14.3.5 Fatigue Strength of CFRP Strand

Fatigue tensile tests of CFRP strands were conducted at four temperatures 25, 120, 135, and 150 °C, with a frequency of $f = 2$ Hz and a strain ratio of minimum strain to maximum strain $R = 0.1$. The tensile fatigue strength σ_f of the CFRP strands was obtained using Eq. (14.18). The tensile fatigue stress levels were selected as 90% and 85% of the static strength scale parameter σ_{f0} at time $t = 1/(2f)$ and each temperature T. The number of cycles to failure N_f was measured throughout the test. Figure 14.18 shows the relationship between fatigue strength and number of cycles to failure for

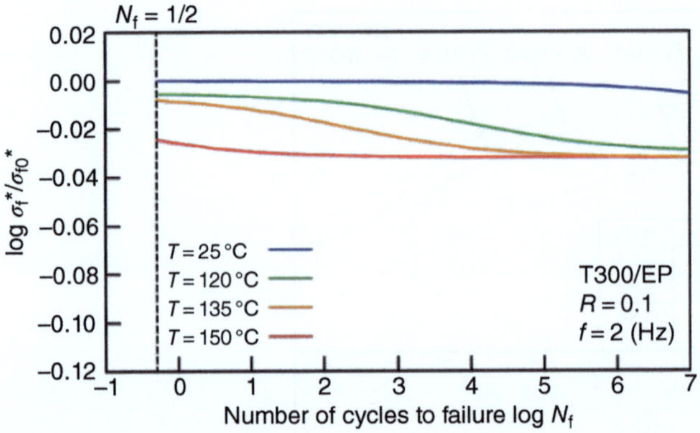

Figure 14.19 Relations between the strength reduction caused by the viscoelasticity of matrix resin and the number of cycles to failure.

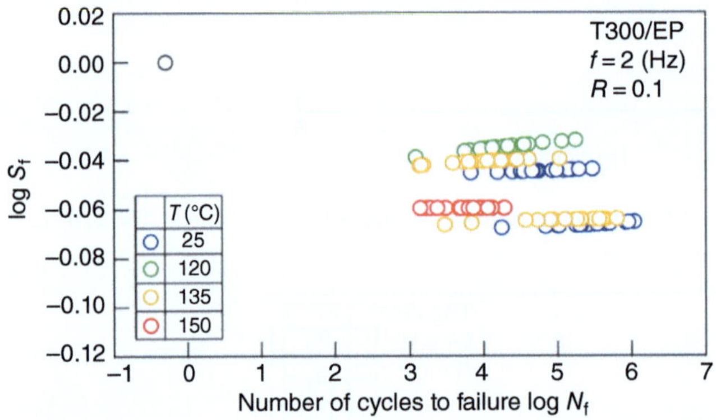

Figure 14.20 Non-dimensional fatigue strength against number of cycles to failure for CFRP strands.

CFRP strands at various maximum loads and temperatures. The fatigue strength of CFRP strands decreases significantly with increasing number of cycles to failure and temperature.

Figure 14.19 shows the relationship between the strength reduction due to the viscoelasticity σ_f^* of the matrix resin and the number of cycles to failure N_f at various temperatures under the conditions of $f = 2$ Hz and $R = 0.1$. These relationships were obtained by substituting the parameters in Tables 14.3 and 14.4 into Eq. (14.12).

Figure 14.20 shows the relationship between the non-dimensional fatigue strength S_f and the number of cycles to failure N_f at each temperature under the conditions of $f = 2$ Hz and $R = 0.1$. These relationships were obtained by substituting the measured values of the fatigue strength σ_f of the CFRP strand, the number of cycles to failure N_f and the viscoelastic parameters of the matrix resin into the right-hand side of Eq. (14.13). This relationship does not form a single curve, but is temperature-dependent. A single curve for the non-dimensional fatigue strength S_{f0} given by Eq. (14.17) can be obtained as shown in Figure 14.22 by determining the parameter a_f as the slope of Figure 14.21. The parameter a_f is listed in Table 14.5.

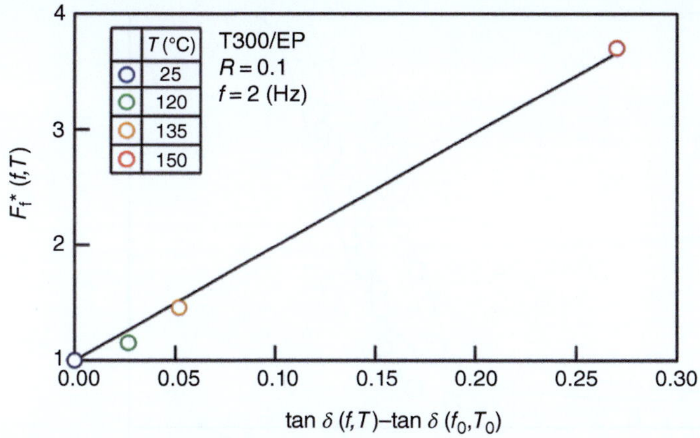

Figure 14.21 Relationship between frequency–temperature parameter against the loss tangent of matrix resin.

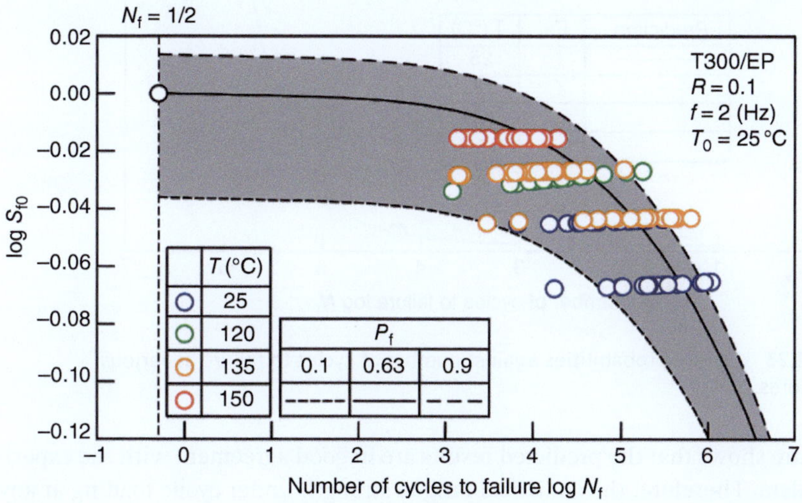

Figure 14.22 Master curve of non-dimensional fatigue strength against number of cycles to failure (S_{f0} – N_f master curve) for CFRP strands.

Table 14.5 Parameters of non-dimensional fatigue strength for CFRP strand.

a_f	$n_{f1}(\times 10^{-4})$	$n_{f2}(\times 10^{-4})$	$n_{f3}(\times 10^{-4})$	$n_{f4}(\times 10^{-4})$
3.7	7.79	2.69	−3.13	1.08

The curve in Figure 14.22 can be defined as the S_{f0} – N_f master curve for a CFRP strand at a reference temperature of $T_0 = 25\,°C$. This curve is expressed by the polynomial function in Eq. (14.16). These parameters $n_{fi}, i = 1, 2, 3, 4$, are listed in Table 14.5. Figure 14.23 also shows the failure probability P_f and number of cycles to failure N_f at various temperatures at $f = 2$ Hz for $\sigma_f/\sigma_{f0} = 0.90$ and 0.85.

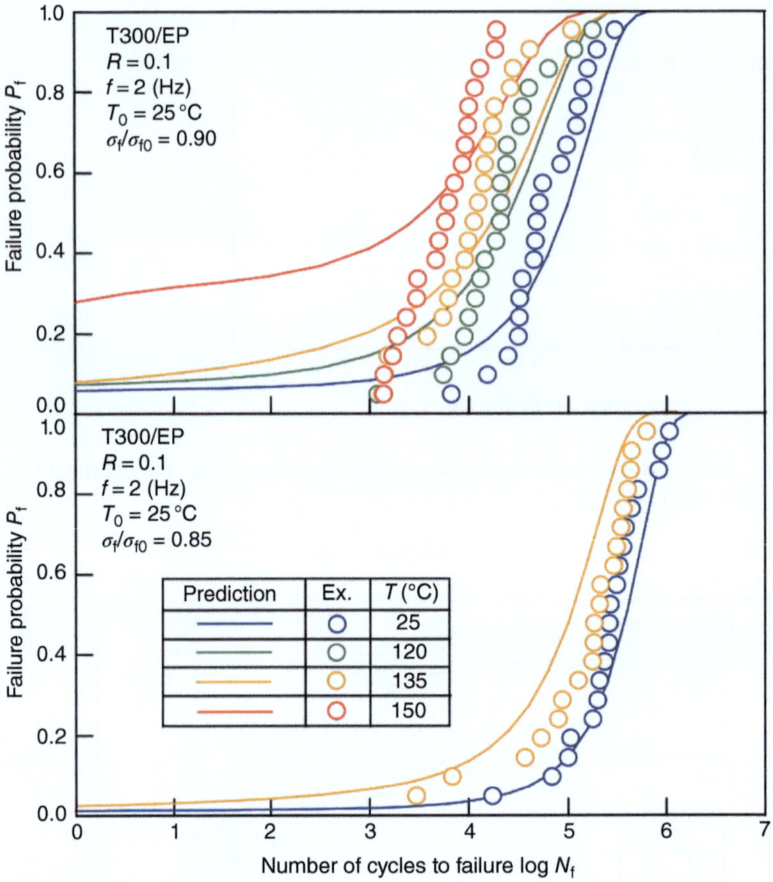

Figure 14.23 Failure probabilities against number of cycles to failure at various temperatures.

This figure shows that the predicted results are in good agreement with the experimental data. Therefore, the statistical fatigue strength under cyclic loading at any time, temperature, frequency, and strain ratio can be clearly predicted from the relaxation modulus of matrix resin listed in Table 14.3 and the statistical static and fatigue strength parameters of CFRP strand listed in Tables 14.4 and 14.5.

14.4 Statistical Long-term Life Prediction of CFRP Strand

Figure 14.24 shows the relationship between the relaxation modulus of matrix resin and the elapsed time, as well as the relationship between the creep and fatigue strengths and the time to failure at temperatures $T = 25\,°C$ and $60\,°C$. The fatigue strength conditions are a frequency of $f = 1$ cycle/min and a strain ratio of $R = 0.1$.

As shown in the upper part of Figure 14.24, the relaxation modulus of the matrix resin decreases significantly over a long period of 10 years even at room temperature.

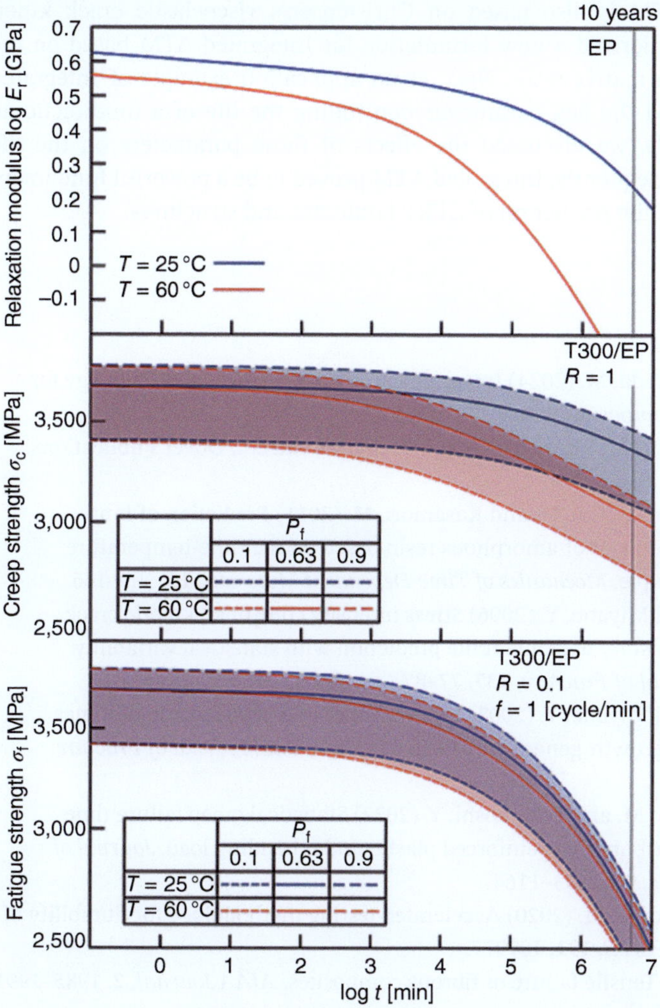

Figure 14.24 Long-term predictions of relaxation modulus of matrix resin and tensile creep and fatigue strengths of CFRP strands.

This decrease is accelerated by increasing temperature. This tendency of the matrix resin is clearly evident in the tensile creep strength of CFRP strand, but is only slightly evident in the fatigue strength.

14.5 Conclusion

In this chapter, we proposed an Integrated ATM for predicting the durability of CFRPs and structures. The unique feature of Integrated ATM is that it can express the life of CFRPs and structures under any environmental temperature and any strain ratio R from constant load ($R = 1$) to cyclic load ($R = 0$) using a single formulation incorporating multiple scientifically based parameters. Like conventional

ATM, Integrated ATM is also based on Christensen's viscoelastic crack kinetics theory. First, we derived a new formulation for Integrated ATM based on the viscoelasticity of the matrix resin. Next, as an application example of Integrated ATM, we determined the key parameters controlling the life of a unidirectional CFRP strand. Finally, we discussed the effects of these parameters on the life of CFRP strand. Consequently, Integrated ATM proved to be a powerful fundamental tool for long-term life prediction of CFRP laminates and structures.

References

1 Miyano, Y. and Nakada, M. (2024) Integrated accelerated testing methodology for CFRP durability. *Composites Part B*, **281**, 111527.

2 Christensen, R.M. (1982) *Theory of Viscoelasticity*, second ed., Dover Publications, Inc., Mineola, New York.

3 Nakada, M., Miyano, Y., Cai, H. and Kasamori, M. (2011) Prediction of long-term viscoelastic behavior of amorphous resin based on the time-temperature superposition principle. *Mechanics of Time-Dependent Materials*, **15**, 309–146.

4 Christensen, R. and Miyano, Y. (2006) Stress intensity controlled kinetic crack growth and stress history dependent life prediction with statistical variability. *International Journal of Fracture*, **137**, 77–87.

5 Christensen, R. and Miyano, Y. (2007) Deterministic and probabilistic lifetimes from kinetic crack growth-generalized forms. *International Journal of Fracture*, **143**, 35–39.

6 Miyano, Y., Nakada, M. and Yoshikoshi, Y. (2022) Statistical creep failure time of unidirectional carbon fiber reinforced plastic under bending load. *Journal of Composite Materials*, **56**, 1153–1164.

7 Miyano, Y. and Nakada, M. (2020) Accelerated testing methodology for durability of CFRP. *Composites Part B*, **191**, 107977.

8 Rosen, B.W. (1964) Tensile failure of fibrous composites. *AIAA Journal*, **2**, 1985–1991.

15

Application 1 of Integrated Accelerated Testing Methodology: Statistical Creep and Fatigue Lives of Unidirectional Carbon-fiber-reinforced Polymer Laminates Under Bending Load

15.1 Introduction

In this chapter [1, 2], our developed integrated accelerated testing methodology (Integrated ATM) [3] is applied for creep and fatigue life prediction under longitudinal bending of unidirectional carbon fiber-reinforced polymer (CFRP) laminates. In this case, the fracture mode of unidirectional CFRP laminates is the microbuckling of carbon fibers in the compression side of the bending specimen. First, the viscoelastic coefficients of the matrix resin are characterized by the viscoelastic bending tests based on the modified time–temperature superposition principle (modified TTSP). Second, the static flexural strengths of CFRP laminates are measured statistically at various constant temperatures under a constant strain rate. Third, the statistical creep failure times under bending loading for CFRP laminates are predicted at a constant temperature using the statistical static strengths and the relaxation modulus of the matrix resin. The validity of predicted results is clarified by comparison with the creep failure times measured statistically using creep tests for CFRP laminates. Fourth, the flexural fatigue strengths of CFRP laminates are measured statistically at various frequencies and temperatures, and the characteristics of the fatigue life of CFRP laminates subjected to bending loads are revealed.

15.2 Experiments

Unidirectional CFRP laminates were made by laminating and molding the prepreg using the autoclave method under the curing conditions presented in Table 15.1.

The relaxation moduli of the matrix resin at various temperatures were determined by dynamic mechanical analysis (DMA) and creep tests for the transverse direction of unidirectional CFRP laminates, as shown in Figure 15.1.

DMA tests were conducted at the cyclic deformation of maximum strain 0.06%, various frequencies f from 0.01 to 10 Hz, and various temperatures T from 25 to 120 °C, and the load responses, were measured. The storage moduli $E'(f,T)$ for the transverse direction of unidirectional CFRP laminates obtained were converted to

Properties and Testing of Fiber-Reinforced Polymers, First Edition. Yasushi Miyano and Masayuki Nakada. © 2026 Wiley-VCH GmbH. All rights reserved, including rights for text and data mining and training of artificial intelligence technologies or similar technologies.
Published 2026 by Wiley-VCH GmbH.

Table 15.1 Prepreg and curing conditions.

Fiber/resin of prepreg (Torayca prepreg)	T800SC/2592
Thickness of prepreg	0.125 mm
Laminate configuration	$[0_8]$
Laminate thickness	2.0 mm
Plate size	200 × 200 mm
Curing method	Autoclave
Curing temperature	130 °C
Curing pressure	0.3 MPa
Curing time	2 h

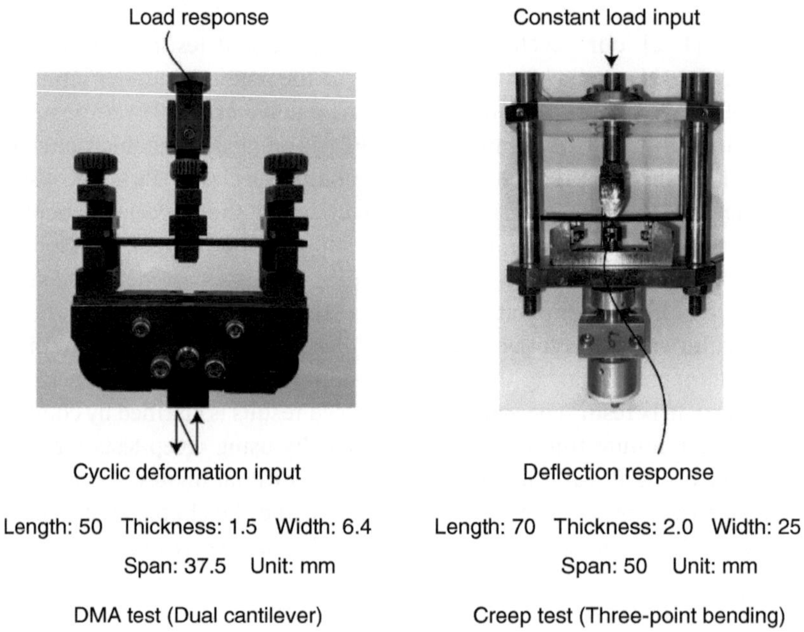

Figure 15.1 DMA and creep tests for the transverse direction of unidirectional CFRP laminates.

the storage moduli E' of the matrix resin by using Chamis's rule of mixture [4] and the mechanical properties of carbon fiber T800S, and the E' of the matrix resin was converted to the relaxation moduli $E_r(t,T)$ based on Eq. (14.1) in Chapter 14.

Creep tests were performed as the confirmation of reliability of the long-term relaxation modulus determined from the results of DMA tests measured in the short-time range. Creep tests were conducted under arbitrary constant loads from 5 to 50 N at various temperatures T from room temperature to 90 °C, and the deflection responses were measured from 1 to 1000 minutes. The creep compliances $D_c(t,T)$ for the transverse direction of unidirectional CFRP laminates obtained were converted to the relaxation moduli $E_r(t,T)$ based on Eq. (14.2). The long-term reliability for the relaxation moduli at various temperatures in the wide range of time determined from the results of DMA tests was confirmed by combining the results of creep tests.

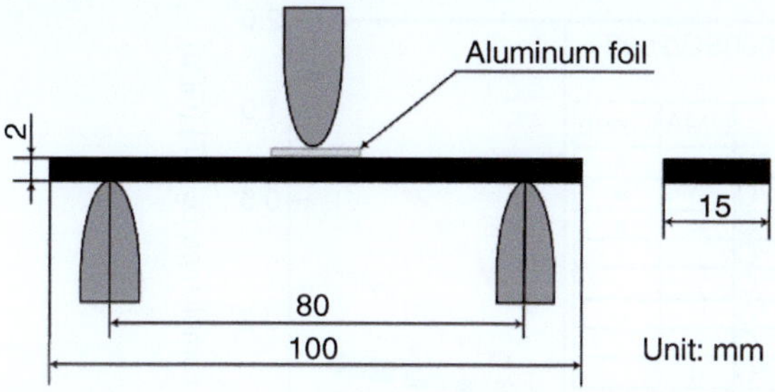

Figure 15.2 Three-point bending test for unidirectional CFRP laminates.

Table 15.2 Effects of aluminum foil cushion sheet number on flexural static strength.

Temperature (°C)	Cushion sheet number	Flexural static strength (MPa)	Average (MPa)	s_p/s_a
25	0	1816, 1758, 1818, 1739, 1583	1681 $(= s_a)$	1.00
	8	1829, 1761, 1771, 1714, 1619	1739 $(= s_p)$	1.03
	16	2393, 2245, 2190, 2257, 2265	2270 $(= s_p)$	1.35
	32	2769, 1865, 2234, 2494, 2859	2444 $(= s_p)$	1.45
80	0	1270, 1288, 1332, 1352, 1332	1315 $(= s_a)$	1.00
	8	1457, 1506, 1527, 1432, 1505	1486 $(= s_p)$	1.13
	16	1589, 1485, 1554, 1651, 1435	1543 $(= s_p)$	1.17

Beam specimens were cut from the laminated plates, as shown in Figure 15.2. Static and creep bending tests were performed with three-point bending in the fiber direction of the unidirectional CFRP laminates with aluminum foil as the cushion material at the loading point, as shown in Figure 15.2. Flexural static strength was found by setting the test speed to 2 mm/min at six temperature levels in the range of room temperature to 120 °C. The number of specimens at each temperature was either 20 specimens or 10 specimens, respectively. The bending creep tests were conducted at a constant temperature of 80 °C and three load levels using the same testing machine for static test. The number of specimens at each load level was 20. Table 15.2 presents the effects of aluminum foil cushion sheet number on the flexural static strength. This table clarifies that the effects of stress concentration at loading points become negligibly small if the number of sheets is selected as 16 for aluminum foil.

15.3 Results and Discussion

15.3.1 Relaxation Modulus and Loss Tangent of the Matrix Resin

DMA and creep tests were conducted at various constant temperatures for the fiber transverse direction of unidirectional CFRP laminates. The storage modulus E' versus the period (the inverse of frequency) t by DMA test and the creep compliance D_c versus the elapsed time t by creep test were measured at various temperatures T for

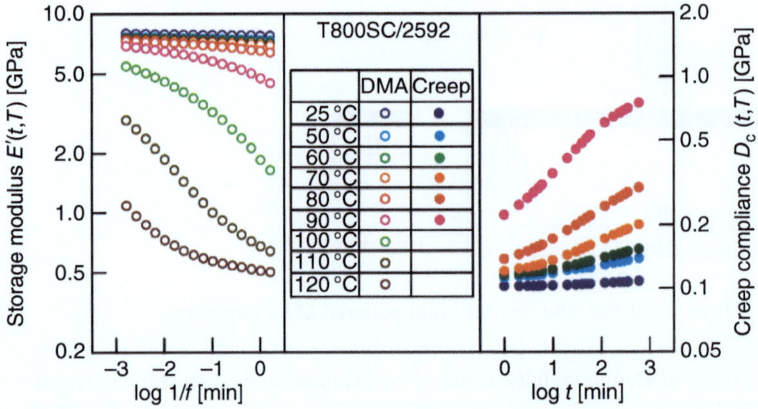

Figure 15.3 Storage modulus and creep compliance versus time at various temperatures for the transverse direction of unidirectional CFRP laminates.

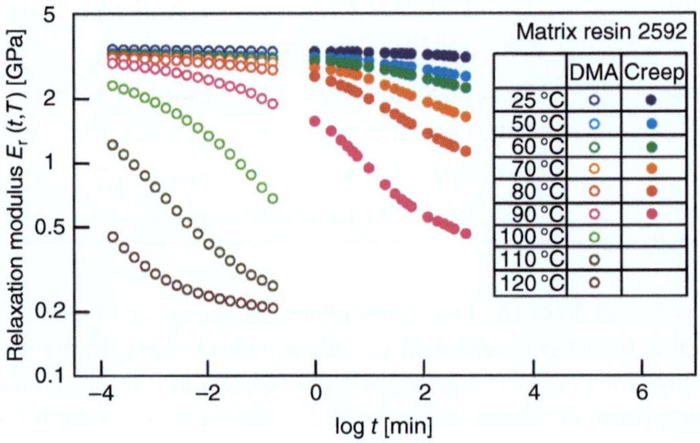

Figure 15.4 Relaxation moduli of the matrix resin at various temperatures.

the transverse direction of unidirectional CFRP laminates as shown in Figure 15.3. The relaxation moduli of the matrix resin in the wide range of time over 7 decades from 10^{-4} to 10^{3} min at various temperatures were obtained using the linear viscoelasticity and rule of mixture, as shown in Figure 15.4.

The master curve of relaxation modulus at reference temperature $T_0 = 25\,°C$ shown in Figure 15.5 was obtained by horizontal and vertical shifting of the relaxation moduli at various temperatures shown in Figure 15.4 with respect to the logarithmic time axis and the logarithmic relaxation modulus axis based on the modified TTSP by which the reliable long-term prediction of viscoelastic coefficient is available. Figure 15.6 shows the time–temperature (horizontal) shift factor $a_{T_0}(T)$ and the temperature (vertical) shift factor $b_{T_0}(T)$, which are obtained by constructing the master curve of the relaxation modulus. The parameters of relaxation modulus of the matrix resin in Eqs. (14.3–14.5) in Chapter 14 are shown in Table 15.3.

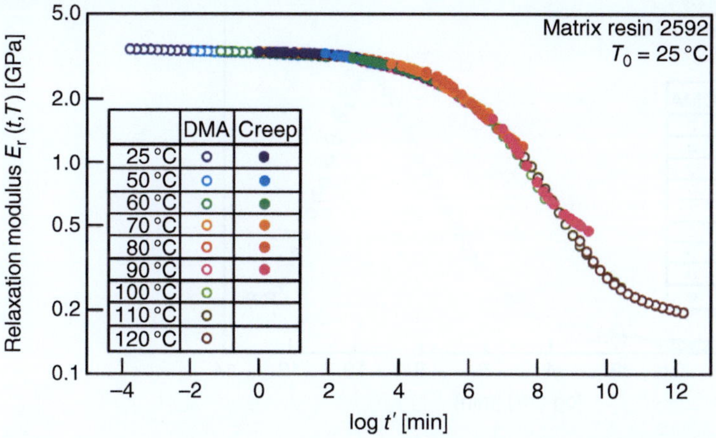

Figure 15.5 Master curve of relaxation modulus of the matrix resin at $T_0 = 25\,°C$.

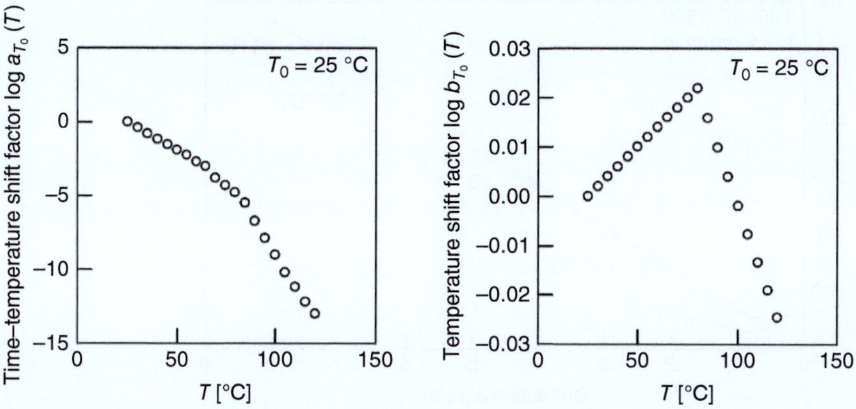

Figure 15.6 Horizontal and vertical shift factors for the relaxation modulus of the matrix resin.

Table 15.3 All parameters expressing the master curve of the relaxation modulus.

$E_r(t_0, T_0)$(GPa)	3.30
m_1	0.007
m_2	0.330
t_g (min)	4.0×10^6
ΔH_1 (kJ/mol)	145
ΔH_2 (kJ/mol)	600
T_{gh} (°C)	80
b_1	0.0004
b_2	−0.0012
T_{gv} (°C)	80

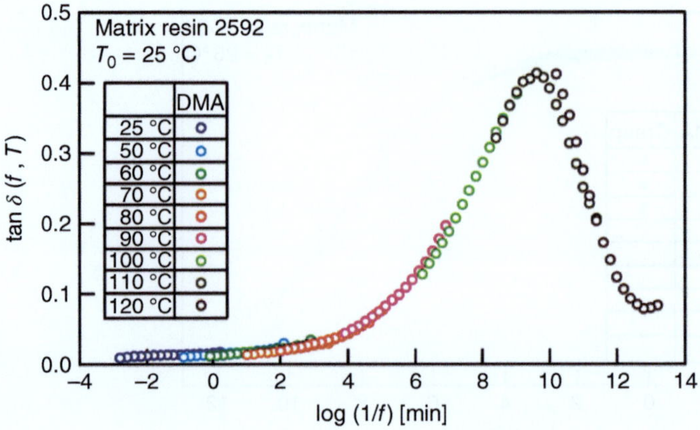

Figure 15.7 Master curve of loss tangent of the matrix resin at $T_0 = 25\,°C$.

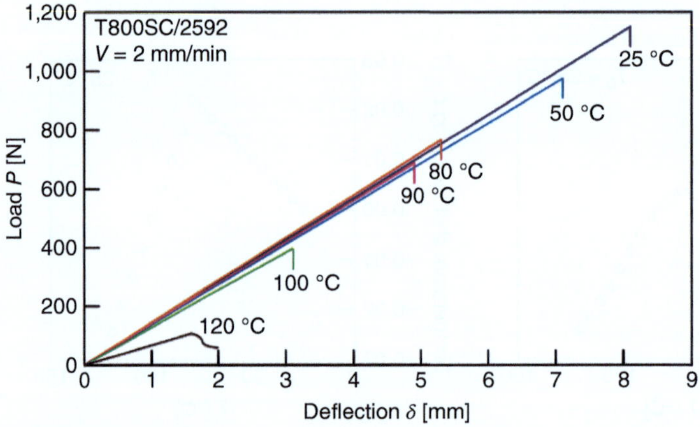

Figure 15.8 Load–deflection diagrams of unidirectional CFRP laminates at various temperatures.

The master curve of loss tangent tan δ of the matrix resin shown in Figure 15.7 was obtained by horizontal shifting of those at various temperatures. This figure is used in the discussion of fatigue strength.

15.3.2 Statistical Flexural Static Strength of Unidirectional CFRP Laminates

Load–deflection diagrams of unidirectional CFRP laminates under static bending tests at various temperatures are presented in Figure 15.8. These diagrams at various temperatures indicate a single straight line with the same slope, except at 120 °C. Flexural strengths obtained from the maximum loads are shown against temperature in Figure 15.9.

Weibull distributions for static strength at various temperatures are depicted in Figure 15.10 for the CFRP laminates. In this figure, α_s is the shape parameter and β_s is the scale parameter of the CFRP laminates. Although the scale

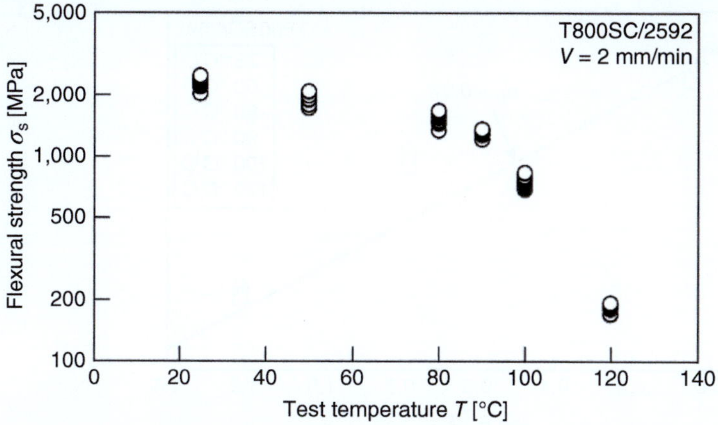

Figure 15.9 Flexural static strength versus temperature for unidirectional CFRP laminates.

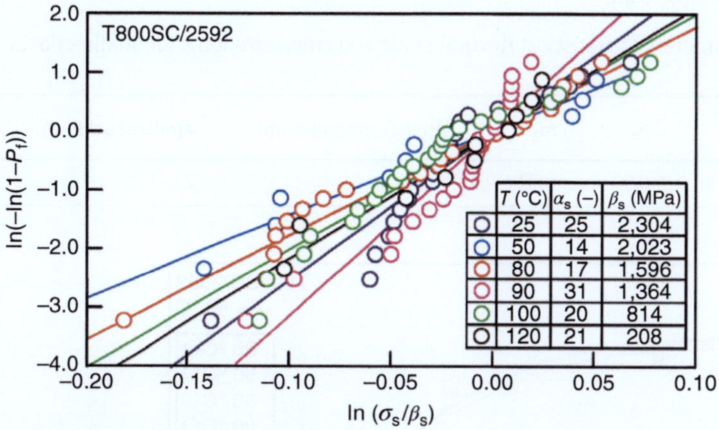

Figure 15.10 Weibull distributions of flexural static strengths at various temperatures for unidirectional CFRP laminates.

parameter decreases according to the temperature rise, the shape parameter maintains an almost constant value for the CFRP laminates to the temperature rise. Shape parameter α_s and scale parameter β_s at temperature $T = 25\,°C$ in this figure can be inferred as the shape parameter α and scale parameter σ_0 of the static strength at the reference temperature $T_0 = 25\,°C$, and the reference failure time $t_0 = 1$ min in Eq. (14.7) in Chapter 14.

15.3.3 Relation Between Flexural Static Strength of CFRP Laminates and Viscoelastic Modulus of the Matrix Resin

Figure 15.11 shows the relation between the flexural static strength of unidirectional CFRP laminates and the viscoelastic modulus of the matrix resin. Linear approximation was made based on data of 25, 50, and 80 °C, which are temperatures below the glass transition temperature of the matrix resin. The viscoelastic parameter $n_R = 0.97$ was obtained.

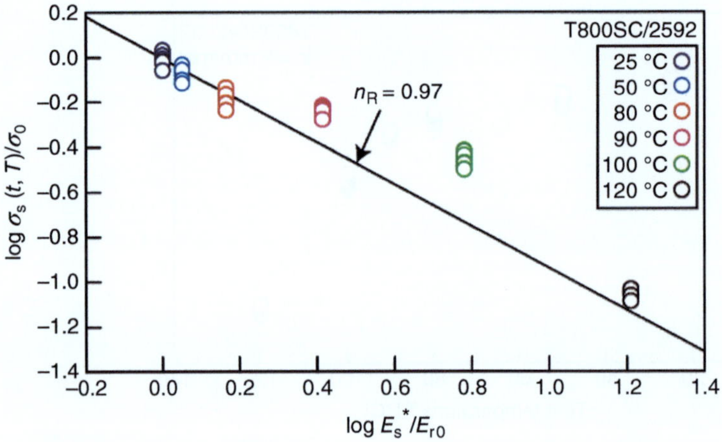

Figure 15.11 Flexural static strength of unidirectional CFRP laminates versus viscoelastic modulus of the matrix resin.

Table 15.4 Parameters of statistical flexural static and creep strengths for unidirectional CFRP laminates.

σ_0 (MPa)	α	n_R	m_R	A(crack propagation)	A(microbuckling)
2304	25.5	0.97	0.220	5.54	2

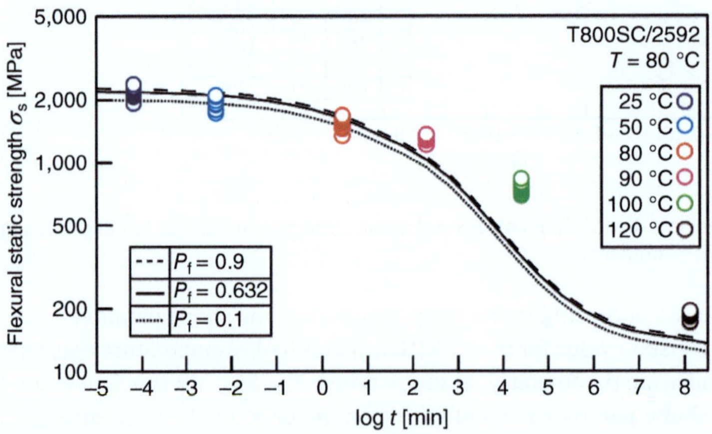

Figure 15.12 Predicted statistical flexural static strengths versus failure time and the measured ones.

All parameters for predicting statistical flexural static and creep strengths of unidirectional CFRP laminates were obtained as presented in Table 15.4.

15.3.4 Statistical Flexural Static Strength Versus Failure Time

Figure 15.12 shows the prediction of statistical flexural static strength versus failure time at temperature $T = 80\,°C$ obtained by substituting the data of Tables 15.3 and 15.4 into Eq. (14.7). The plots presented in this figure show the measured data. The predicted values agree well with the measured data in the short-time region, and the predicted values deviate slightly from the measured data in the long-term region.

15.3.5 Statistical Flexural Creep Strength Versus Failure Time

Figure 15.13 presents the prediction of the statistical flexural creep strength versus failure time at temperature $T = 80\,°C$, as obtained by substituting the data of Tables 15.3 and 15.4 into Eq. (14.8). The flexural creep failure times measured experimentally at three levels of creep stress are presented in this figure. The measured creep failure times agree well with the predicted curve in the case of $A = 2.00$, for which the failure mode is the microbuckling as explained in Chapter 14. Figure 15.14 presents the prediction of

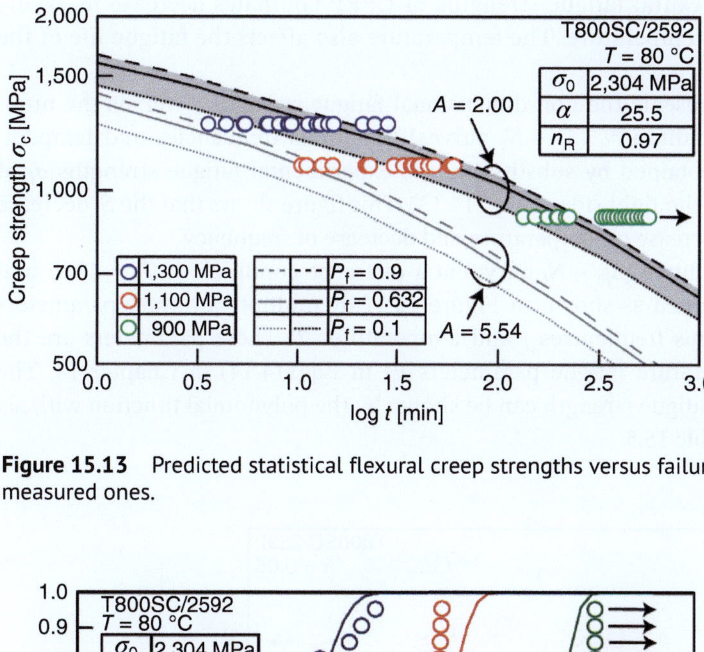

Figure 15.13 Predicted statistical flexural creep strengths versus failure time and measured ones.

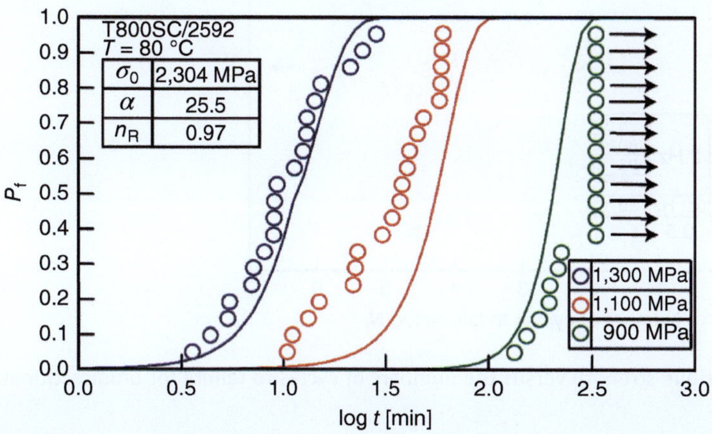

Figure 15.14 Predicted and experimental failure probability versus creep failure time.

failure probability P_f versus failure time t at three levels of creep stress. The plots presented in this figure show the measured data. This figure also indicates that the experimental data agree well with the predicted curves in the case of $A = 2.00$, for which the failure mode is microbuckling.

15.3.6 Statistical Flexural Fatigue Strength Against Number of Cycles to Failure for CFRP Laminates

Fatigue bending tests for unidirectional CFRP laminates were conducted at four temperatures with frequency $f = 2$Hz and the strain ratio $R = 0.05$, and additionally at temperature $T = 80\,°C$ with frequency $f = 0.2$ Hz. Flexural fatigue strength σ_f of CFRP laminates was measured against the number of cycles to failure N_f through testing. Figure 15.15 portrays the flexural fatigue strength versus the number of cycles to failure for CFRP laminates at various frequencies and temperatures. The flexural fatigue strengths of CFRP laminates decrease markedly with increase of temperature. The temperature also affects the fatigue life of the CFRP laminates.

Figure 15.16 presents the non-dimensional fatigue strength S_f versus the number of cycles to failure N_f ($S_f - N_f$ curves) at various frequencies and temperatures, which is obtained by substituting the experimental fatigue strengths σ_f of Figure 15.15 into the right side of Eq. (14.13). This figure shows that the S_f decrease markedly with increase of temperature and decrease of frequency.

Master $S_f - N_f$ curve ($S_{f0} - N_f$ curve) at a reference condition of $T_0 = 25\,°C$ and $f_0 = 2$ Hz is obtained as shown in Figure 15.17 by multiply arbitrary parameters to the S_f at various frequencies f and temperatures T. These parameters are the frequency–temperature fatigue parameters F_f^* in Eq. (14.14) in Chapter 14. The master curve of fatigue strength can be shown by the polynomial function with six parameters in Table 15.5.

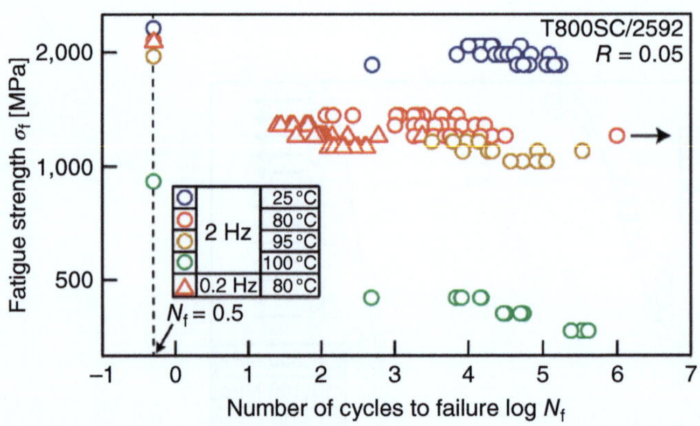

Figure 15.15 Fatigue strength versus the numbers of cycles to failure for unidirectional CFRP laminates.

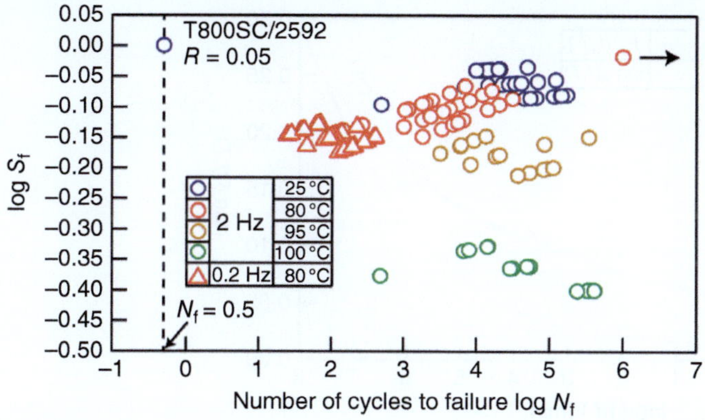

Figure 15.16 S–N curves at various conditions for unidirectional CFRP laminates.

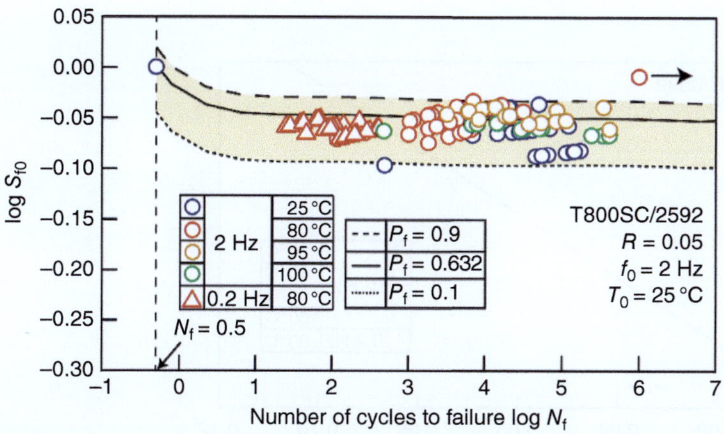

Figure 15.17 S–N master curve at reference condition for unidirectional CFRP laminates.

Table 15.5 Parameters for fatigue strength of unidirectional CFRP laminates.

a_f	$n_{f1} (\times 10^{-4})$	$n_{f2} (\times 10^{-4})$	$n_{f3} (\times 10^{-4})$	$n_{f4} (\times 10^{-4})$	$n_{f5} (\times 10^{-4})$	$n_{f6} (\times 10^{-4})$
52.3	0.25	−6.01	56.7	258.5	599.6	−676.3

The frequency–temperature fatigue parameter $F_f^*(f,T)$ and the loss tangent tan δ versus the inverse of reduced frequency $1/f'$ at $T_0 = 25\,°C$ are presented in Figure 15.18. This figure indicates that the frequency–temperature fatigue parameter $F_f^*(f,T)$ increases monotonically with the inverse of reduced frequency $1/f'$ in a similar manner to that of the loss tangent, which is one of the expressions for the matrix resin viscoelasticity. Figure 15.19 shows the relationship between F_f^* and tan δ at the same frequency f and temperature T, which is approximately linear.

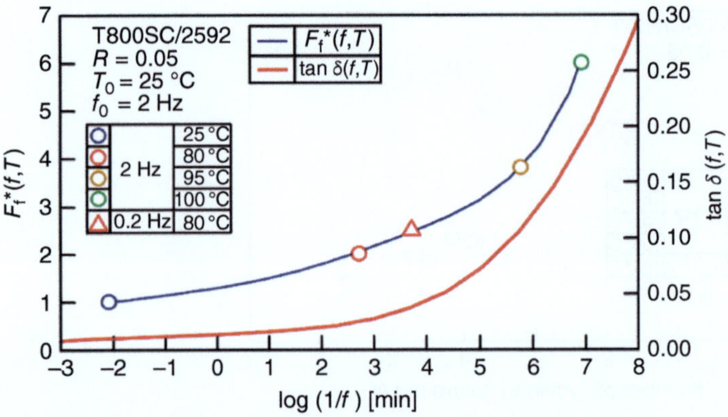

Figure 15.18 Frequency–temperature fatigue parameter and loss tangent versus inverse of frequency at a reference condition of $T_0 = 25\,°C$ and $f_0 = 2$ Hz.

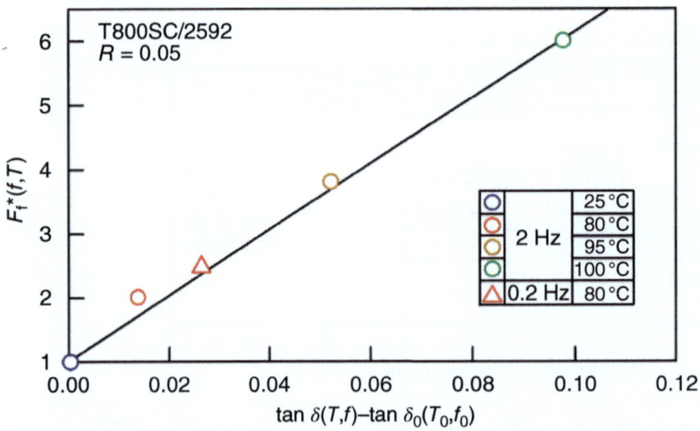

Figure 15.19 Relationship between F_f^* and tan δ at the same frequency f and temperature T.

Therefore, the F_f^* is shown as Eq. (14.15) controlled by the matrix resin viscoelasticity and the parameter a_f, which is the slope of F_f^* against tan δ can be expressed in Table 15.5.

15.3.7 Fractographies After Static, Constant, and Cyclic Bending Loads

Figures 15.20–15.22 show the fractographies at the load point after static, constant, and cyclic bending loads, respectively. These figures clarify that all fractures occur near the load point at the compression surface of the specimen. It can be presumed easily that the failure mode is microbuckling under compressive stress in the longitudinal direction of unidirectional CFRP laminates. Therefore, it can be inferred that the flexural static, creep, and fatigue strengths for unidirectional CFRP laminates described above are the compressive static, creep, and fatigue strengths.

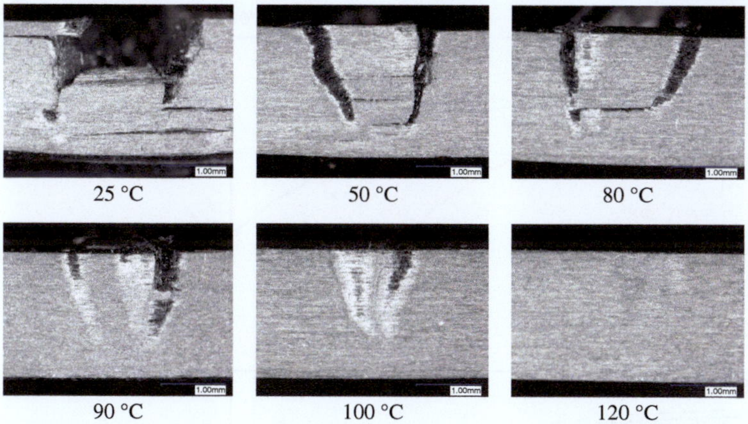

Figure 15.20 Fractographies at the load point after static bending loads.

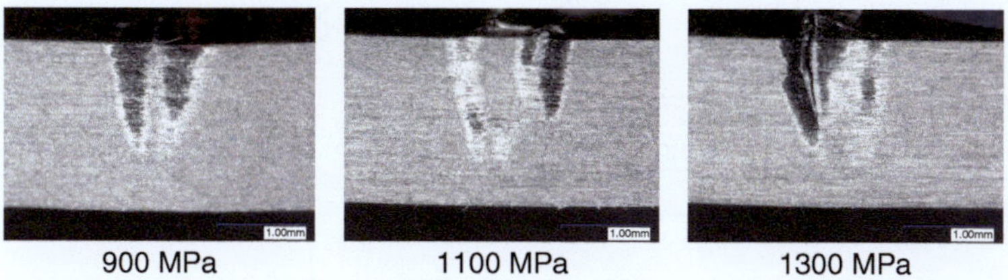

Figure 15.21 Fractographies at the load point after constant bending loads at $T = 80\,°C$.

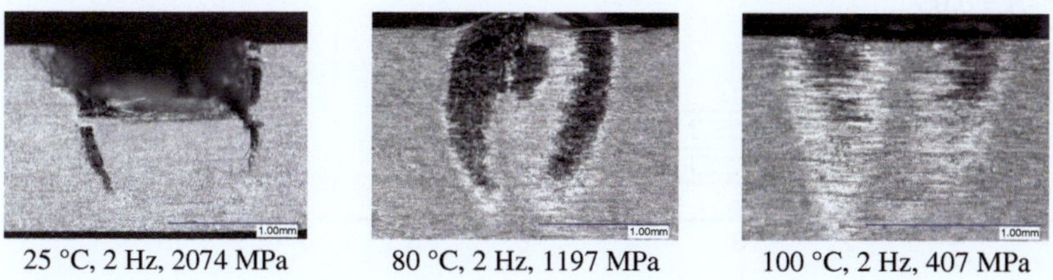

Figure 15.22 Fractographies at the load point after cyclic bending loads under various conditions.

15.3.8 Long-term Prediction of Flexural Creep and Fatigue Strengths of Unidirectional CFRP Laminates

Figure 15.23 portrays long-term predictions for the relaxation modulus of the matrix resin and the flexural creep and fatigue strengths of unidirectional CFRP laminates. This figure clarifies that the long-term flexural creep strength of unidirectional CFRP laminates is strongly controlled by the relaxation modulus of the matrix resin. The flexural creep strength decreases drastically with time, with similar behavior of the relaxation modulus of the matrix resin, and this decrease accelerates with increasing temperature. However, the flexural fatigue strength decreases only slightly with

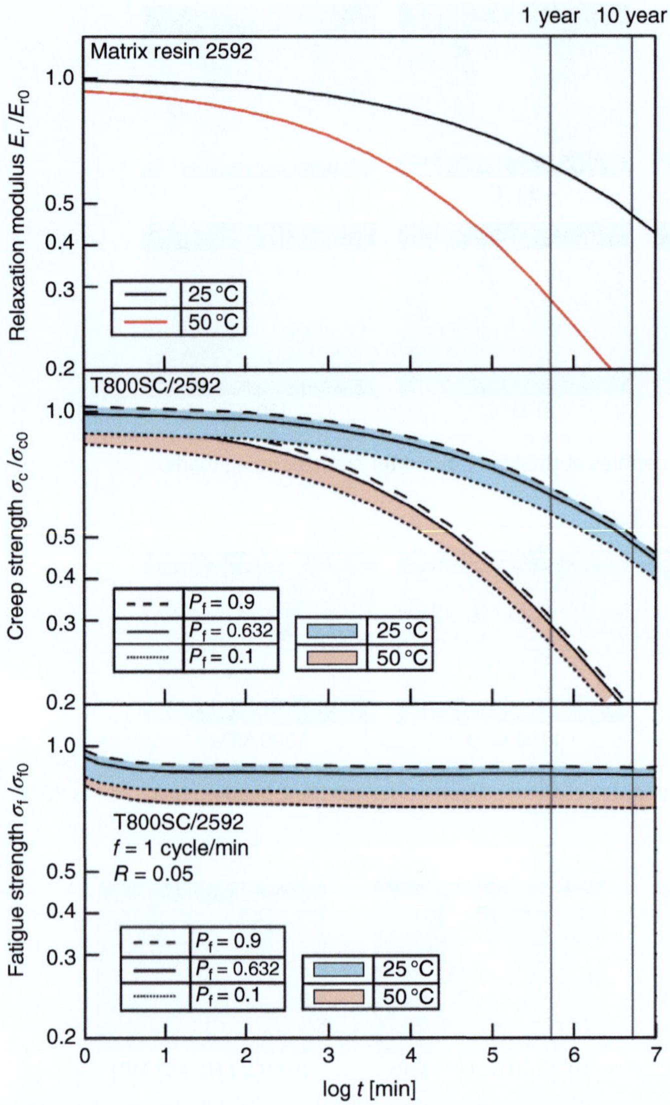

Figure 15.23 Long-term prediction of relaxation modulus of the matrix resin and flexural creep and fatigue strengths of unidirectional CFRP laminates.

increasing of time, temperature, and the number of cycles to failure N_f. Especially, the fatigue strength remains constant with increasing N_f in the region of numerous cycles to failure, which is considered as the fatigue less. This behavior is attributable to the fact that the trigger of failure is the microbuckling of carbon fibers in the matrix resin.

15.4 Conclusion

Our developed Integrated ATM based on the matrix resin viscoelasticity for the creep and fatigue failure lives of CFRP laminates was applied to the statistical prediction of long-term creep and fatigue failure lives for the longitudinal bending of unidirectional CFRP laminates, which is an important basic item for the durability design of CFRP structures used for aircraft and others. The validity of Integrated ATM for this creep and fatigue life prediction under the longitudinal bending of unidirectional CFRP laminates was discussed herein. From that discussion, the following results were obtained.

1. It was clarified that the Integrated ATM presented in Chapter 14 can be applied to longitudinal bending of unidirectional CFRP laminates. As a result, Integrated ATM makes it possible to predict long-term creep and fatigue lives for longitudinal bending of unidirectional CFRP laminates. The fracture mode is microbuckling in all static, creep, and fatigue fractures.
2. The long-term flexural creep strength of unidirectional CFRP laminates is strongly controlled by the relaxation modulus of the matrix resin. The flexural creep strength decreases drastically with time, with similar behavior of the relaxation modulus of the matrix resin, and this decrease accelerates with increasing temperature.
3. The flexural fatigue strength decreases only slightly with increasing time, temperature, and the number of cycles to failure. Especially, the fatigue strength remains constant with increasing N_f in the region of numerous cycles to failure, which is considered as fatigue less. This behavior is attributable to the fact that the trigger of failure is the microbuckling of carbon fibers in the matrix resin.

References

1 Miyano, Y., Nakada, M. and Yoshikoshi Y. (2022) Statistical creep failure time of unidirectional carbon fiber reinforced plastic under bending load. *Journal of Composite Materials*, **56**, 1153–1164.

2 Miyano, Y., Nakada, M. and Morisawa, Y. (2022) Statistical fatigue failure life of unidirectional CFRP under bending load. *Journal of Composite Materials*, **56**, 4173–4181.

3 Miyano, Y. and Nakada, M. (2024) Integrated accelerated testing methodology for CFRP durability. *Composites Part B*, **281**, 111527.

4 Chamis, C.C. (1984) Simplified composites micromechanics equations for hygral, thermal and mechanical properties. *SAMPE Q*, **15**, 14–23.

16

Application 2 of Integrated Accelerated Testing Methodology: Carbon Fiber and Matrix Resin Mechanical Properties Controlling Statistical Tensile Fatigue Life of Unidirectional Carbon-fiber-reinforced Polymer

16.1 Introduction

In this chapter [1], the integrated accelerated testing method (Integrated ATM) [2] is applied to evaluate the effects of the mechanical properties of carbon fiber and matrix resin on the statistical fatigue life of unidirectional carbon fiber-reinforced polymers (CFRP) under cyclic tensile loading. First, a statistical fatigue strength formulation for Integrated ATM is proposed by considering the mechanical properties of carbon fiber, matrix resin and the strain ratio. Next, the static and fatigue tensile tests are conducted on the epoxy resin-impregnated carbon fiber strands (CF/EP strands) using three types of carbon fiber as unidirectional CFRPs, and the parameters for these CF/EP strands in the formulation are determined from the results. Finally, the effects of the mechanical properties of carbon fiber and matrix resin on the statistical fatigue life of the CF/EP strands are discussed.

16.2 Formulations

In this section, the mechanisms governing the strength degradation of unidirectional CFRP under longitudinal cyclic loading are presented and formulated.

16.2.1 Formulations of Fatigue Strength of Unidirectional CFRP

The formulation of statistical fatigue strength σ_f in the longitudinal direction of unidirectional CFRP with fatigue degradation parameter F_f based on the viscoelasticity of the matrix resin is given by the following equation.

$$\begin{aligned} \log \sigma_f(t,T,R) &= \log\sigma_0(t_0,T_0) + \frac{1}{\alpha}\log[-\ln(1-P_f)] \\ &+ n_R \log\left[\frac{E_f(t,T,R)}{E_r(t_0,T_0)}\right] - F_f(t,f,T,R) \\ N_f &= \frac{t}{f}, \quad N_f \geq \frac{1}{2} \end{aligned} \tag{16.1}$$

Properties and Testing of Fiber-Reinforced Polymers, First Edition. Yasushi Miyano and Masayuki Nakada. © 2026 Wiley-VCH GmbH. All rights reserved, including rights for text and data mining and training of artificial intelligence technologies or similar technologies.
Published 2026 by Wiley-VCH GmbH.

t:	Failure time
t_0:	Reference time
T:	Temperature
T_0:	Reference temperature
R:	Strain ratio
P_f:	Failure probability
f:	Frequency
N_f:	Number of cycles to failure
σ_0 and α:	Scale and shape parameters of static strength at t_0 and T_0
n_R:	Viscoelastic parameter determined by failure mode
E_f:	Viscoelastic modulus of matrix resin under cyclic loading
E_r:	Relaxation modulus of matrix resin
F_f:	Fatigue degradation parameter

The statistical fatigue strength formulation can be expressed using two elements: the strength reduction caused by the viscoelasticity of matrix resin, and the non-dimensional fatigue strength reduction caused by the energy loss in the matrix resin due to cyclic loading. The strength reduction amount σ_f^* caused by the viscoelasticity of the matrix resin can be calculated as follows.

$$\log \sigma_f^*(t,T,R) = \log \sigma_0(t_0,T_0) + n_R \log\left[\frac{E_f(t,T,R)}{E_r(t_0,T_0)}\right]$$

$$E_f(t,f,T,R) = \frac{1}{2}\left[(1-R)E_r\left(1/(\pi^2 f),T\right) + (1+R)E_r(At/2,T)\right], \quad t = N_f/f$$

$$\log A_i = \log(1 + 1/k_{Ri}), \quad k_{Ri} = m_i \times n_R, \quad i = 1,2 \tag{16.2}$$

where m_i represents the slope of the logarithmic relaxation modulus of matrix resin versus the logarithmic time curve. All parameters of σ_f^* are determined using the static strength test results of CFRP and the viscoelastic measurement results of matrix resin conducted at various temperatures.

The non-dimensional fatigue strength, including the variation can be expressed by the following equation:

$$\begin{aligned}\log S_f(N_f,f,T,R) &= \frac{1}{\alpha}\log\left[-\ln\left(1-P_f\right)\right] - F_f(N_f,f,T,R)\\ &= \log\frac{\sigma_f(N_f,f,T,R)}{\sigma_0(t_0,T_0)} - n_R\log\left[\frac{E_f(N_f,f,T,R)}{E_r(t_0,T_0)}\right]\end{aligned} \tag{16.3}$$

The non-dimensional fatigue strength can be calculated by substituting the fatigue strength and number of cycles to failure measured at each frequency, temperature, and strain ratio into the right-hand side of Eq. (16.3).

16.2.2 Fatigue Degradation Parameter for Unidirectional CFRP

The fatigue degradation parameter F_f in the middle part of Eq. (16.3) is expressed by the following equation for the tension load of the longitudinal direction of unidirectional CFRP.

$$F_f(N_f, f, T, R) = F_f^R(R) F_f^*(f, T) F_f^{**} F_f'(N_f), \tag{16.4}$$

where F_f^R is the strain ratio parameter, F_f^* is the frequency–temperature parameter, F_f^{**} is the mechanical parameter of the fiber, and F_f' is the fatigue parameter as the function of number of cycles to failure. In this chapter, we will present, discuss, and attempt to formulate, based on scientific evidence, how these four parameters are determined by the carbon fiber and matrix resin.

We propose a longitudinal mechanical model for unidirectional CFRP, as shown in Figure 16.1. The elastic modulus of this model is determined by the carbon fiber's elastic modulus E_c, because it is much larger than the matrix resin's elastic modulus E_m. The static strength of this model is determined by the statistical strength of carbon fiber, whose shape and scale parameters follow a Weibull distribution. The energy loss during cyclic loading is determined by the viscoelasticity of the matrix resin, whose elastic modulus E_m and viscosity η vary with frequency and temperature. Incidentally, since carbon fiber is completely elastic, it does not contribute to the energy loss by cyclic loading. As the number of loading cycles increases, the resin's energy loss accumulates, degrading the interface between the carbon fiber and the resin, leading to a decrease in the strength of the CFRP. This is considered CFRP fatigue.

Figure 16.2 shows the relationship between tensile stress and tensile strain when cyclic loading is applied to unidirectional CFRP in the longitudinal direction. The shaded area indicates the energy loss due to the viscoelasticity of the matrix resin. Using this model, we consider the four parameters on the right side of Eq. (16.4). Energy loss deteriorates the interface between the carbon fiber and resin, resulting in a decrease in the strength of the CFRP. In this chapter, we focus on the loading conditions expressed in Eq. (16.4) and the virgin properties of the constituent materials, and evaluate their effects on fatigue strength.

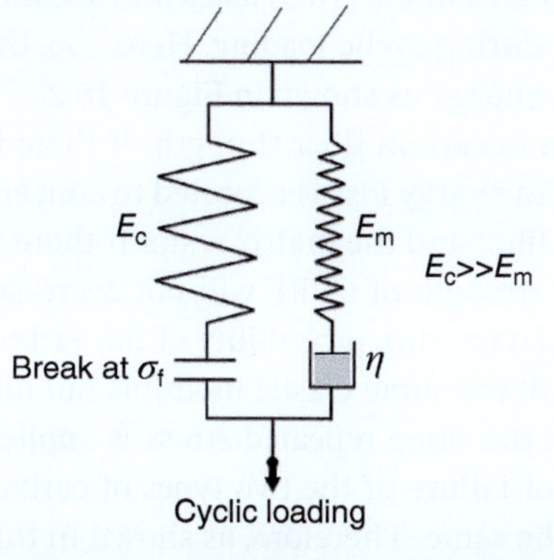

Figure 16.1 Mechanical model of unidirectional CFRP under longitudinal tension loading.

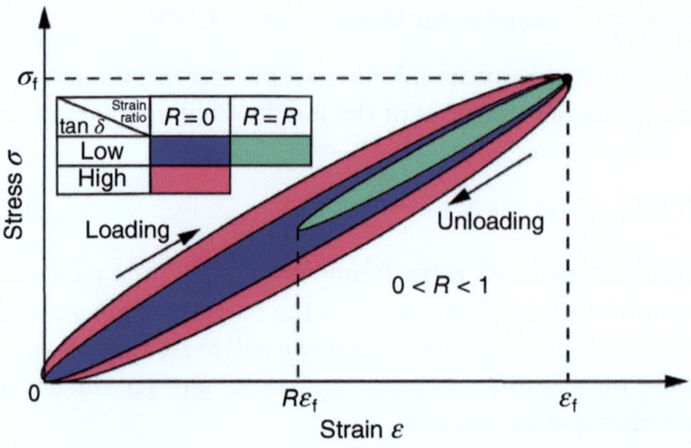

Figure 16.2 Stress–strain relations of unidirectional CFRP with various strain ratios and various loss tangents of matrix resin.

The strain ratio parameter F_f^R in Eq. (16.4) can be expressed as the effect of strain energy generated in the matrix resin during cyclic loading by the following function of strain ratio R as presented in Figure 16.2.

$$F_f^R(R) = (1 - R)^2 \tag{16.5}$$

The frequency–temperature parameter F_f^* in Eq. (16.4) can be expressed as the effect of viscoelasticity of matrix resin during cyclic loading by the following function of the loss tangent $\tan\delta$ of matrix resin, as shown in Figure 16.2.

$$F_f^*(f,T) = 1 + a_f[\tan\delta(f,T) - \tan\delta(f_0,T_0)] \tag{16.6}$$

In this equation, f_0 and T_0 represent the reference frequency and reference temperature, respectively. The modified time–temperature superposition principle (modified TTSP) for the viscoelasticity of the matrix resin also applies to the frequency–temperature fatigue parameter $F_f^*(f,T)$.

Next, we consider how the mechanical properties of carbon fibers affect the fatigue strength of CFRP. Specifically, we clarify how the fiber mechanical parameter F_f^{**} in Eq. (16.4) is expressed as a function of the mechanical properties of the carbon fiber. The strain ratio parameter F_f^R expressed in Eq. (16.5) takes into account the strain energy generated in the matrix resin during cyclic loading. Here, too, the primary effect to consider is the effect of strain energy, as shown in Figure 16.2.

The second effect to consider is the variation in carbon fiber strength. If there is variation in carbon fiber strength, damage due to energy loss is expected to concentrate at the weak interface between the carbon fiber and the matrix resin. If there is no scatter in carbon fiber strength, the fatigue strength of CFRP will not decrease. Furthermore, at point A in Figure 16.3, the initial fracture probability of the carbon fiber is the same for the two types of CFRP with the same elastic modulus but different strength shape parameters. In this case, the same repeated stress is applied to the two types of CFRP, and the probability of failure of the two types of carbon fiber when the stress reaches its maximum is the same. Therefore, as shown in this figure, it can be inferred that the number of cycles until failure of the two types

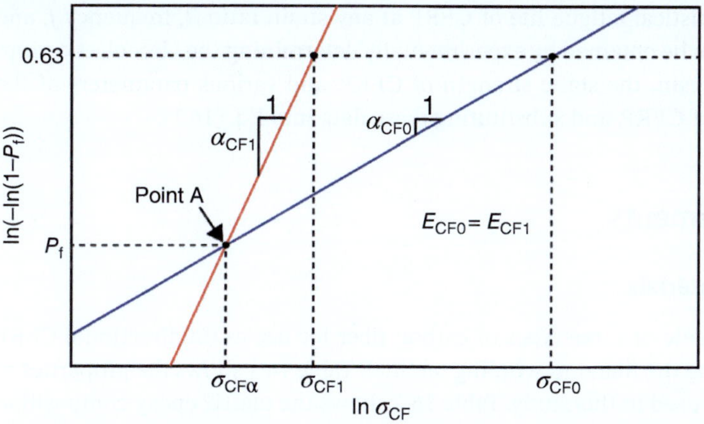

Figure 16.3 Weibull distributions of carbon fiber strengths with two kinds of shape and scale parameters having the same elastic modulus.

of CFRP is approximately the same. From these considerations, it can be inferred that the mechanical parameter of the fiber F_f^{**}, which is affected by the mechanical properties of the carbon fiber, can be obtained by the following equation.

$$F_f^{**} = \left(\frac{\alpha_{cf0}}{\alpha_{cf}}\right)\left(\frac{\varepsilon_s}{\varepsilon_{s0}}\right)^2 \tag{16.7}$$

where α_{cf} and α_{cf0} are the shape parameters of the strength of the arbitrary carbon fiber and the reference carbon fiber, respectively, and ε_s and ε_{s0} are the scale parameters of the static ultimate strain at the reference time and temperature for the arbitrary unidirectional CFRP and the reference unidirectional CFRP, respectively.

The fatigue parameter F_f' can be expressed as a function of the number of cycles to failure N_f as follows:

$$F_f'(N_f) = \sum_{i=1}^{n} n_{fi} \log(2N_f)^i, \quad N_f \geq 1/2 \tag{16.8}$$

Here, the fatigue strength at $N_f = 1/2$ is equal to the static strength when the time to failure t is equal to $1/(2f)$.

Furthermore, S_{f0} defined as follows is the master curve of the non-dimensional fatigue strength at the reference frequency, temperature, and strain ratio when statistically generating one curve for various frequencies, temperatures, and strain ratios.

$$\log S_{f0}(N_f) = \log S_f + F_f'(N_f)\left[F_f^R(R)\, F_f^*(f,T) F_f^{**} - F_f^R(R_0)\right] \tag{16.9}$$

The coefficients n_{fi} in Eq. (16.8) are determined from the fatigue strength versus the number of cycles to failure curve (S–N curve) obtained by applying the least squares method to experimental values measured at a reference strain ratio R_0, a reference frequency f_0, and a reference temperature T_0. The coefficient a_f in Eq. (16.6) is similarly determined from fatigue tests performed at various frequencies and temperatures.

Finally, the statistical fatigue life of CFRP at any strain ratio R, frequency f, and temperature T can be obtained by experimentally determining the viscoelastic properties of matrix resin, the static strength of CFRP, and various parameters of the fatigue strength of CFRP, and substituting these data into Eq. (16.1).

16.3 Experiments

16.3.1 Test Materials

CF/EP strands made of three types of carbon fiber for use as unidirectional CFRP were molded using the filament winding method. Table 16.1 shows the properties of the carbon fibers used in this study. Table 16.2 shows the matrix epoxy composition and curing conditions of the CF/EP strands.

16.3.2 Testing Method and Test Conditions

The configuration of the CF/EP strand specimen is shown in Figure 16.4. The diameter and gauge length of the CF/EP strand are approximately 1.0 and 180 mm for T700/EP and M40J/EP, respectively, and approximately 0.8 and 200 mm for T300/EP. The glass transition temperature T_g of the epoxy was determined to be 157 °C from the peak of the loss tangent versus temperature at 1 Hz using a dynamic mechanical analyzer (DMA). Static and fatigue tension tests of the CF/EP strand were reliably performed using the test system developed by us, as shown in Figure 16.4.

Table 16.1 Carbon fiber specifications.

Carbon fiber	Density ρ_{CF} (g/cm^3)	Tex t_e (g/km)	Elastic modulus E_{CF} (GPa)	Tensile strength σ_{CF} (MPa)	Ultimate strain ε_{CF} (%)
T700-6000	1.80	400	230	4900	2.1
T300-3000	1.76	198	230	3530	1.5
M40J-6000	1.77	225	377	4400	1.2

Table 16.2 Compositions and cure schedules of CF/EP strands.

CF/EP strand	Carbon fiber	Composition of resin (weight ratio)	Product name of resin (company)	Cure schedule
T700/EP T300/EP M40J/EP	T700-6000 T300-3000 M40J-6000	Matrix resin (100) + Hardener (103.6) + Cure accelerator (1.0)	jER 828 (Mitsubishi Chemical Analytec Co. Ltd.) MHAC-P (Resonac Holdings Corp.) 2E4MZ (Shikoku Chemicals Corp.)	70 °C × 12 h + 150 °C × 4 h + 190 °C × 2 h

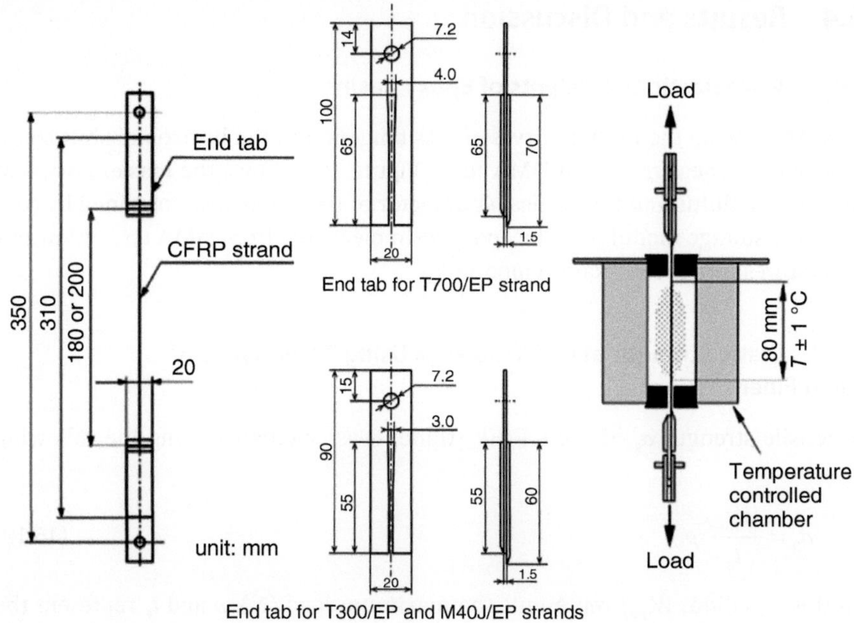

Figure 16.4 Specimen configuration and temperature-controlled chamber.

Table 16.3 Number of tests for each temperature condition on static tension tests.

	25 (°C)	120 (°C)	135 (°C)	150 (°C)
T700/EP	20	20	20	20
T300/EP	30	30	30	30
M40J/EP	30	–	30	30

Table 16.4 Number of tests for each temperature condition on fatigue tension tests.

	Strain ratio R (–)	25 (°C)	120 (°C)	135 (°C)	150 (°C)
T700/EP	0.1	25	10	21	5
T300/EP		24	20	40	20
M40J/EP		21	–	–	–
T700/EP	0.5	17	–	–	–

Static tension tests of CF/EP strands were conducted at various constant temperatures with a crosshead speed of $V = 2$ mm/min using the number of specimens shown in Table 16.3. Fatigue tension tests of CF/EP strands were conducted at various conditions with a frequency of $f = 2$ Hz using the number of specimens shown in Table 16.4.

16.4 Results and Discussion

16.4.1 Viscoelastic Coefficients of Epoxy Resin

Figure 16.5 shows the master curves of loss tangent and the horizontal shift factor of epoxy resin measured from DMA tests. Figure 16.6 shows the master curves of relaxation modulus and the vertical shift factor of the epoxy resin obtained by converting the storage modulus of the epoxy resin measured from DMA tests at various temperatures into the relaxation modulus.

16.4.2 Static Strength of CF/EP Strands Using Three Types of Carbon Fiber

The tensile strength σ_s of the CF/EP strands was calculated using the following equation.

$$\sigma_s = \frac{W_{max}}{t_e} \rho \tag{16.10}$$

In this equation, W_{max} represents the maximum load (N). ρ and t_e represent the density of the carbon fiber (kg/m^3) and the tex (strand weight per unit length) of the carbon fiber strand (g/km), respectively. Figure 16.7 shows the relationship between

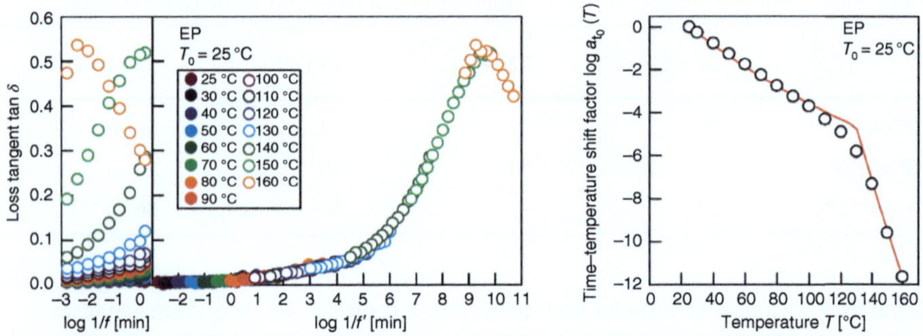

Figure 16.5 Master curve of loss tangent and horizontal shift factor for epoxy resin.

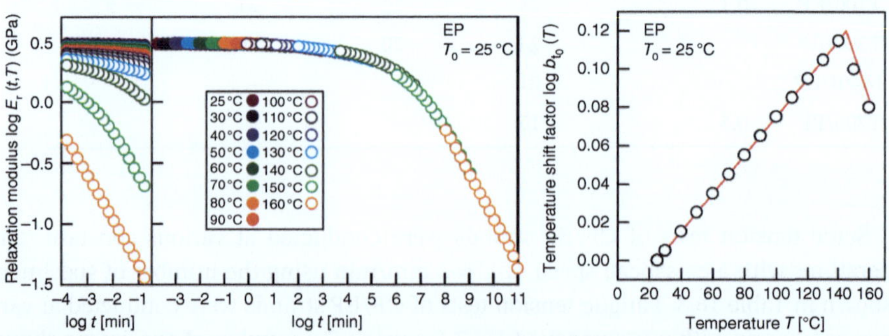

Figure 16.6 Master curve of relaxation modulus and vertical shift factor for epoxy resin.

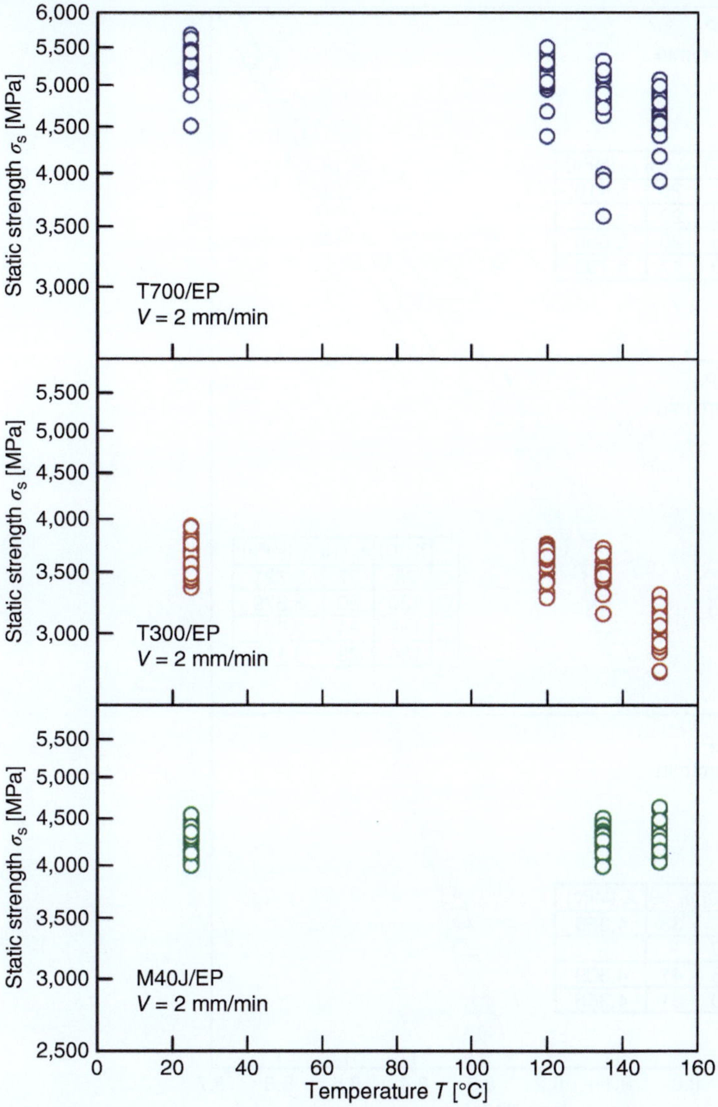

Figure 16.7 Static strengths of CF/EP strands against temperature.

static strength and temperature for three types of CF/EP strands. The static tensile strength of CF/EP strands shows unique temperature dependencies depending on the type of carbon fiber.

Figure 16.8 shows the Weibull distribution of static strength at various temperatures for CF/EP strands. The parameters were estimated from the regression lines. α_s is the shape parameter and β_s is the scale parameter of the CF/EP strand in this figure. The scale parameter decreases with increasing temperature, while the shape parameter maintains a nearly constant value for the CF/EP strand during the temperature increase. The shape parameter α_s and the scale parameter β_s in this figure at the temperature $T = 25\,°C$ can be used as the shape parameter α and the scale

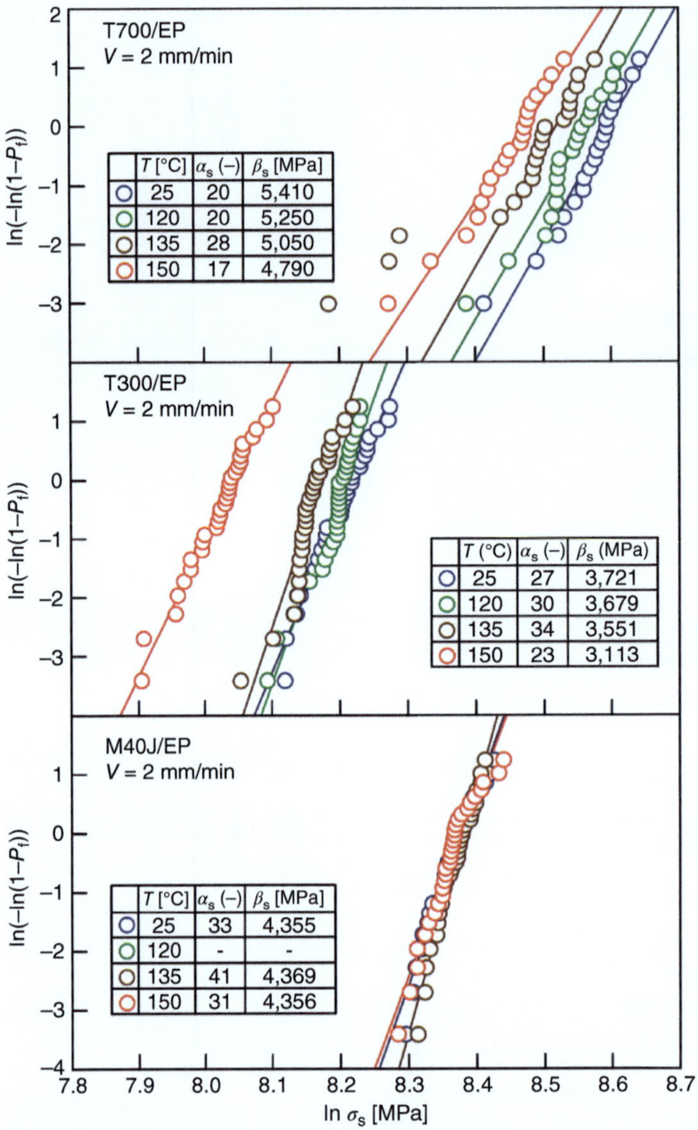

Figure 16.8 Weibull distributions of the static strengths of CF/EP strands at various temperatures.

parameter σ_0 at the reference time to failure $t_0 = 1$ min for the static strength at the reference temperature $T_0 = 25\,°\text{C}$ in Eq. (16.1).

Figure 16.9 shows the relationship between the non-dimensional static strength σ_s/σ_0 of the CF/EP strand and the non-dimensional relaxation modulus E_s^*/E_{r0} of the matrix resin. As can be seen from the regression line, the relationship between σ_s/σ_0 and E_s^*/E_{r0} can be expressed as a straight line with the slope of the viscoelastic parameter n_R in Eq. (16.1). The parameters σ_0, α, and n_R of the CF/EP strand are shown in Table 16.5.

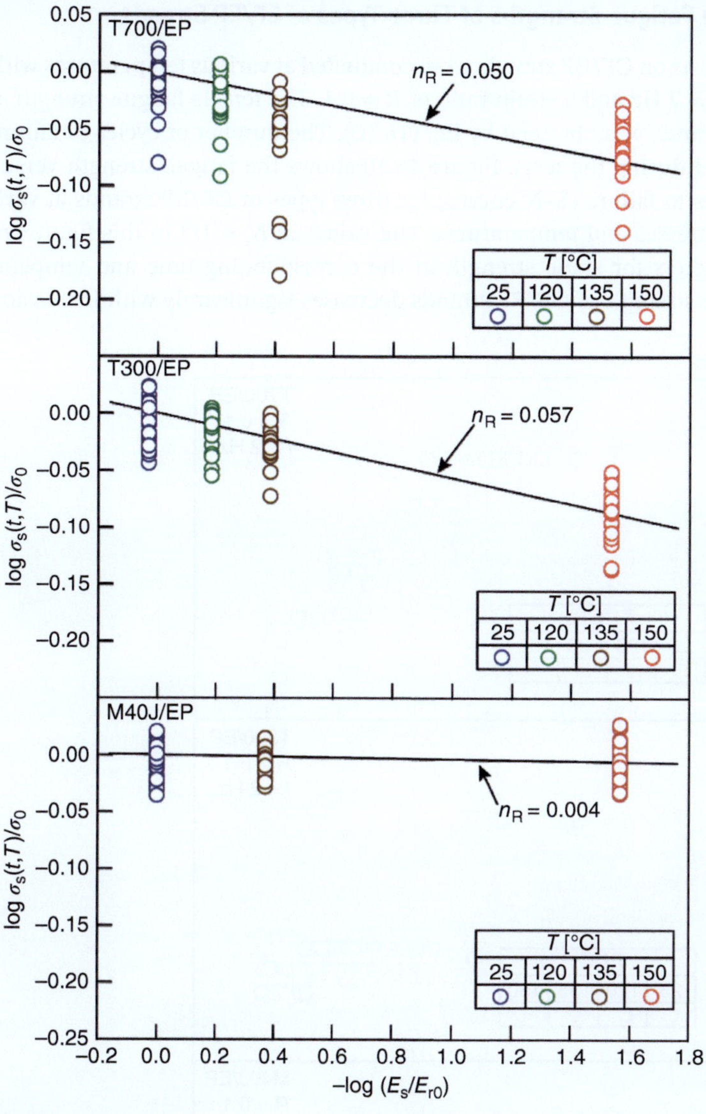

Figure 16.9 Static strengths of CF/EP strand of three kinds against the viscoelastic modulus of the matrix resin.

Table 16.5 Parameters of statistically inferred tensile static strengths of CF/EP strands.

CFRP strand	σ_0 (MPa)	α	n_R
T700/EP	5410	20	0.050
T300/EP	3720	27	0.057
M40J/EP	4355	33	0.004

16.4.3 Tensile Fatigue Strengths of Three Types of CF/EP Strands

Fatigue tensile tests on CF/EP strands were conducted at various temperatures with a frequency of $f = 2$ Hz and a strain ratio of $R = 0.1$. The tensile fatigue strength σ_f of the CF/EP strands was obtained by Eq. (16.10). The number of cycles to failure N_f was measured during the tests. Figure 16.10 shows the fatigue strength versus number of cycles to failure (S–N curves) for three types of CF/EP strands at various maximum stresses and temperatures. The values at $N_f = 1/2$ in this figure are the scale parameters for static strength at the corresponding time and temperature. The fatigue strength of CF/EP strands decreases significantly with increasing

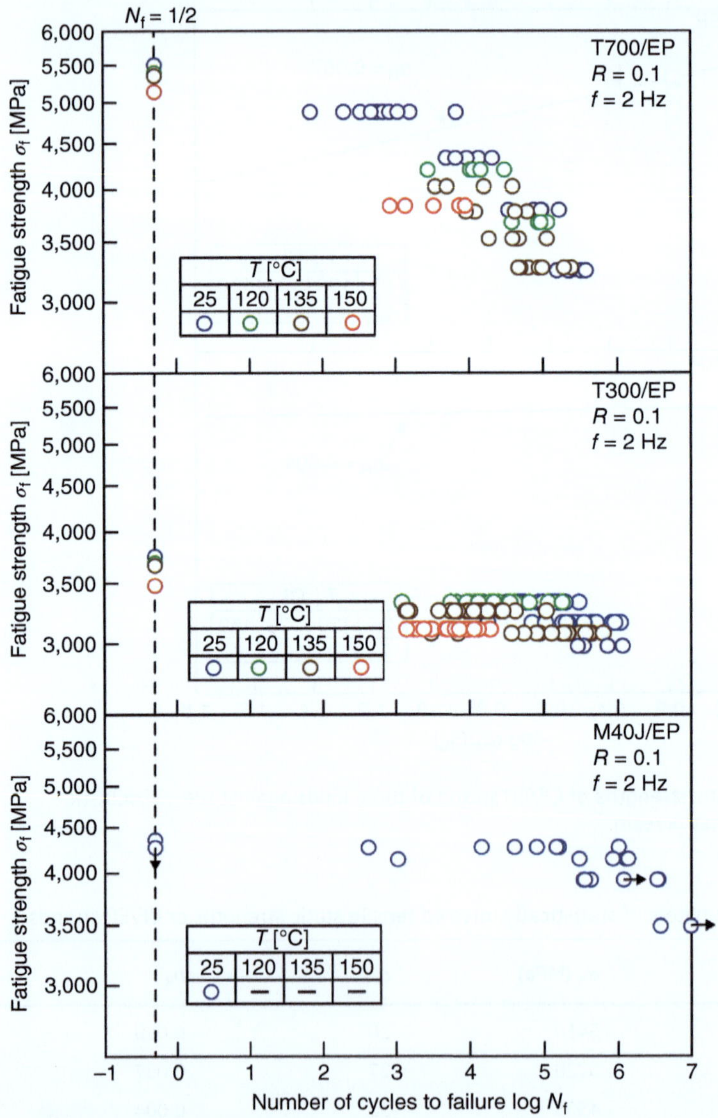

Figure 16.10 Fatigue strengths against the number of cycles to failure for CF/EP strand of three kinds.

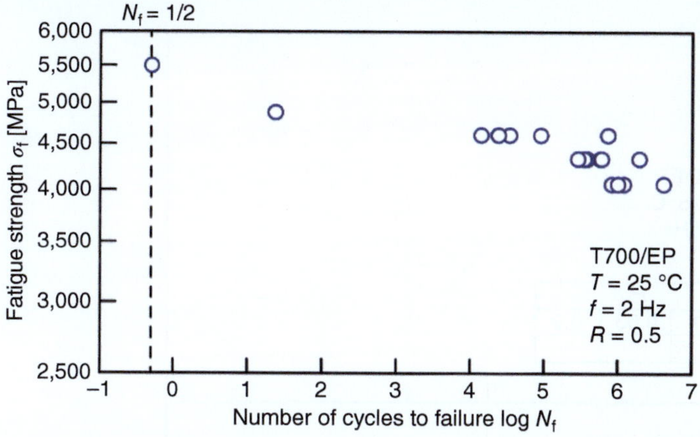

Figure 16.11 Fatigue strengths against the number of cycles to failure at 25 °C under $R = 0.5$ for T700/EP strand.

temperature and number of cycles to failure. Figure 16.11 shows the S–N curve at strain ratio $R = 0.5$ for T700/EP at $T = 25\,°C$.

In the following sections, the effects of strain ratio, temperature, and mechanical properties of carbon fibers on the tensile fatigue strength of CF/EP strands are discussed.

16.4.4 Influence of Strain Ratio on Tensile Fatigue Strength of CF/EP Strands

Figure 16.12 shows the results of tensile fatigue tests performed on T700/EP strand at 25 °C and 2 Hz for strain ratios $R = 1$, 0.5, and 0.1. The top graph in this figure shows the S–N curve for strain ratio $R = 1$, i.e. the creep failure strength versus time to failure, predicted using the viscoelastic properties of the epoxy resin shown in Figures 16.5 and 16.6 and the static strength parameters for T700/EP in Table 16.5. The bottom graph in this figure shows the experimental points for T700/EP at 25 °C and $R = 0.1$ shown in the top graph in Figure 16.10 and their approximations using the least-squares method. The parameter n_{fi} in Eq. (16.8) determined using this process is presented in the next section. The predicted line for $R = 0.5$ shown in the center of Figure 16.12 was obtained from the predicted lines for $R = 0.1$ and 1 based on the equation introduced in Section 16.2.2. The predicted values and the measured values are in good agreement. Therefore, we can infer that the concept shown in Eq. (16.5) is correct.

16.4.5 Influence of Matrix Resin Viscoelasticity on Tensile Fatigue Strength of CF/EP Strands

Figure 16.13 shows the relationship between the strength reduction due to the viscoelasticity σ_f^* of the matrix resin and the number of cycles to failure N_f at various temperatures under the conditions of $f = 2$ Hz and $R = 0.1$. These relationships are obtained by substituting the parameters in Table 16.5 into Eq. (16.2).

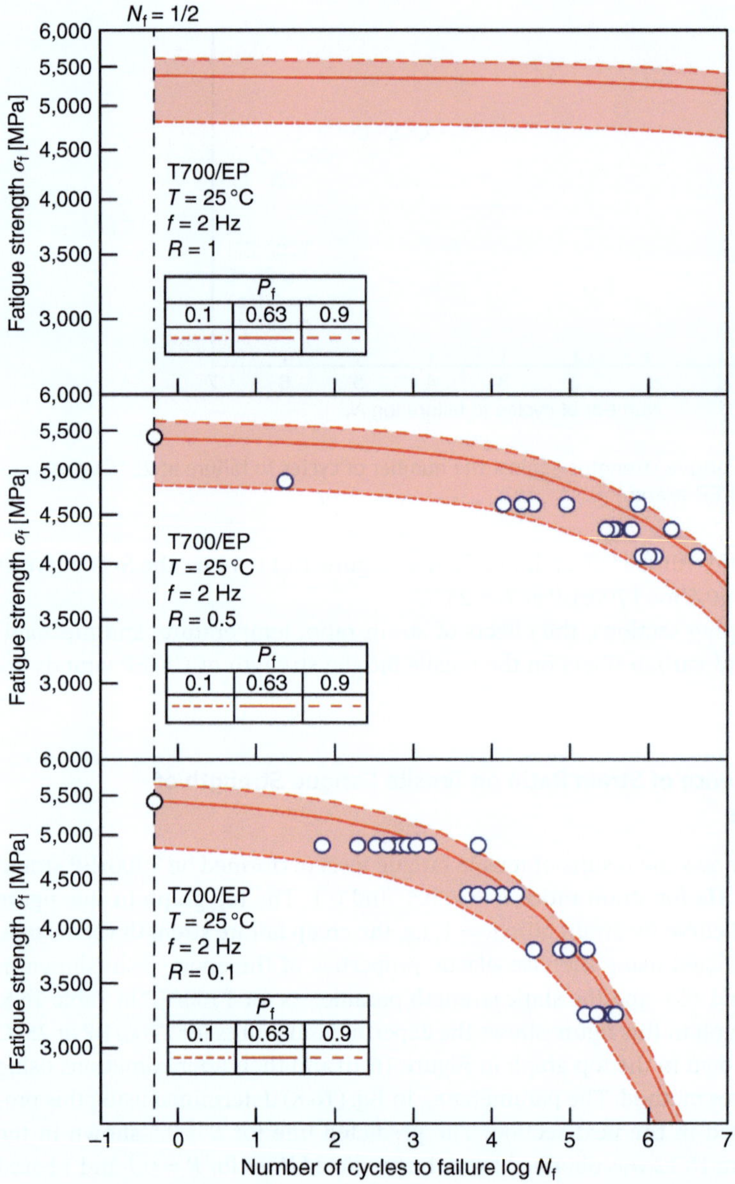

Figure 16.12 Fatigue strengths against the number of cycles to failure for T700/EP at 25 °C under $R = 1$, 0.5, and 0.1.

Figure 16.14 shows the relationship between the non-dimensional fatigue strength S_f and the number of cycles to failure N_f at various temperatures under the conditions of $f = 2$ Hz and $R = 0.1$. These relationships are obtained by substituting the measured fatigue strength σ_f versus N_f of the CF/EP strand and the viscoelastic properties of the matrix resin shown in Figures 16.5 and 16.6 into the right-hand side of Eq. (16.3). This relationship does not produce a single curve; it depends on frequency and temperature. A single curve for the non-dimensional fatigue strength S_{f0} given by Eq. (16.9) can be obtained as shown in Figure 16.15 by determining

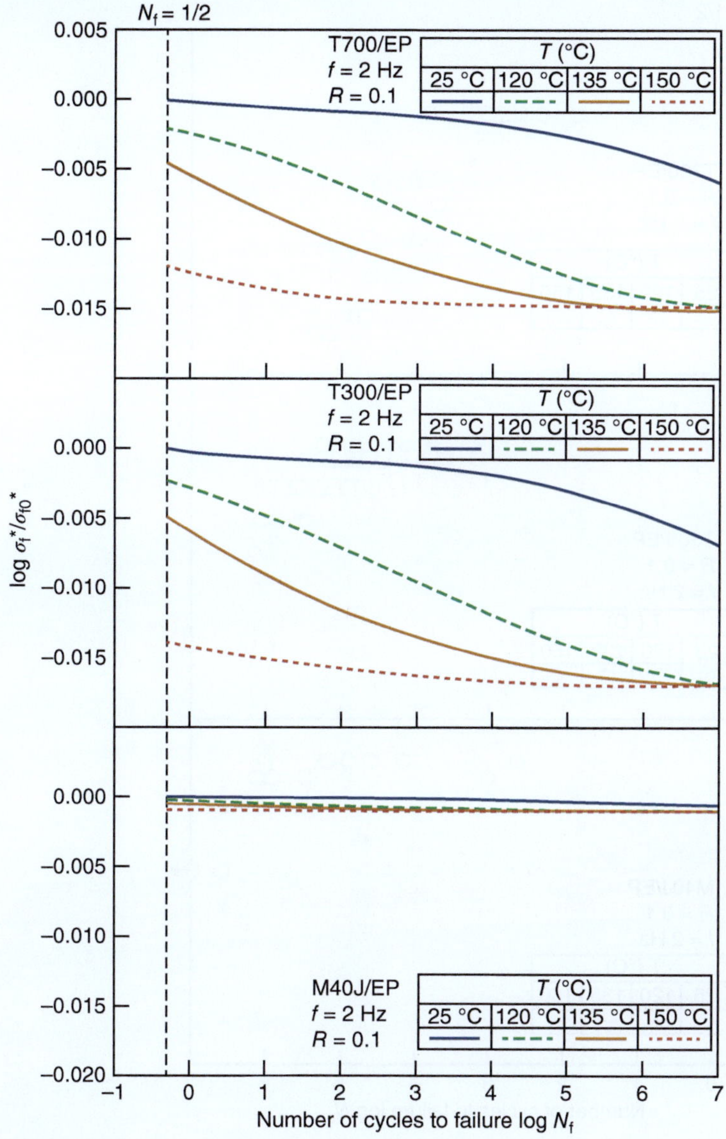

Figure 16.13 Strength reduction attributable to matrix resin viscoelasticity against the number of cycles to failure for CF/EP strands.

the parameter a_f as the slope of Figure 16.16. Figure 16.16 shows the relationship between the frequency and temperature parameters shown in Figure 16.5 and the loss tangent of the matrix resin. The parameter a_f is listed in Table 16.6. The parameters n_{fi} in this table are those at the reference conditions of $T = 25\,°C$, $f = 2$ Hz, and $R = 0.1$.

Comparing Figures 16.13 and 16.14, it can be seen that the decrease in strength due to the viscoelasticity σ_f^* of the matrix resin relative to the number of cycles to failure N_f is extremely small compared to the decrease in the non-dimensional fatigue strength S_f relative to the number of cycles to failure N_f.

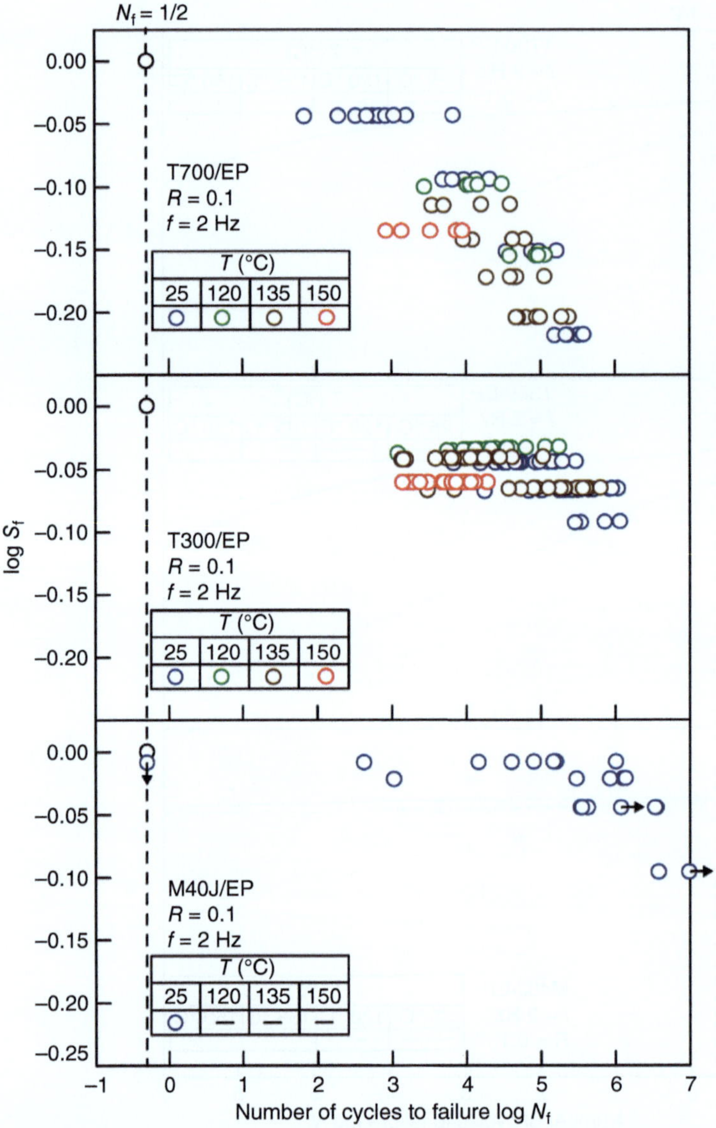

Figure 16.14 Non-dimensional fatigue strength against the number of cycles to failure for CF/EP strands.

16.4.6 Influence of Mechanical Properties of Carbon Fibers on CF/EP Strand Fatigue Strengths

Figure 16.17 shows the Weibull distribution of the strength of 25 mm long single fiber for three types of carbon fiber: T700-6000, T300-3000, and M40J-6000. From this figure, the strength shape parameter α_{cf} for the three types of carbon fiber can be determined. The ultimate strains of the three types of CF/EP strands at the reference temperature $T_0 = 25\,°C$ and the reference time $t_0 = 1$ min can be easily

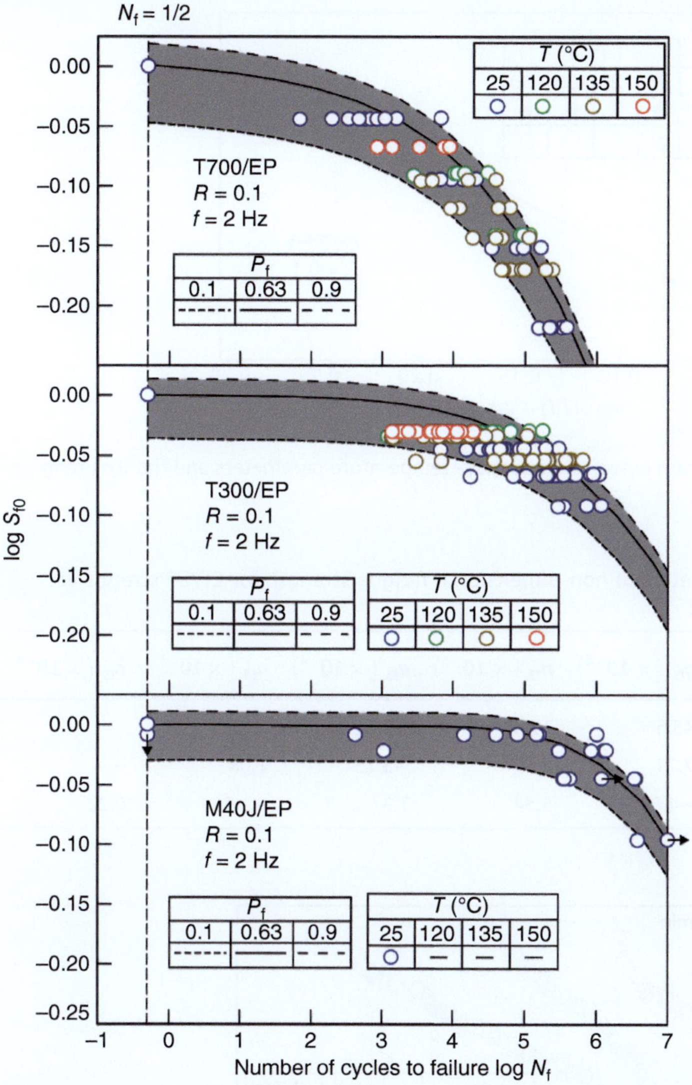

Figure 16.15 Master curves of non-dimensional fatigue strength against the number of cycles to failure for CF/EP strands.

obtained from Tables 16.1 and 16.5. Therefore, the mechanical parameter of the fiber F_f^{**} shown by Eq. (16.7) can be determined.

Figure 16.18 shows the non-dimensional fatigue strength—the number of cycles to failure curves of T300/EP and M40J/EP predicted using Eq. (16.7) with T700/EP as the reference when the strain ratio R is 0.1, the temperature T is 25 °C, and the frequency f is 2 Hz. Figure 16.19 shows the results of Figure 16.18 superimposed on Figure 16.15. The non-dimensional fatigue strength—the number of cycles to failure curves of T300/EP and M40J/EP predicted from that of T700/EP using Eq. (16.7) are in close agreement with the measured values. This figure indicates

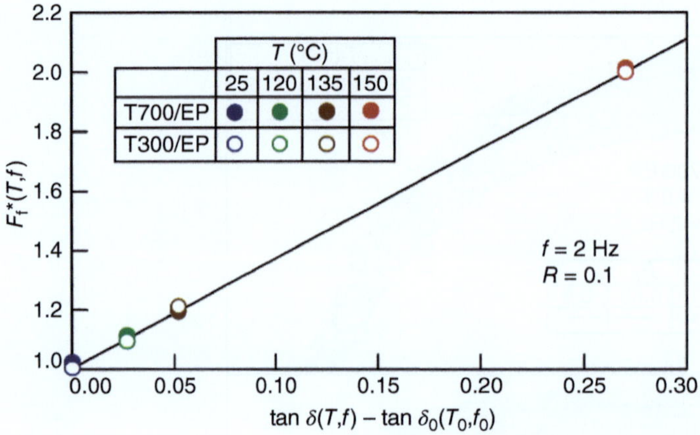

Figure 16.16 Relation between frequency–temperature parameters and the loss tangent of the matrix resin.

Table 16.6 Parameters of non-dimensional fatigue strength for CF/EP strands at $R = 0.1$ and $f = 2$ Hz.

	a_f	n_{f1} $(\times 10^{-4})$	n_{f2} $(\times 10^{-4})$	n_{f3} $(\times 10^{-4})$	n_{f4} $(\times 10^{-4})$	n_{f5} $(\times 10^{-4})$
T700/EP	3.7	45.6	19.8	−3.21	1.85	–
T300/EP	3.7	7.79	2.69	−3.13	1.08	–
M40J/EP	–	−4.09	5.54	1.33	−1.46	0.22

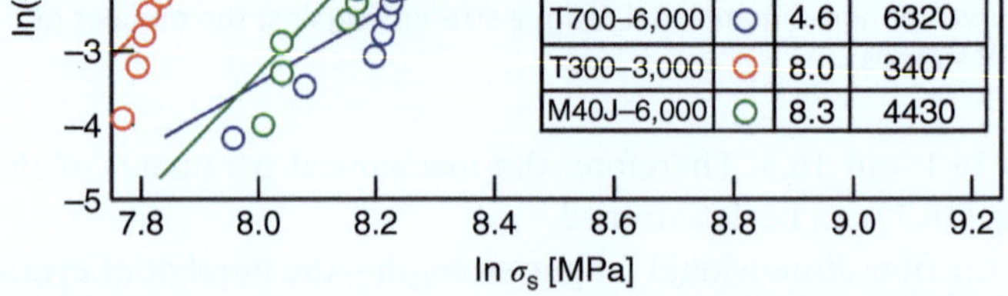

Figure 16.17 Weibull distributions for tensile strengths for carbon fibers of three types.

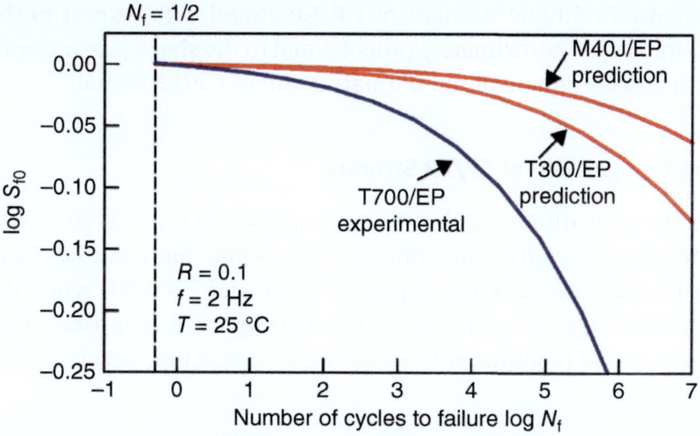

Figure 16.18 Non-dimensional fatigue strength against the number of cycles to failure for T300/EP and M40J/EP predicted from that for T700/EP.

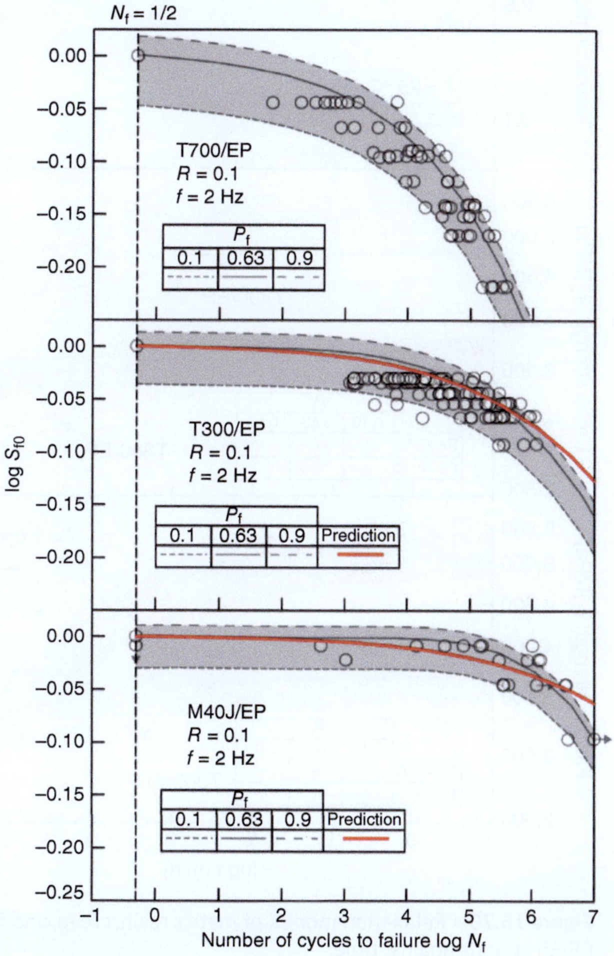

Figure 16.19 Comparing prediction with experiments for non-dimensional fatigue strength against the number of cycles to failure curves of T300/EP and M40J/EP.

that the rate of decrease in fatigue strength of CF/EP strand with respect to the number of cycles to failure is approximately proportional to the shape parameter of carbon fiber strength and the square of the ultimate strain of CF/EP strand.

16.4.7 Long-term Fatigue Life of CF/EP Strands

Figure 16.20 shows the relaxation modulus of matrix resin at 25 °C and 70 °C and the decrease in fatigue strength with time at the same temperatures and $f = 1$ cycle/min for CF/EP strands of two types, T700/EP and T300/EP, when the strain ratios of the CF/EP strands are 1 and 0. From this figure, it is apparent that the decrease of the relaxation modulus of the resin with elapsed time accelerates as

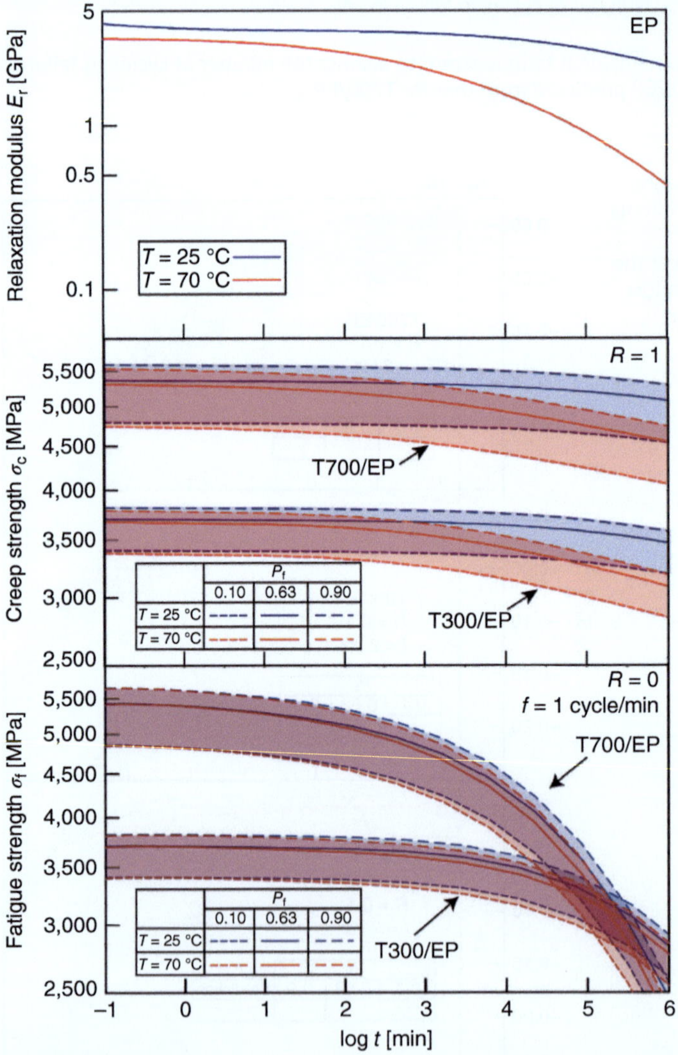

Figure 16.20 Relaxation moduli of matrix resin, creep, and fatigue strengths of CF/EP strands against time.

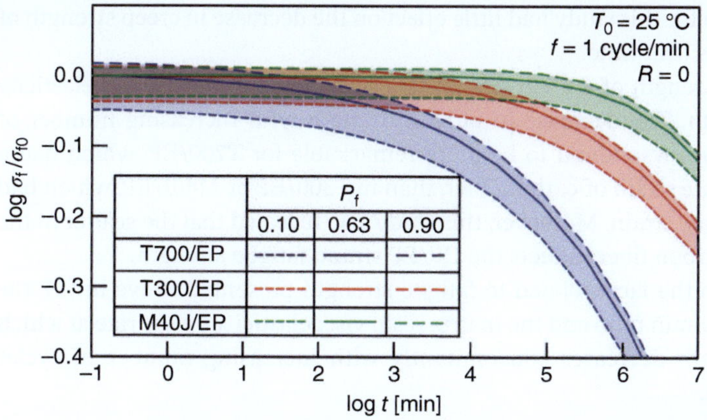

Figure 16.21 Non-dimensional fatigue strengths against elapsed time of CF/EP strands of three kinds.

the temperature increases. The decrease in the creep strength of the CF/EP strand also accelerates similarly. For fatigue strength, such temperature dependence is observed only slightly. The difference in the decrease attributable to the difference in carbon fiber becomes clear.

These characteristics of fatigue strength are indicated clearly in Figure 16.21. The rate at which fatigue strength decreases with increasing number of cycles to failure clearly depends on the fiber. The mechanism of this decrease is explained in the preceding section.

16.5 Conclusion

The influence of carbon fiber and matrix resin on the statistical life of CF/EP strands under constant and cyclic tension loadings was evaluated using Integrated ATM. The main results are described below.

1. The role of carbon fiber and resin in the generation of energy loss during cyclic loading was clarified for the longitudinal direction of unidirectional CFRP. Especially, it was found that the fatigue degradation parameter F_f representing the decrease in fatigue strength caused by cyclic loading is expressed as the product of four functions. They are the strain ratio parameter F_f^R determined by the strain ratio, the frequency–temperature parameter F_f^* determined by matrix resin viscoelasticity, the fiber mechanical parameter F_f^{**} determined by carbon fiber mechanical properties and the fatigue parameter F_f' determined by the number of cycles to failure. The three parameters of F_f^R, F_f^*, and F_f^{**} were formulated based on the roles of carbon fiber and matrix resin. Moreover, the validity of the formulations was clarified through experimentation.
2. The creep strength of CF/EP strands, which depends on the matrix resin viscoelasticity, decreases markedly with increasing the elapsed time and increasing temperature. Differences in mechanical properties in carbon fiber of three

kinds selected for this study had little effect on the decrease in creep strength of CF/EP strands over time.

3. The fatigue strength of CF/EP strands is less dependent on the viscoelasticity of matrix resin. The decrease in fatigue strength with increasing number of cycles to failure was found to be more remarkable for T700/EP, which had a greater ultimate strain of carbon fiber, than in T300/EP or M40J/EP, which had a lower ultimate strain. Moreover, this study also revealed that the scatter in the strength of carbon fibers affects the CF/EP strand fatigue strength.
4. In addition to the facts related to fatigue strength presented above in (3), the effects of the strain ratio and the matrix resin viscoelasticity on the rate at which fatigue strength decreases concomitantly with increasing number of cycles were clarified.

References

1 Miyano, Y., Nakada, M., Takeuchi, Y. and Morisawa, Y. (2025) Carbon fiber and matrix resin mechanical properties controlling statistical tensile fatigue life of unidirectional CFRP. *Journal of Composite Materials*. https://doi.org/10.1177/00219983251371581.

2 Miyano, Y. and Nakada, M. (2024) Integrated accelerated testing methodology for CFRP durability. *Composites Part B*, **281**, 111527.

17

Application 3 of Integrated Accelerated Testing Methodology: Influence of Mechanical Properties of Carbon Fiber on Statistical Creep and Fatigue Lives of Carbon-fiber-reinforced Polymer Strands with Thermoplastic Epoxy Resin as Matrix

17.1 Introduction

This chapter evaluates the durability of thermoplastic epoxy resin-impregnated carbon fiber strands (CF/TPEP strands) industrially molded by the pultrusion method [1, 2]. Specifically, the creep and fatigue lives of CF/TPEP strands using two types of carbon fibers, T700-6000 and T300-3000, are evaluated using the integrated accelerated testing methodology (Integrated ATM) [3], as shown in Chapter 14, and the influence of the mechanical properties of the carbon fibers on the lives of the CF/TPEP strands is examined. Furthermore, the creep and fatigue lives of CF/TPEP strands are compared with those of epoxy resin-impregnated carbon fiber strands (CF/EP strands) using a thermosetting epoxy matrix evaluated in Chapter 16.

17.2 Experimental Procedure

17.2.1 Specimen Preparation

CF/TPEP strands consisting of high-strength type carbon fibers of two types, T700-6000 and T300-3000, and thermoplastic epoxy were molded using the pultrusion method. Table 17.1 presents the characteristics of the two types of carbon fibers used for this study and their mechanical properties. Table 17.2 shows the matrix epoxy composition and the curing conditions for the CF/TPEP strands.

17.2.2 Static, Creep, and Fatigue Tests of CF/TPEP Strands

The specimen configurations of T700/TPEP and T300/TPEP strands are just the same to those of T700/EP and T300/EP strands presented in Chapter 16. The glass transition temperature T_g of the thermoplastic epoxy was 102 °C, as ascertained from the peak of loss tangent against temperature at 1 Hz using dynamic mechanical analysis (DMA). Static, creep, and fatigue tension tests for two types of CF/TPEP strands were conducted reliably using our developed testing system, depicted in Chapter 16. The number of tests for each condition on static, creep, and fatigue tension tests is shown in Table 17.3.

Properties and Testing of Fiber-Reinforced Polymers, First Edition. Yasushi Miyano and Masayuki Nakada. © 2026 Wiley-VCH GmbH. All rights reserved, including rights for text and data mining and training of artificial intelligence technologies or similar technologies.
Published 2026 by Wiley-VCH GmbH.

Table 17.1 Specifications of carbon fibers.

Carbon fiber	Density (g/cm^3)	Tex (g/km)	Elastic modulus (GPa)	Tensile strength (MPa)
T700-6000	1.80	400	230	4900
T300-3000	1.76	198	230	3530

Table 17.2 Compositions and cure schedules of CF/TPEP strands.

CFRTP strand	Carbon fiber	Composition of resin (weight ratio)	Product name of resin (company)	Cure schedule
T700/TPEP	T700-6000	Thermoplastic epoxy (100) + Cure accelerator (6.5)	XNR6850V (Nagase ChemteX Corp.)	150 °C × 0.5 h
T300/TPEP	T300-3000		XNH6850V (Nagase ChemteX Corp.)	

Table 17.3 Number of tests for each condition on static, creep, and fatigue tension tests.

(a) Static

	25 °C	80 °C	87 °C	90 °C
T700/TPEP	20	22	20	20
T300/TPEP	20	20	–	20

(b) Creep

	Temperature (°C)	$\sigma_c/\sigma_0 = 0.9$	$\sigma_c/\sigma_0 = 0.85$
T700/TPEP	70	20	20
T300/TPEP		19	20

(c) Fatigue

	Temperature (°C)	σ_f/σ_0									
		0.9	0.87	0.85	0.82	0.8	0.72	0.65	0.57	0.5	0.47
T700/TPEP	25	13	–	10	–	10	12	3	2	2	2
	80	5	–	–	–	5	–	–	–	–	–
	87	–	–	–	–	5	–	–	–	–	–
	90	–	–	–	–	3	–	–	–	–	–
T300/TPEP	25	11	21	–	12	–	–	–	–	–	–
	75	–	10	–	21	–	–	–	–	–	–
	80	–	–	–	20	–	–	–	–	–	–
	90	–	–	–	20	–	–	–	–	–	–

17.3 Results and Discussion

The following evaluations will be carried out according to the procedures of Integrated ATM described in Chapter 14.

17.3.1 Relaxation Modulus of TPEP Resin

The relaxation moduli E_r for thermoplastic epoxy (TPEP) neat resin at various temperatures were obtained approximately as the inverse of creep compliance, as shown in the left panel in Figure 17.1. Long-term values of E_r at $T = 25\,°C$ were obtained by shifting these values horizontally and vertically at various temperatures based on the modified time–temperature superposition principle (modified TTSP), as shown in the right panel of Figure 17.1. The time–temperature (horizontal) shift and temperature (vertical) shift factors are presented in Figure 17.2. The parameters expressing the master curve of the relaxation modulus of the matrix resin defined in Chapter 14 are shown in Table 17.4 for TPEP resin.

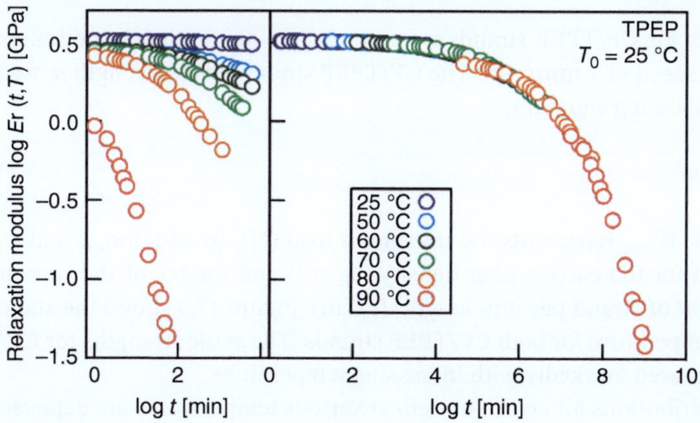

Figure 17.1 Master curve of relaxation modulus for thermoplastic epoxy resin.

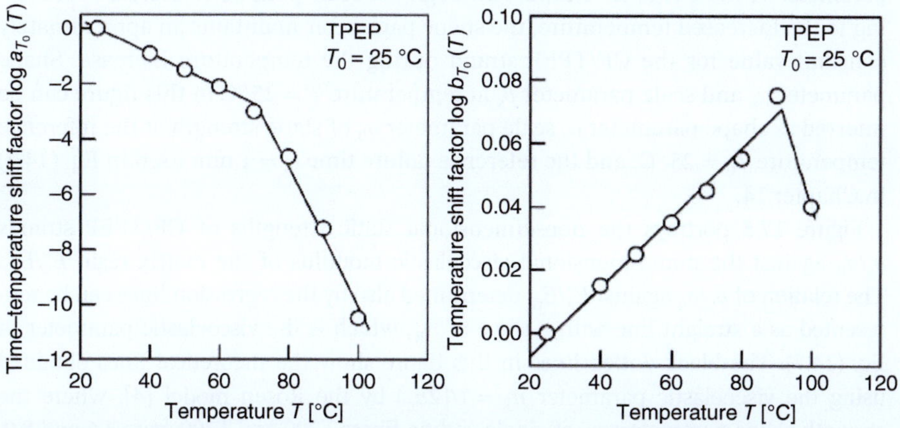

Figure 17.2 Horizontal and vertical shift factors of the relaxation modulus for thermoplastic epoxy resin.

Table 17.4 All parameters expressing the master curve of the relaxation modulus of TPEP resin.

$E_r\,(t_0, T_0\,)$ (GPa)	3.17
m_1	0.02
m_2	0.77
t_g (min)	5.6×10^6
ΔH_1 (kJ/mol)	132
ΔH_2 (kJ/mol)	764
T_{gh} (°C)	75
b_1	0.001
b_2	−0.001
T_{gv} (°C)	75

17.3.2 Static Strength of CF/TPEP Strands with Two Types of Carbon Fibers

Static tension tests for CF/TPEP strands were conducted at various temperatures with a cross-head speed of 2 mm/min. The CF/TPEP strand tensile strength σ_s was found using the following equation.

$$\sigma_s = \frac{W_{max}}{t_e}\rho \tag{17.1}$$

In this equation, W_{max} represents the maximum load (N). In addition, ρ and t_e, respectively, stand for the carbon fiber density (kg/m^3) and the tex of the carbon fiber strand (weight of strand per unit length) (g/km). Figure 17.3 shows the static strength versus temperature for both CF/TPEP strands. The static strengths for CF/TPEP strands decreased markedly with increasing temperature.

The Weibull distributions for static strength at various temperatures are depicted in Figure 17.4 for the CF/TPEP strands. The parameters were determined from the regression lines. In these figures, α_s is the shape parameter and β_s is the scale parameter of the CF/TPEP strand. Although the scale parameter decreases according to the increased temperature, the shape parameter maintains an approximately constant value for the CF/TPEP strand during the temperature increase. Shape parameter α_s and scale parameter β_s at temperature $T = 25\,°C$ in this figure can be inferred as shape parameter α, scale parameter σ_0 of static strength at the reference temperature $T_0 = 25\,°C$, and the reference failure time $t_0 = 1$ min used in Eq. (14.7) in Chapter 14.

Figure 17.5 portrays the non-dimensional static strengths of CF/TPEP strands σ_s/σ_0 against the non-dimensional viscoelastic modulus of the matrix resin E_s^*/E_{r0}. The relation of σ_s/σ_0 against E_s^*/E_{r0} determined also by the regression lines can be represented as a straight line with a slope of n_R, which is the viscoelastic parameter in Eq. (14.7). The black dotted lines in this figure show the theoretical lines obtained using the viscoelastic parameter $n_R = 1/(2\alpha_c)$ by the Rosen model [4], where the strength shape parameters α_c of single carbon fibers T700 and T300 were 4.6 and 8.0, as shown in Figure 17.6. All measured static strengths agree well with the theoretical

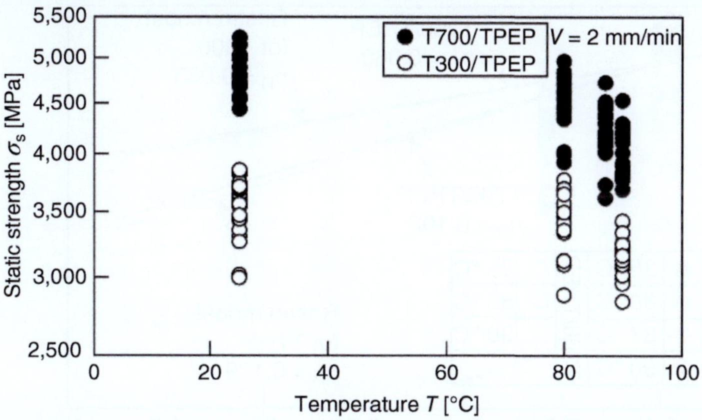

Figure 17.3 Static strengths of CF/TPEP strands against temperature.

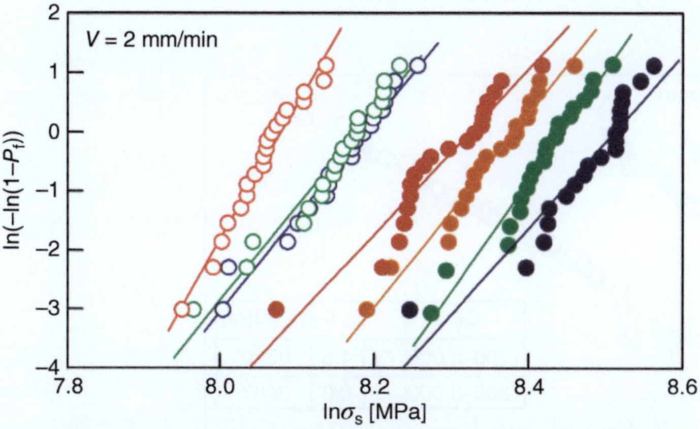

		T [°C]	α_s (–)	β_s [MPa]
T300/TPEP	○	25	16	3,600
	○	80	16	3,570
	○	90	23	3,230
T700/TPEP	●	25	15	4,970
	●	80	19	4,650
	●	87	16	4,360
	●	90	14	4,120

Figure 17.4 Weibull distributions of the static strength of CF/TPEP strands at various temperatures.

lines. Therefore, the viscoelastic parameter n_R can be found using the Rosen model in the cases of combinations of carbon fibers and matrix resin selected in this chapter.

Parameters σ_0, α, and n_R for CF/TPEP strands are presented in Table 17.5. Parameters n_R of CF/TPEP strands representing the influence of viscoelasticity clearly depend on the carbon fiber type.

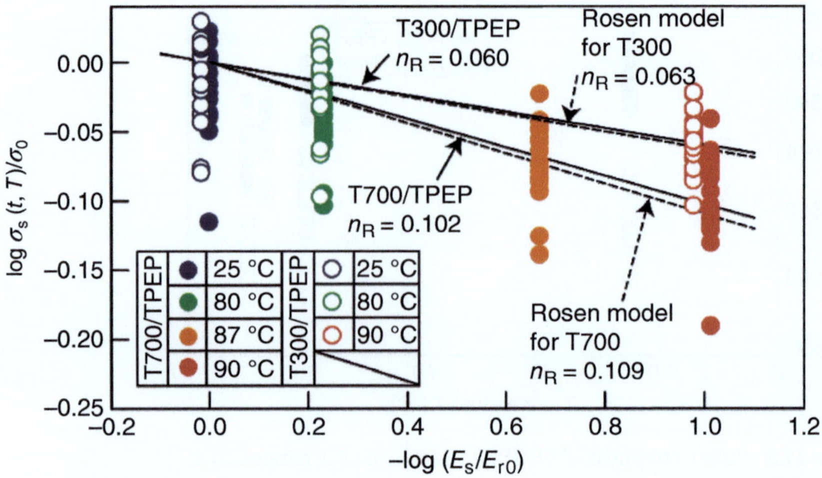

Figure 17.5 Static strength of CF/TPEP strands against the viscoelastic modulus of the matrix resin.

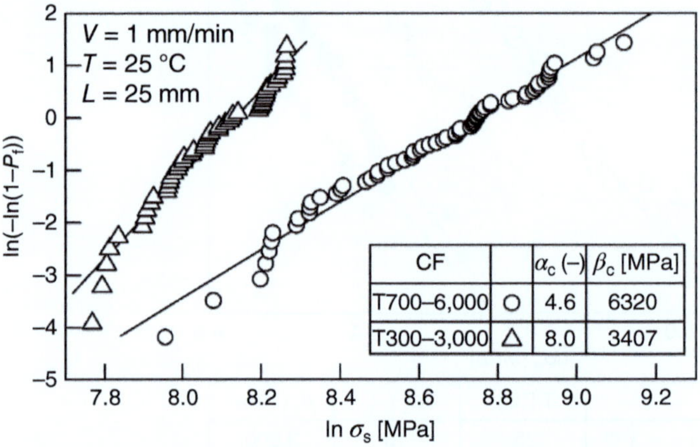

Figure 17.6 Weibull distributions for tensile strengths of T700 and T300 single fibers.

Table 17.5 Parameters of statistically determined tensile static strength for CF/TPEP strands.

	σ_0 (MPa)	α	n_R
T700/TPEP	4970	15	0.102
T300/TPEP	3600	16	0.060

17.3.3 Creep Strength of CF/TPEP Strands with Two Types of Carbon Fibers

Creep failure tests of CF/TPEP strands were conducted at arbitrary constant stress levels and at a constant temperature. The results are presented in Figure 17.7. Tensile creep strengths σ_c of CF/TPEP strands are also predicted by substituting the parameters of Tables 17.4 and 17.5 into Eq. (14.8) in Chapter 14. The predicted results

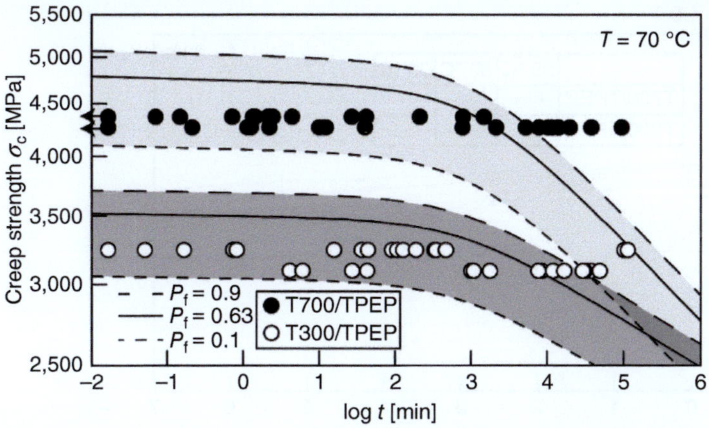

Figure 17.7 Predicted and experimentally obtained creep strengths.

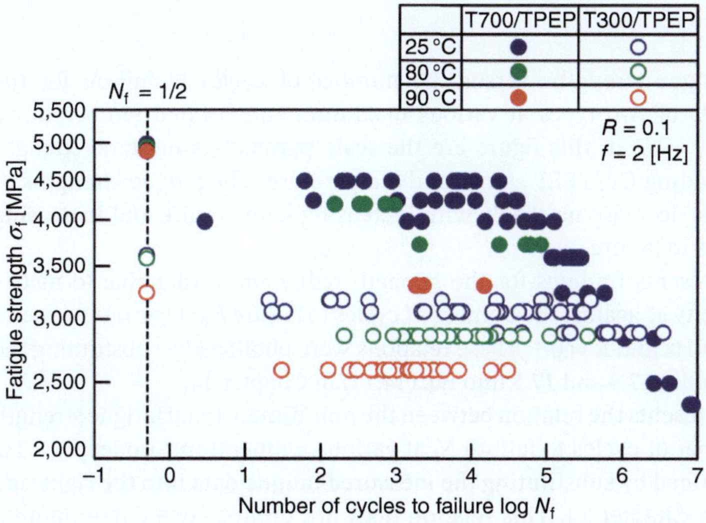

Figure 17.8 Fatigue strengths of CF/TPEP strands against the number of cycles to failure.

are presented in Figure 17.7. The predicted statistical creep failure time agrees well with the experimentally obtained data. Therefore, the results clarified that the statistical creep failure time can be predicted from the relaxation modulus of the matrix resin and the static strengths of CF/TPEP strands measured at various temperatures.

17.3.4 Fatigue Strength of CF/TPEP Strands with Two Types of Carbon Fibers

Fatigue tension tests for CF/TPEP strands were conducted with frequency $f = 2$ Hz and the strain ratio $R = 0.1$ at the three temperatures of 25 °C, 80 °C, and 90 °C. Tensile fatigue strengths σ_f of CF/TPEP strands were obtained using Eq. (17.1). The numbers of cycles to failure N_f were measured through testing. Figure 17.8

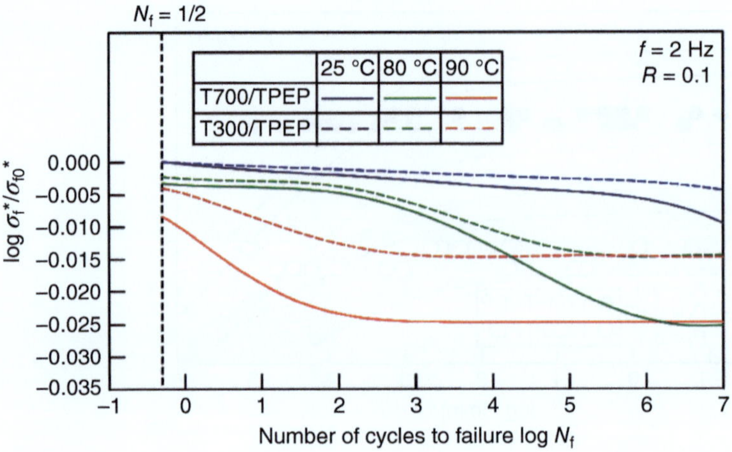

Figure 17.9 Strength reduction attributable to matrix resin viscoelasticity against the number of cycles to failure for CF/TPEP strands.

presents the fatigue strengths versus the number of cycles to failure for the CF/TPEP strands of two types at various maximum stresses and temperatures. The values at $N_f = 1/2$ in this figure are the scale parameters of static strength for the corresponding CF/TPEP strand and temperature. The fatigue strengths of CF/TPEP strands decrease markedly with increasing temperature and increasing number of cycles to failure.

Figure 17.9 presents findings for the strength reduction attributable to matrix resin viscoelasticity σ_f^* against the number of cycles to failure N_f at various temperatures under $f = 2$ Hz and $R = 0.1$. These relations were obtained by substituting the parameters of Tables 17.4 and 17.5 into Eq. (14.12) in Chapter 14.

Figure 17.10 presents the relation between the non-dimensional fatigue strength S_f and the number of cycles to failure N_f at various temperatures under $f = 2$ Hz and $R = 0.1$ obtained by substituting the measured fatigue data into the right side of Eq. (14.13) in Chapter 14. This relation does not produce one curve; moreover, it is frequency–temperature dependent. The single curve of non-dimensional fatigue strength S_{f0} shown by Eq. (14.17) is obtainable as presented in Figure 17.11 by determining parameter a_f as the slope of Figure 17.12. Figure 17.12 depicts the relation between the frequency–temperature parameter and the loss tangent of the matrix resin presented in Figure 17.13. Parameter a_f is shown in Table 17.6.

Comparing Figures 17.9 and 17.10, the decrease in the strength reduction is attributable to the matrix resin viscoelasticity σ_f^* against the number of cycles to failure N_f is extremely small compared to the decrease in the non-dimensional fatigue strength S_f against the number of cycles to failure N_f. Figure 17.11 shows that the decrease in the master curves of non-dimensional fatigue strength S_{f0} varies greatly depending on the type of carbon fiber. This will be discussed in the next section together with the case of CF/EP strands.

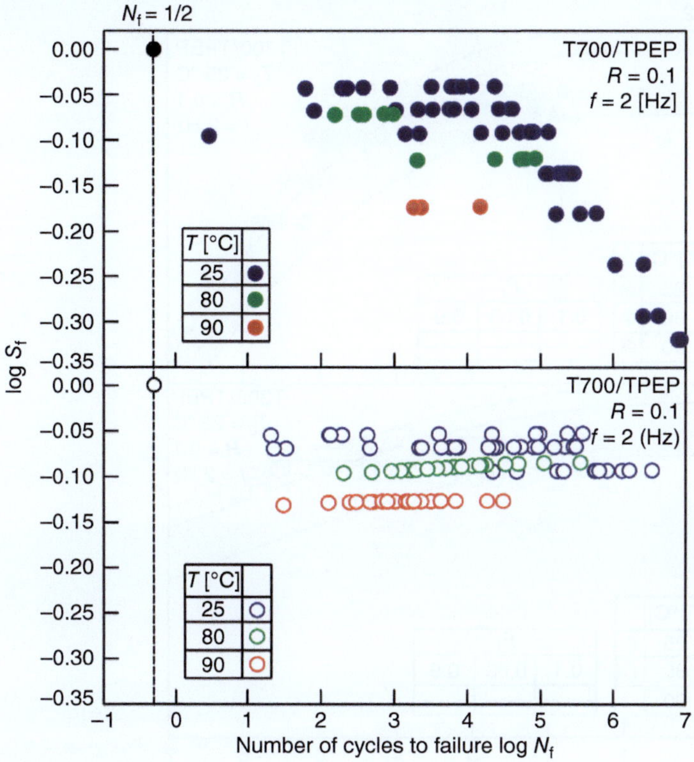

Figure 17.10 Non-dimensional fatigue strength against the number of cycles to failure for CF/TPEP strands.

17.3.5 Influence of Mechanical Properties of Carbon Fibers on Creep and Fatigue Strengths of CF/TPEP Strands

Figure 17.14 shows the relaxation moduli of TPEP resin, the creep strengths, and the fatigue strengths at strain ratio $R = 0$ and frequency $f = 1$ cycle/min of CF/TPEP strands, respectively, against the elapsed time at 25 °C and 70 °C, which are obtained by the procedure of Integrated ATM described Chapter 14 using the parameters of Tables 17.4–17.6. The upper part of this figure shows that the relaxation modulus of the matrix resin decreases markedly over time. The degree of this decrease is accelerated by an increase in temperature. The middle and bottom parts of this figure show that both of creep and fatigue strengths of CF/TPEP strands using high-strength T700 carbon fibers are higher than those using low-strength T300 carbon fibers. The creep strength of CF/TPEP strands decreases remarkably over time and temperature in the similar behavior of the matrix resin, as shown in the middle part of this figure. By contrast, the fatigue strength of CF/TPEP strands changes slightly with changes in time and temperature, with a slight effect of the matrix viscoelasticity, as shown in the bottom part of this figure.

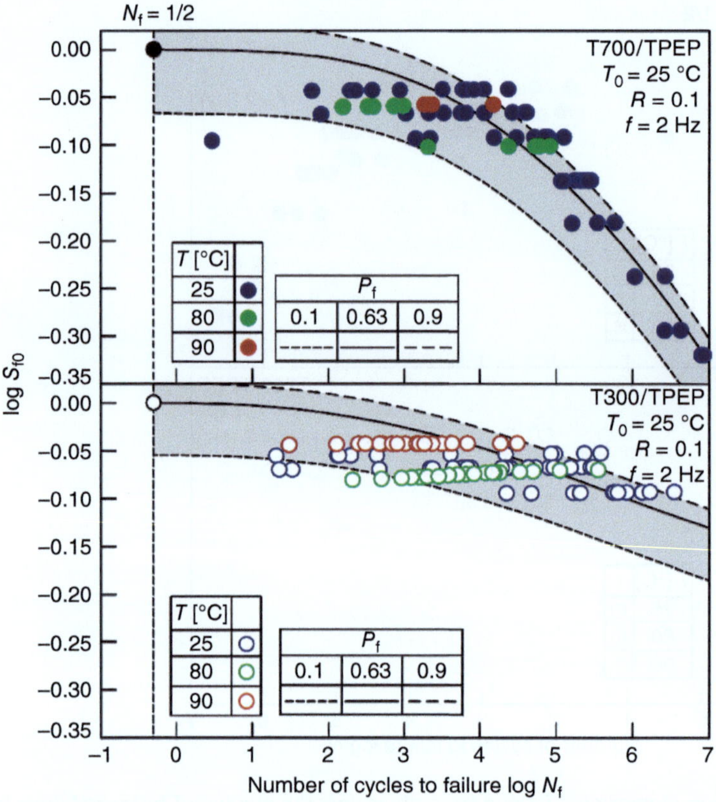

Figure 17.11 Master curve of non-dimensional fatigue strength against the number of cycles to failure for CF/TPEP strands.

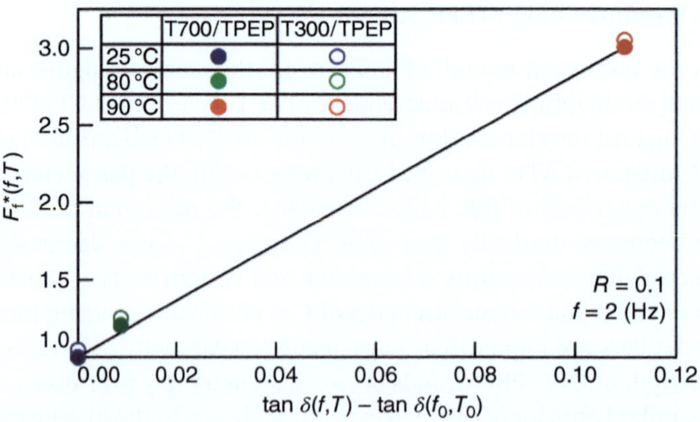

Figure 17.12 Relation between frequency–temperature parameter against the loss tangent of the matrix resin.

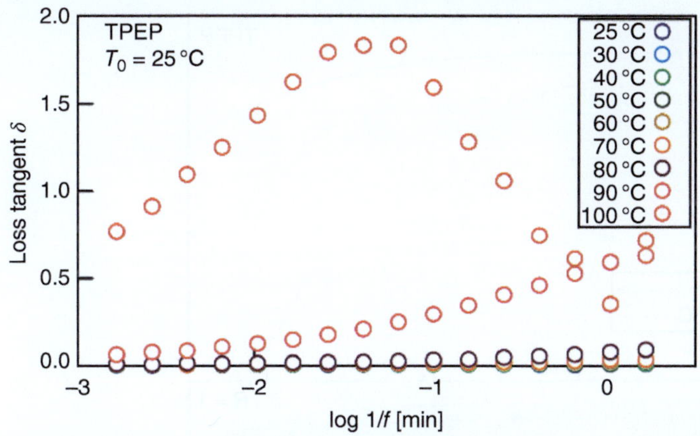

Figure 17.13 Loss tangent against frequency for matrix resin at various temperatures.

Table 17.6 Parameters of non-dimensional fatigue strength for CF/TPEP strand.

	a_f	n_{f1} $(\times 10^{-4})$	n_{f2} $(\times 10^{-4})$	n_{f3} $(\times 10^{-4})$	n_{f4} $(\times 10^{-4})$
T700/TPEP	18.2	14.2	17.1	17.3	0.81
T300/TPEP	18.2	3.0	18.8	4.0	0.42

Figure 17.15 shows the non-dimensional creep and fatigue strengths against elapsed time of two types of CF/TPEP strands, with T300 and T700 carbon fibers at 25 °C. The degradation rate of creep strength against time of CF/TPEP strands using T700 carbon fibers having a low shape parameter for the strength is higher than that using T300 carbon fibers having a high shape parameter for strength, as shown in the upper panel of this figure. The rate of fatigue strength degradation against elapsed time or the number of cycles to failure of CF/TPEP strands using T700 carbon fibers having a high ultimate strain is higher than that obtained using T300 carbon fibers having a low ultimate strain, as the bottom panel shows.

17.3.6 Creep and Fatigue Lives of CF/EP Strands and Their Comparison with CF/TPEP Strands

In Chapter 16, we predicted the creep and fatigue lives of CF/EP strands that use the same carbon fibers, T700-6000 and T300-3000, as the reinforcing fibers and a thermosetting epoxy resin as the matrix resin.

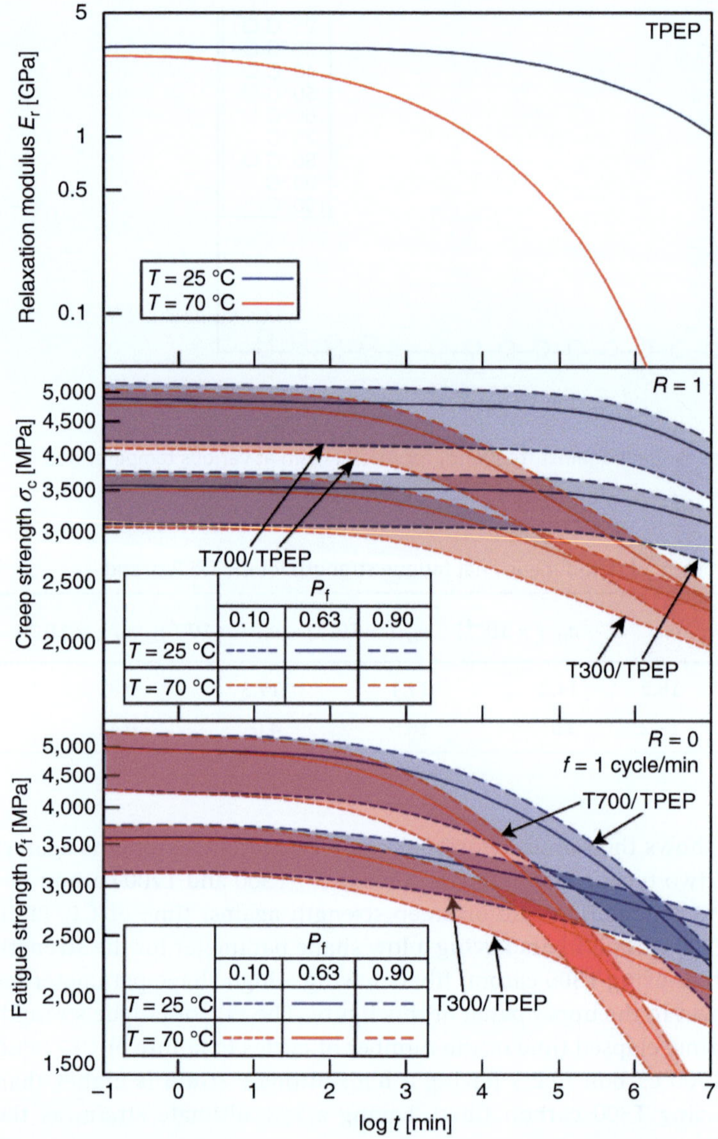

Figure 17.14 Relaxation moduli of the matrix resin, creep, and fatigue strengths of CF/TPEP strands against time.

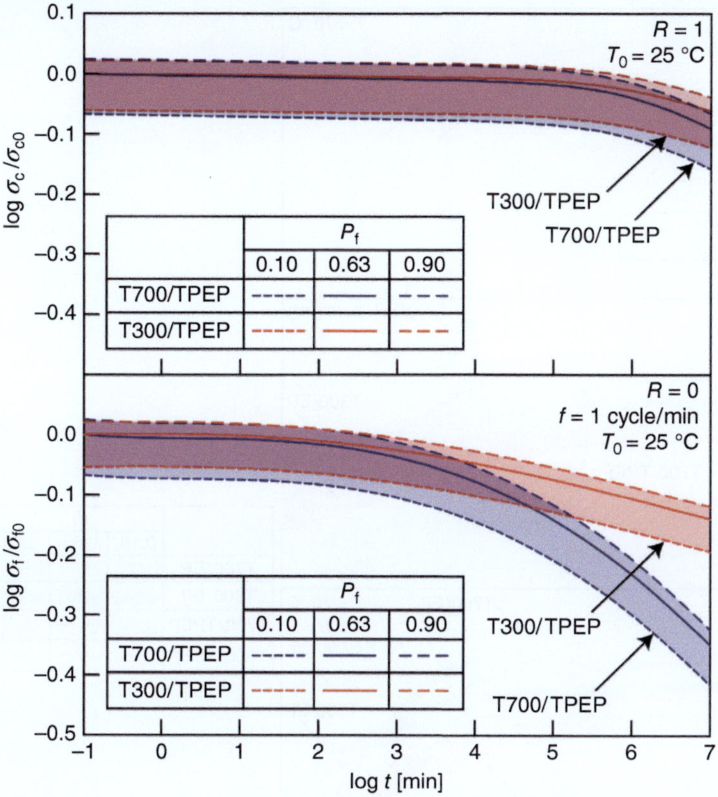

Figure 17.15 Non-dimensional creep and fatigue strengths against elapsed time of CF/TPEP strands of two types.

Figure 17.16 shows the relaxation moduli of EP and TPEP resins, the creep strengths, and the fatigue strengths at strain ratio $R = 0$ and frequency $f = 1$ cycle/min of CF/EP and CF/TPEP strands, respectively, against the elapsed time at 70 °C. The upper part of this figure shows that the relaxation modulus of TPEP resin decreases markedly over time compared with that of EP resin. The middle and bottom parts of this figure show the creep and fatigue strengths of CF/EP and CF/TPEP strands. The creep and fatigue strengths of CF/TPEP strands are lower than those using EP resin. The value is low over the entire time range. This phenomenon in the long-term range is due to the difference in the viscoelastic properties of the matrix resin, while the phenomenon

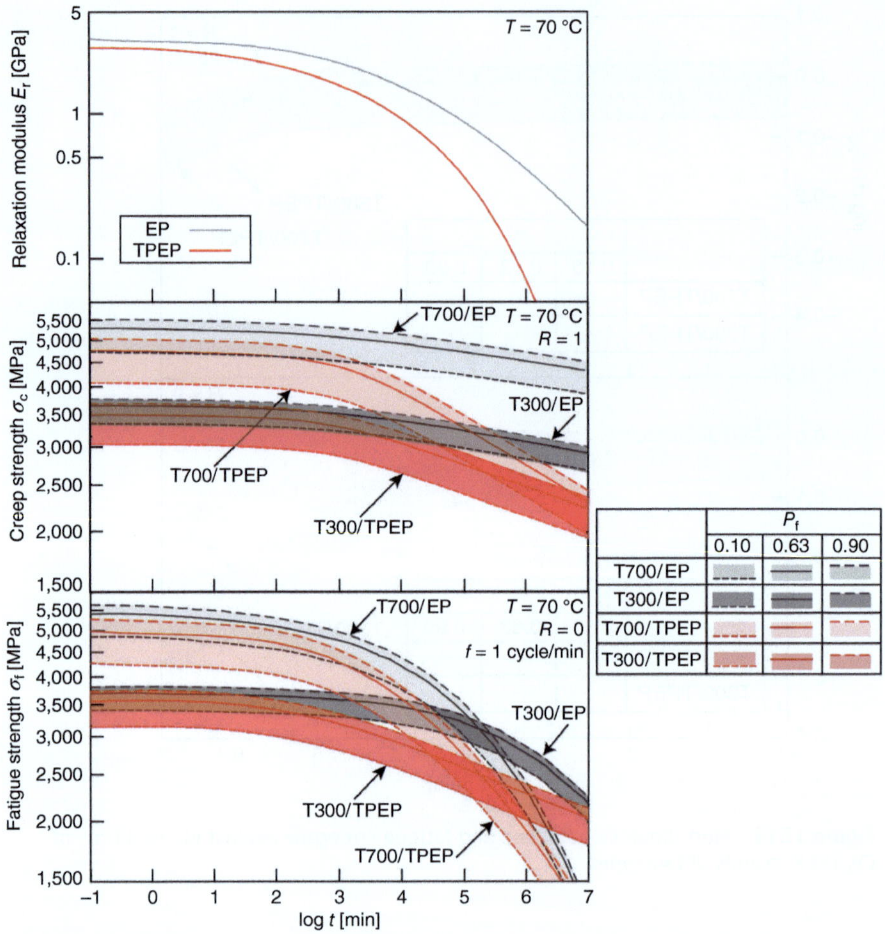

Figure 17.16 Relaxation moduli of the matrix resin, creep, and fatigue strengths against time at 70 °C for CF/EP and CF/TPEP strands with two types of carbon fiber.

in the short-term range is presumably due to the fact that the CF/TPEP strands are industrially molded by extrusion molding, causing disorder in the fibers.

Figure 17.17 shows the non-dimensional creep and fatigue strengths against elapsed time of CF/EP and CF/TPEP strands at 70 °C. The degradation rate of creep strength against time of CF/TPEP strands is higher than that of CF/EP strands, as shown in the upper panel of this figure because of the difference in viscoelasticity of the matrix resin. The rate of fatigue strength degradation against elapsed time or the number of cycles to failure for both of CF/EP and CF/TPEP strands using T700 carbon fibers having a high ultimate strain is higher than that obtained using T300 carbon fibers having a low ultimate strain, as the bottom panel shows.

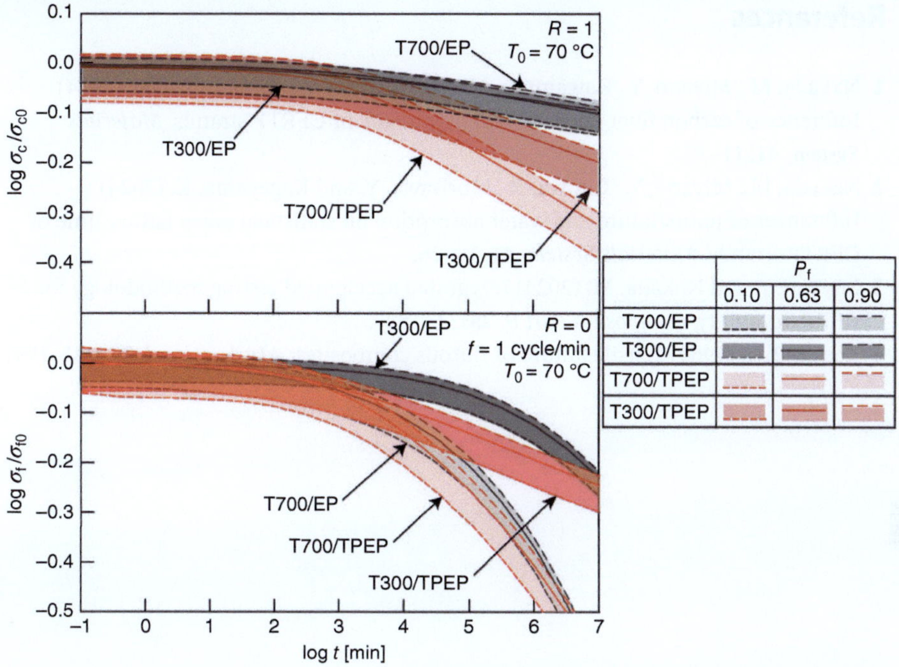

Figure 17.17 Non-dimensional creep and fatigue strengths against elapsed time at 70 °C for CF/EP and CF/TPEP strands with two types of carbon fiber.

17.4 Conclusion

The influence of carbon fiber on the statistical life of CF/TPEP strands under constant and cyclic tension loadings was evaluated using our developed Integrated ATM based on the modified TTSP, which holds for matrix resin viscoelasticity. The salient results are presented below.

1. The creep strength of CF/TPEP strand depends on the viscoelasticity of the matrix resin and decreases markedly with increased passage of time and a rise in temperature, but the rate of decrease is greater when the shape parameter in the strength of the carbon fibers is small.
2. The fatigue strength is less dependent on the viscoelasticity of the matrix resin. Moreover, greater ultimate strain of carbon fiber is associated with a greater degradation of fatigue strength with respect to the number of cycles to failure.
3. When the creep and fatigue lives of a CF/EP strand using the same carbon fibers and a thermoset epoxy matrix resin were compared with those of a CF/TPEP strand, it was found that the creep life was significantly affected by the viscoelasticity of the matrix resin, and the fatigue life was significantly affected by the mechanical properties of the carbon fibers.

References

1 Nakada, M., Miyano, Y., Kawamura, M., Morisawa, Y. and Kageyama, K. (2024) Influence of carbon fiber on statistical fatigue life of CFRTP strands. *Materials System*, **41**, 11–17.

2 Nakada, M., Miyano, Y., Tomiya, R., Morisawa, Y. and Kageyama, K. (2024) Influences of temperature and water absorption on statistical creep failure time of CFRTP strands. *Materials System*, **41**, 19–26.

3 Miyano, Y. and Nakada, M. (2024) Integrated accelerated testing methodology for CFRP durability. *Composites Part B*, **281**, 111527.

4 Rosen, B.W. (1964) Tensile failure of fibrous composites. *AIAA Journal*, **2**, 1985–1991.

18

Application 4 of Integrated Accelerated Testing Methodology: Statistical Tensile and Flexural Creep and Fatigue Lives of Unidirectional Carbon-fiber-reinforced Polymer Laminates with Polypropylene as Matrix

18.1 Introduction

As described in [1–3], integrated accelerated testing methodology (Integrated ATM) [4] is applied to the prediction of creep and fatigue strengths of unidirectional carbon fiber-reinforced polypropylene laminates (CF/PP laminates) under tension and bending loadings. Then, the respective effects of time, temperature, frequency, and the number of cycles to failure on the tensile and flexural creep and fatigue strengths are evaluated. Finally, the characteristic behavior of the creep and fatigue strengths of CF/PP laminates depending on the loading direction of tension and bending is discussed.

18.2 Experimental Procedure

18.2.1 Specimen Preparation

Unidirectional CF/PP sheets were prepared using the standard modulus grade CF ($E = 240$ GPa, 7 μm filament diameter) with 137 g/m fiber weight and the volume fraction of fiber $V_f = 50\%$, the sheet thickness was controlled to $t = 0.16$ mm. Neat PP plates (3 mm thick) were also prepared. Thermal treatment for unidirectional CF/PP sheets and neat PP plates was made under 120 °C for 1 hour to stabilize the crystallinity of PP.

18.2.2 Test Methods and Test Conditions for CF/PP Laminates

The dynamic mechanical analysis (DMA) tests were conducted with cyclic deformation of maximum strain 0.06%, at various frequencies f from 0.01 to 10 Hz, and at various temperatures T from 25 °C to 160 °C. Storage moduli E' and the loss tangent tan δ for PP plates are measured. The time–temperature (horizontal) shift factor is determined by making the master curve of tan δ. The temperature (vertical) shift factor is also determined by producing the master curve of E' based on the modified

Properties and Testing of Fiber-Reinforced Polymers, First Edition. Yasushi Miyano and Masayuki Nakada. © 2026 Wiley-VCH GmbH. All rights reserved, including rights for text and data mining and training of artificial intelligence technologies or similar technologies. Published 2026 by Wiley-VCH GmbH.

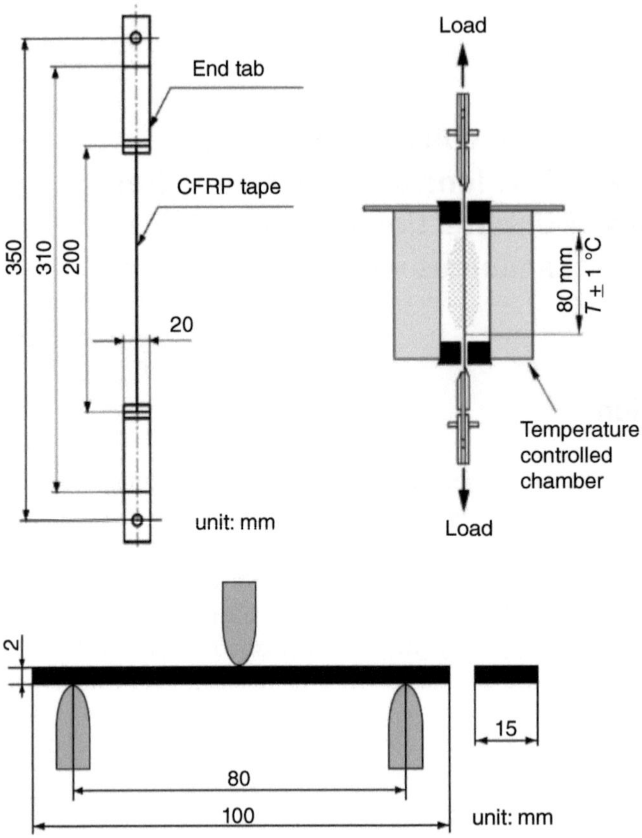

Figure 18.1 Tension and bending test methods.

time–temperature superposition principle (modified TTSP). The storage moduli E' can be converted to relaxation moduli E_r based on the linear viscoelasticity.

For this study, static, creep and fatigue tests under tension and bending loadings for the fiber direction of the CF/PP laminates were conducted using the test methods as shown in Figure 18.1. An Instron-type testing apparatus with a constant temperature chamber was used for measuring the static, creep, and fatigue strengths under tension and bending loadings at various temperatures and frequencies. The test conditions and number of specimens for static, creep, and fatigue strengths under tension and bending loadings are presented in Tables 18.1–18.3.

18.3 Results and Discussion

18.3.1 Relaxation Modulus of Matrix Resin

The storage moduli $E'(f,T)$ and the loss tangent tan $\delta(f,T)$ for PP plates were measured at various temperatures. The storage moduli $E'(f,T)$ were converted to the relaxation moduli $E_r(t,T)$. The details of the procedure are described in

Table 18.1 Test conditions and number of specimens for static tests.

Temperature (°C)	Tension test	Bending test
25	20	20
50	–	10
80	20	20
120	20	20
145	20	–
150	20	20

Note: Total number of specimens 190.

Table 18.2 Test conditions and number of specimens for creep tests.

Creep tension test				Creep bending test			
Temperature (°C)	Creep stress (MPa)	σ_c/σ_0^a (%)	Number of specimens	Temperature (°C)	Creep stress (MPa)	σ_c/σ_0^b (%)	Number of specimens
120	921	67	20	25	600	64	20
				40	600	64	20
					550	59	20
					500	53	20
				45	500	53	20
				50	500	53	20

Note: Total number of specimens 140.
a. Creep stress/scale parameter of tensile static strength at $T_0 = 25\,°C$, $\sigma_0 = 1368$ MPa at 25 °C.
b. Creep stress/scale parameter of flexural static strength at $T_0 = 25\,°C$, $\sigma_0 = 936$ MPa at 25 °C.

Table 18.3 Test conditions and number of specimens for fatigue tests.

Loading direction	Temperature (°C)	Frequency (Hz)	Maximum stress/static strength[a] (%)								
			90	85	80	75	70	65	60	55	50
Tension	25	2.0	20	20	–	–	4	20	2	–	–
	50		–	20	–	–	13	3	–	–	–
		0.5	–	20	–	1	1	–	–	–	–
Bending	25	2.0	20	20	7	1	1	–	1	–	–
	50		–	–	2	–	3	–	19	10	2
		0.2	–	–	–	–	2	–	20	15	1

Note: Total number of specimens: 248.
a. Maximum stress/static strength: Maximum stress/scale parameter of static strength at each temperature and frequency.

Chapter 14, and the results are shown in Figure 18.2 in this chapter. The left part of Figure 18.2 shows the relaxation moduli for PP plates at various temperatures. The right part of Figure 18.2 shows the long-term relaxation modulus for PP plates, which was obtained by shifting measured data at various temperatures along the log scale of time and the log scale of the relaxation modulus until they mutually overlapped based on the modified TTSP.

The left part of Figure 18.3 is the loss tangents for PP plates at various temperatures. The right part of this figure shows the long-term loss tangent for PP plates, which was obtained by shifting measured data at various temperatures along the log scale of time.

Figure 18.4 depicts the time–temperature shift factor (horizontal shift) and temperature shift factor (vertical shift) for the master curves of relaxation modulus and loss tangent.

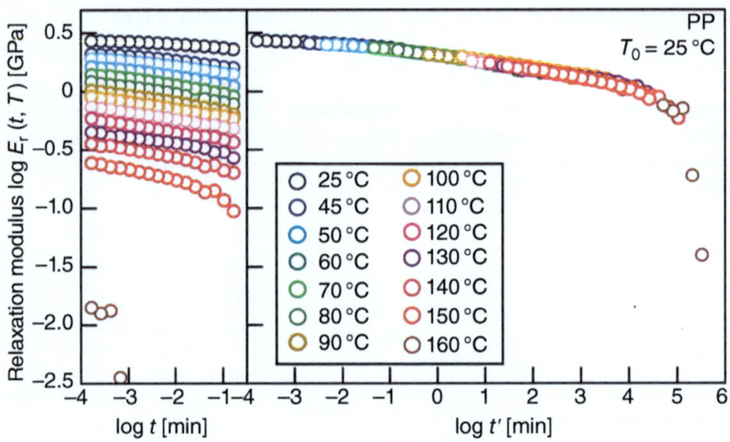

Figure 18.2 Master curve of relaxation modulus versus reduced time at $T_0 = 25$ °C for PP plates.

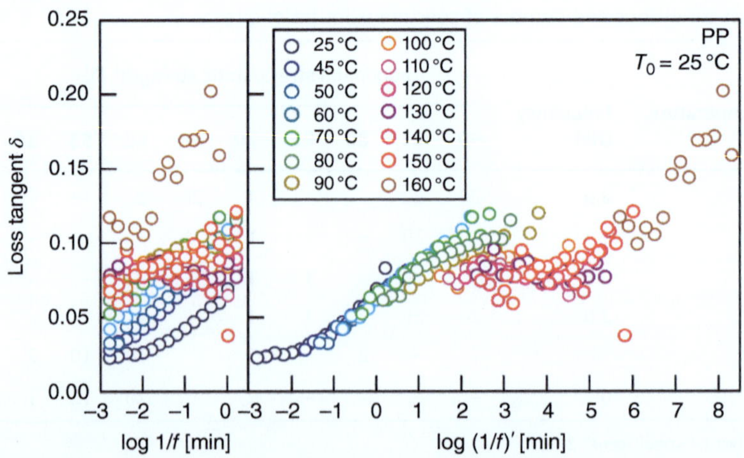

Figure 18.3 Master curve of loss tangent versus reduced time at $T_0 = 25$ °C for PP plates.

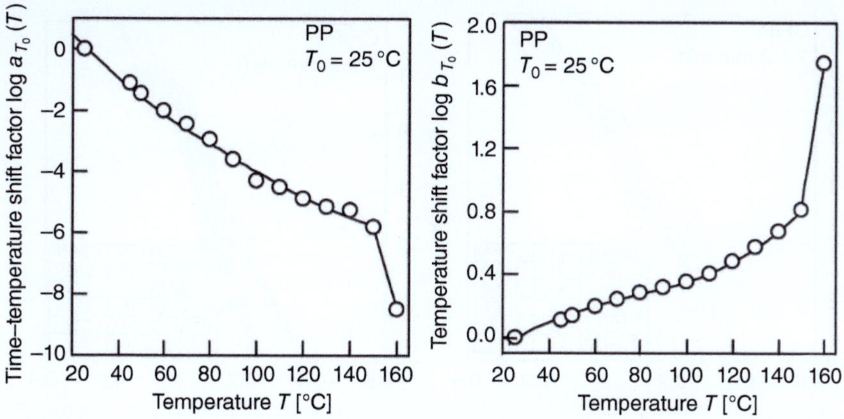

Figure 18.4 Time–temperature (horizontal) and temperature (vertical) shift factors for PP plates.

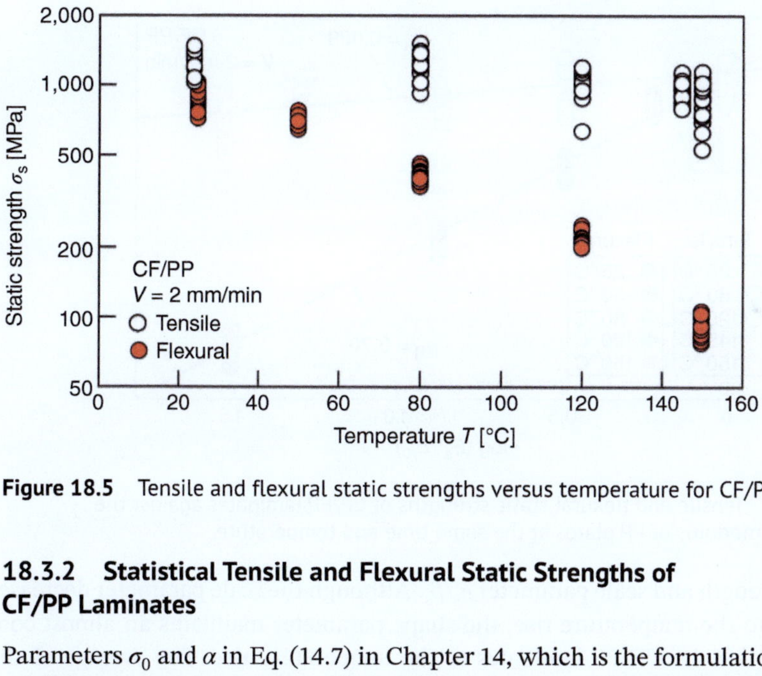

Figure 18.5 Tensile and flexural static strengths versus temperature for CF/PP laminates.

18.3.2 Statistical Tensile and Flexural Static Strengths of CF/PP Laminates

Parameters σ_0 and α in Eq. (14.7) in Chapter 14, which is the formulation of statistical fatigue strength of carbon-fiber-reinforced polymer (CFRP), are determined by measuring the statistical static strength of CFRP at various temperatures.

Statistical tensile and flexural static strengths of CF/PP laminates versus temperature are shown in Figure 18.5. Flexural strength decreases considerably with increasing temperature, but tensile strength does not decrease to any considerable degree.

Weibull distributions for tensile and flexural static strengths at various temperatures are depicted in Figure 18.6 for CF/PP laminates. In the table included in this figure, α_s is the shape parameter and β_s is the scale parameter of CF/PP laminates, which were determined using least squares fitting. The horizontal axis is the ratio

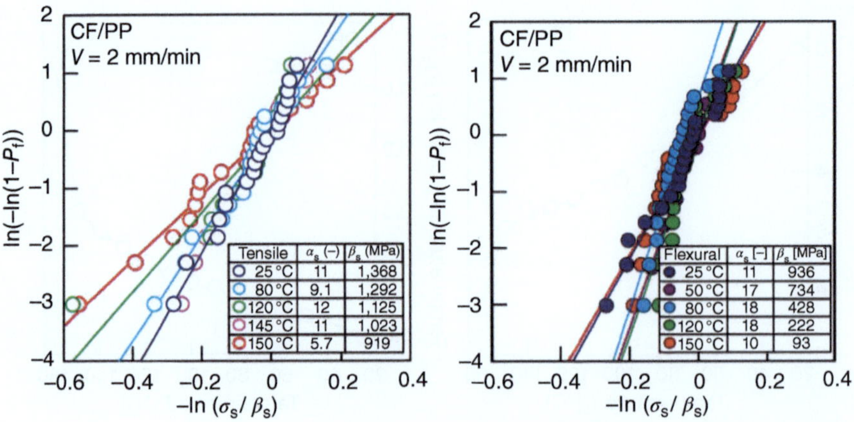

Figure 18.6 Weibull distributions of tensile and flexural static strengths for CF/PP laminates.

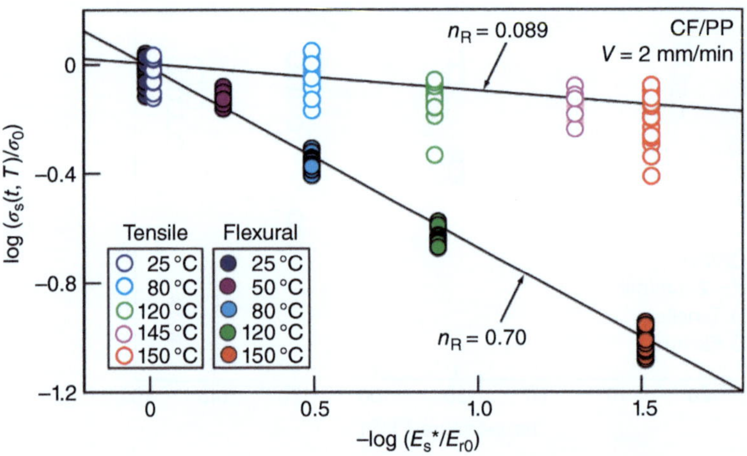

Figure 18.7 Tensile and flexural static strengths of CF/PP laminates against the viscoelastic modulus of PP plates at the same time and temperature.

of static strength and scale parameter σ_s/β_s. Although the scale parameter decreases according to the temperature rise, the shape parameter maintains an almost constant value for the CF/PP laminates to the temperature rise. The average of the shape parameter α_s at six temperatures and the scale parameter β_s at temperature $T = 25\,°\text{C}$ in this figure can be inferred as shape parameter α and scale parameter σ_0 of the tensile and flexural static strengths at reference temperature $T_0 = 25\,°\text{C}$ and reference failure time $t_0 = 1$ min used in Eq. (14.7).

Parameter n_R in Eq. (14.7) is determined according to the relaxation modulus of the matrix resin and the statistical static strength of CF/PP laminates at various temperatures. Figure 18.7 portrays the relation between the tensile and flexural static strengths of CF/PP laminates and the viscoelastic modulus of PP plates. After a linear approximation was made, the viscoelastic parameters $n_R = 0.089$ for tensile strength and $n_R = 0.70$ for flexural strength were obtained.

Table 18.4 Parameters of statistical tensile and flexural static strengths for CF/PP laminates.

	σ_0(MPa)	α (-)	n_R (-)
Tensile	1368	11	0.089
Flexural	936	11	0.70

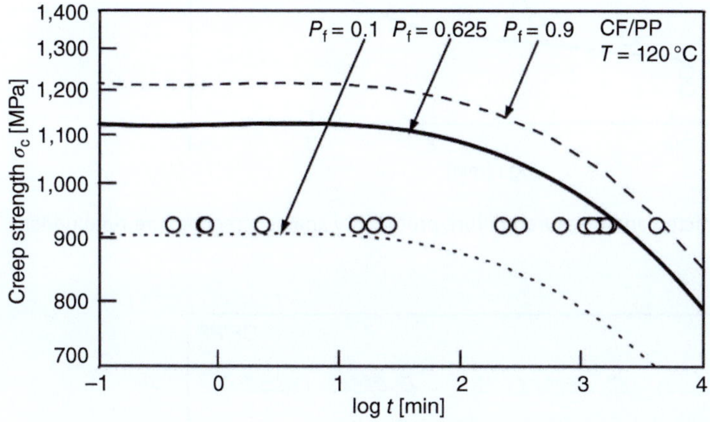

Figure 18.8 Predicted and measured tensile creep strength against failure time for CF/PP laminates.

All parameters in Eq. (14.11), except parameter $F_f(t,f,T,R)$ for predicting statistical tensile and flexural fatigue strengths for CF/PP laminates, were obtained as presented in Table 18.4. Strength reduction attributable to viscoelasticity σ_f^* shown by Eq. (14.12) is also obtainable using the data presented in Table 18.4.

18.3.3 Statistical Tensile and Flexural Creep Strengths of CF/PP Laminates

The statistical tensile creep strength of CF/PP laminates predicted by substituting the parameters of Table 18.4 and the relaxation modulus of matrix resin of Figure 18.2 into Eq. (14.8) is shown in Figure 18.8 with the experimental data measured at 920 MPa and 120 °C. Figure 18.9 shows the predicted and measured failure probability against creep failure time under tension loading. Both figures show that the tensile creep failure of CF/PP laminates can be predicted using the relaxation modulus of matrix resin and the tensile strengths of CF/PP laminates measured statistically at various temperatures based on Integrated ATM.

The statistical flexural creep strengths of CF/PP laminates measured at various load levels and temperatures presented in Table 18.2 are shown in Figure 18.10. This figure clarifies that the flexural creep failure times depend on the load level and temperature with scatter.

Figure 18.11 presents the predicted statistical flexural creep strength versus failure time for CF/PP laminates at 40 °C calculated by substituting the parameters for the CF/PP laminates presented in Table 18.4 and the relaxation modulus of PP plates shown in Figure 18.2 into Eq. (14.8) with the experimentally

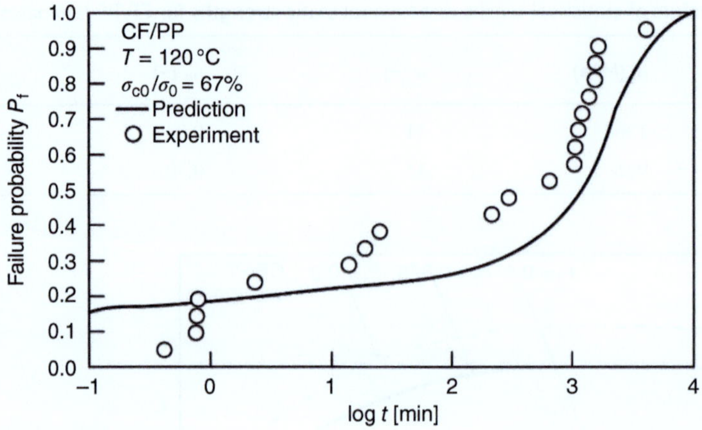

Figure 18.9 Predicted and measured failure probability against creep failure time under tension loading.

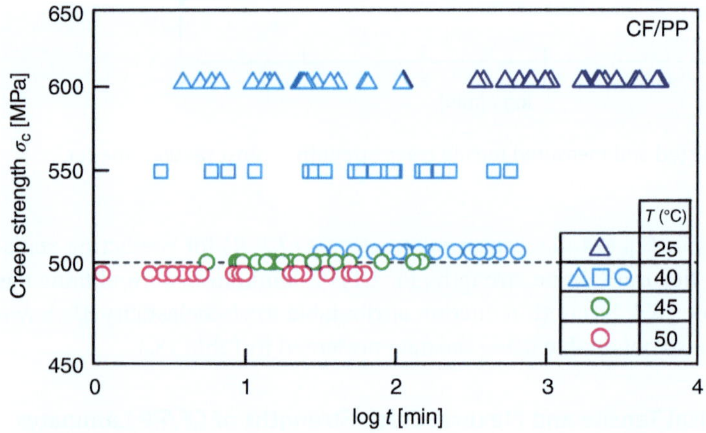

Figure 18.10 Flexural creep strength versus failure time at various constant load levels and temperatures.

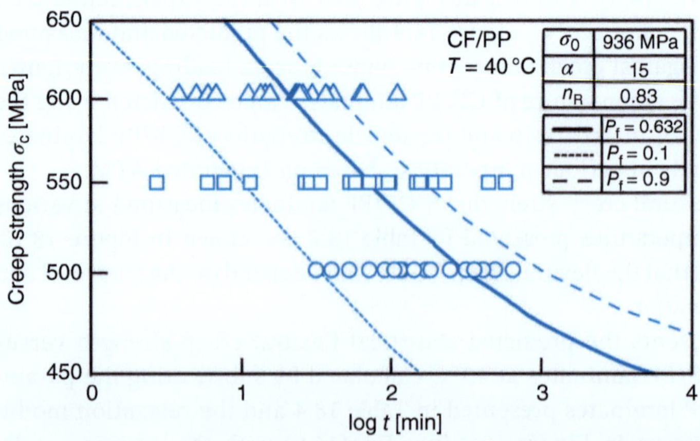

Figure 18.11 Predicted and measured flexural creep strengths versus failure time.

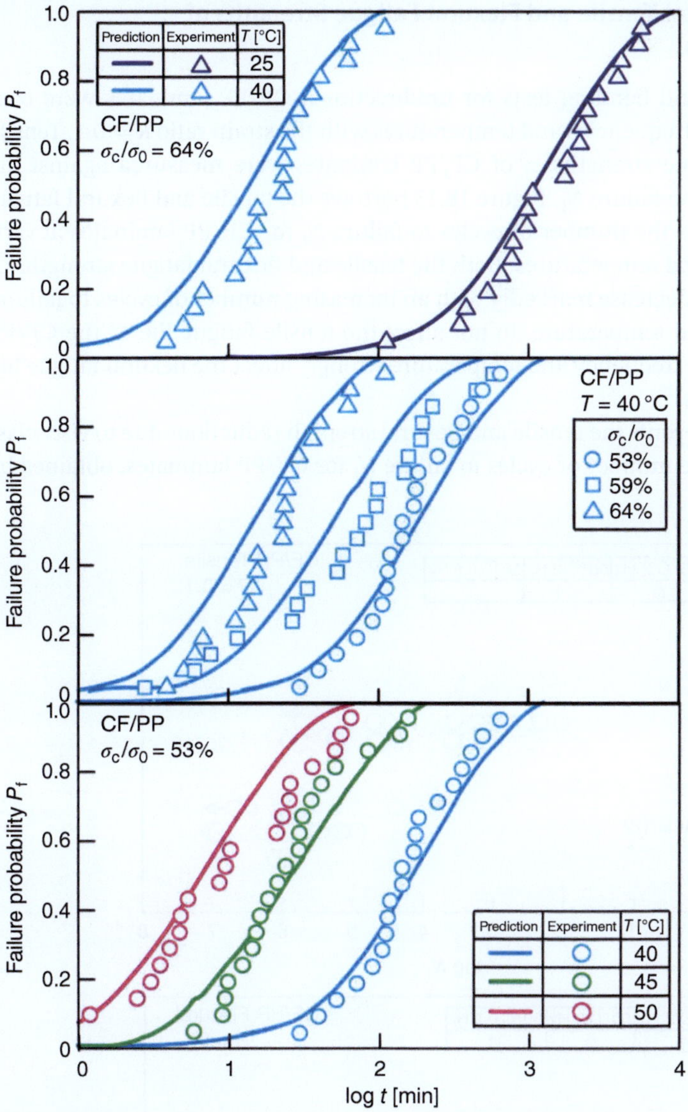

Figure 18.12 Predicted and measured failure probability versus creep failure time under bending load.

obtained data measured at three load levels. Figure 18.12 shows the prediction of failure probability P_f versus failure time t at three cases of load level and temperature. The plots in these figures show the measured data. Figures 18.11 and 18.12 indicate that the experimental data agree well with the predicted curves. Both figures show that the flexural creep failure of CF/PP laminates can also be predicted using the relaxation modulus of the matrix resin and the flexural strengths of CF/PP laminates measured statistically at various temperatures based on Integrated ATM.

18.3.4 Statistical Tensile and Flexural Fatigue Strengths of CF/PP Laminates

Fatigue tension and bending tests for unidirectional CF/PP laminates were conducted at various frequencies and temperatures with the strain ratio $R = 0.1$. Tensile and flexural fatigue strengths σ_f of CF/PP laminates were measured against the number of cycles to failure N_f. Figure 18.13 portrays the tensile and flexural fatigue strengths σ_f versus the number of cycles to failure N_f for CF/PP laminates at various frequencies and temperatures. Both the tensile and flexural fatigue strengths of CF/PP laminates decrease markedly with an increasing number of cycles to failure. The frequency and temperature do not affect the tensile fatigue life of the CF/PP laminates, but the frequency and temperature strongly affect the flexural fatigue life of these laminates.

Figure 18.14 presents the tensile and flexural strength reductions due to viscoelasticity σ_f^* versus the number of cycles to failure N_f for CF/PP laminates, obtained by

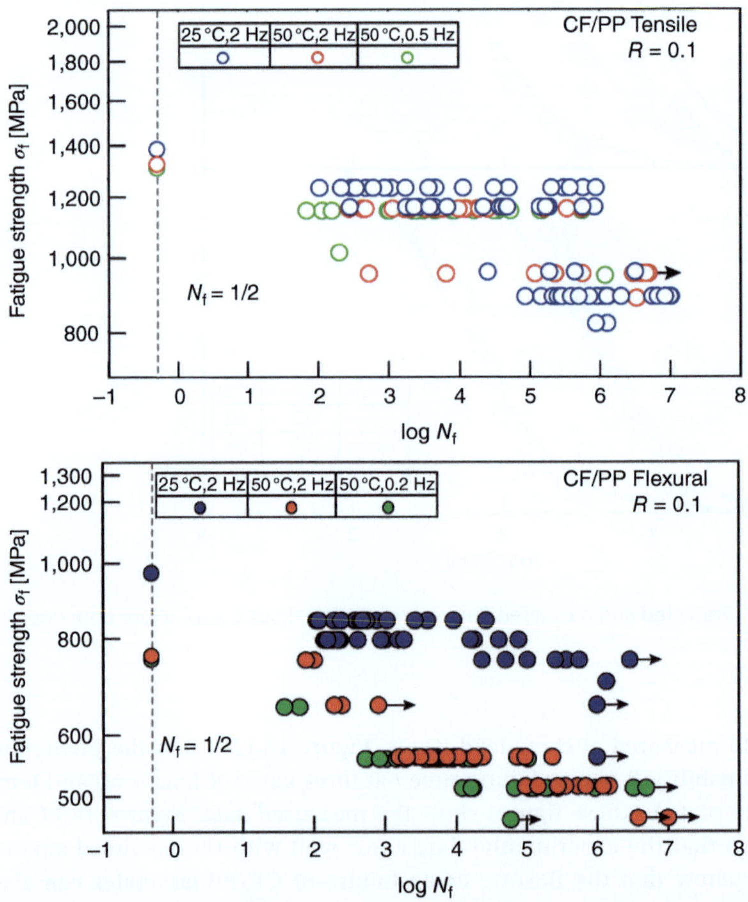

Figure 18.13 Tensile and flexural fatigue strengths versus the number of cycles to failure.

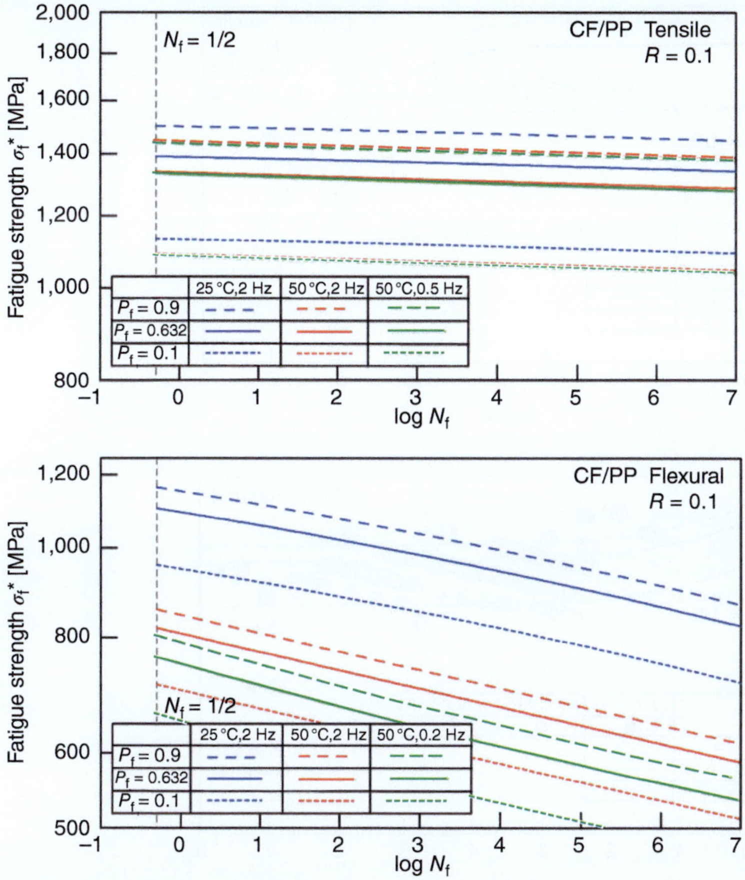

Figure 18.14 Tensile and flexural strength reductions attributable to viscoelasticity versus the number of cycles to failure.

substituting the data in Table 18.4 and the relaxation modulus of PP plates shown in Figure 18.2 into Eq. (14.12). The σ_f^* under tension loading does not so change with the number of cycles to failure N_f or according to the frequency and temperature. The σ_f^* under bending load changes markedly with the number of cycles to failure N_f, just as it does with frequency and temperature.

Figure 18.15 presents the non-dimensional tensile and flexural fatigue strengths versus the number of cycles to failure ($S_{f0} - N_f$ curves) standardized at the reference frequency $f_0 = 2$ Hz, the reference temperature $T_0 = 25\,°C$ and the strain ratio $R = 0.1$ as obtained by substituting the experimental fatigue strengths σ_f of Figure 18.13, the data in Table 18.4, and the relaxation modulus of PP plates shown in Figure 18.2 into the right-hand side of Eq. (14.13). The upper and lower graphs in this figure form one smooth curve with scatter for various frequencies and temperatures, respectively, and then these curves can be considered as the master curves. Therefore, the frequency–temperature fatigue parameter F_f^* defined by Eq. (14.15)

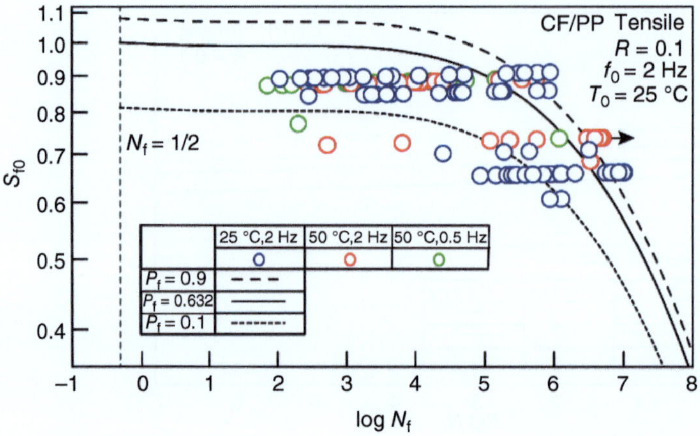

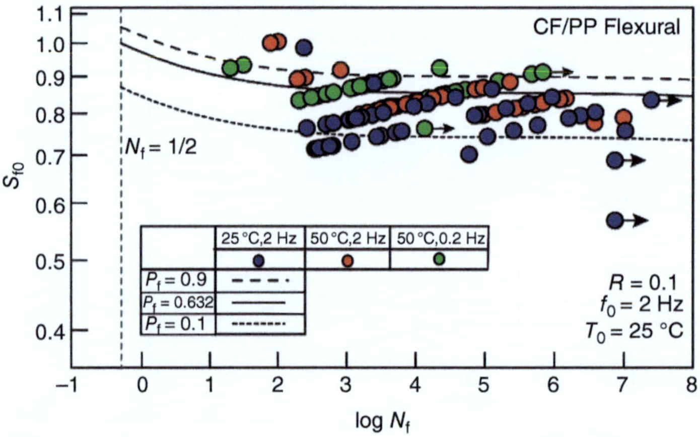

Figure 18.15 Non-dimensional tensile and flexural fatigue strengths versus number of cycles to failure.

Table 18.5 Fatigue coefficient n_{fi} in cyclic fatigue parameter F_f' for tensile and flexural fatigue strengths of CF/PP laminates.

	$n_{f1}\ (\times 10^{-4})$	$n_{f2}\ (\times 10^{-4})$	$n_{f3}\ (\times 10^{-4})$	$n_{f4}\ (\times 10^{-4})$
Tensile	46	−1.0	−9.0	2.0
Flexural	291	−51.5	3.76	−0.09

is equal to 1 for both loads. The solid curves shown in Figure 18.15 were determined by calculating the fatigue coefficients n_{fi} in the cyclic fatigue parameter F_f' shown in Eq. (14.16) by the least squares fitting. The fatigue coefficients n_{fi} for tensile and flexural fatigue strengths of CF/PP laminates are shown in Table 18.5. The S_{f0} under tension load decreases at an accelerated rate as the number of cycles increases. Also, S_{f0} under a bending load tends to converge to the critical value as the number of cycles increases.

18.3.5 Long-term Prediction of Tensile and Flexural, Creep and Fatigue Strengths of CF/PP Laminates

Figure 18.16 shows the long-term predictions for the relaxation modulus of PP plates and the statistical tensile and flexural creep strengths of CF/PP laminates at $T = 25\,°C$. The figure clarifies that the long-term tensile and flexural creep strengths of CF/PP laminates are controlled perfectly by the relaxation modulus of the matrix resin. The flexural creep strength decreases drastically with increasing time, showing similar behavior to the relaxation modulus of the matrix resin. The tensile creep strength decreases moderately with increasing time.

Figure 18.17 portrays the long-term predictions for tensile and flexural fatigue strengths of CF/PP laminates at 25 °C and 80 °C under the conditions of the frequency $f = 1$ cycle/min and the strain ratio $R = 0.1$. This figure clarifies that the long-term tensile fatigue strength of CF/PP laminates decreases at an accelerated rate as the number of cycles increases. The flexural fatigue strength is affected by temperature and frequency rather than by the number of cycles to failure.

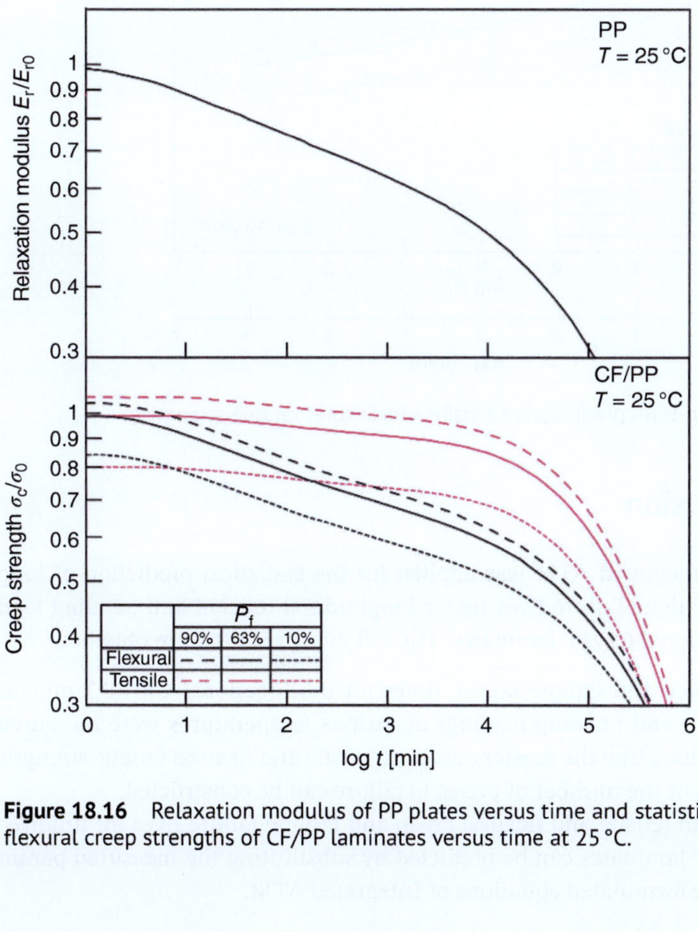

Figure 18.16 Relaxation modulus of PP plates versus time and statistical tensile and flexural creep strengths of CF/PP laminates versus time at 25 °C.

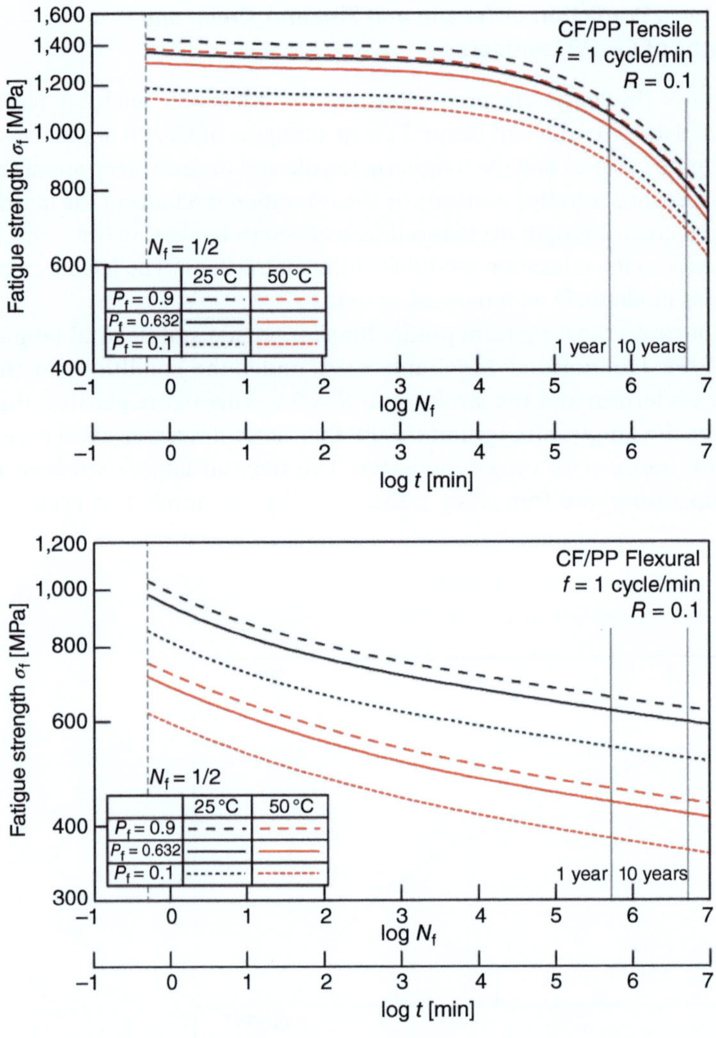

Figure 18.17 Long-term prediction of tensile and flexural fatigue strengths.

18.4 Conclusion

Our developed Integrated ATM was applied for the statistical prediction of long-term creep and fatigue failure lives under longitudinal tension and bending loadings for unidirectional CF/PP laminates. The following results were obtained.

1. Statistical creep and fatigue failure times for unidirectional CF/PP laminates under tension and bending loadings at various temperatures were measured, which confirmed that the master curves of tensile and flexural fatigue strengths as a function of the number of cycles to failure can be constructed.
2. The long-term tensile and flexural creep and fatigue failure lives for unidirectional CF/PP laminates can be predicted by substituting the measured parameters into the formulated equations of Integrated ATM.

3. The long-term tensile and flexural creep strengths of CF/PP laminates are controlled perfectly by the relaxation modulus of the matrix resin. The flexural creep strength decreases drastically with increasing time, with similar behavior to the relaxation modulus of the matrix resin. The tensile creep strength decreases moderately with increasing time.
4. The long-term tensile fatigue strength of CF/PP laminates decreases at an accelerated rate as the number of cycles increases. The flexural fatigue strength is affected by temperature and frequency rather than by the number of cycles to failure.

References

1 Nakada, M., Miyano, Y., Morisawa, Y., Isaki, T., Hirano, T. and Uzawa, K. (2020) Statistical life prediction of unidirectional carbon fiber/polypropylene tape under creep tension load. *Journal of Reinforced Plastics and Composites*, **39**, 278–284.

2 Nakada, M., Miyano, Y., Morisawa, Y., Nonaka, T., Isaki, T., Hirano, T. and Uzawa, K. (2022) Statistical life prediction of a unidirectional carbon fiber/polypropylene laminates under creep bending load. *Journal of Reinforced Plastics and Composites*, **41**, 825–836.

3 Nakada, M., Miyano, Y., Nonaka, T., Morisawa, Y., Isaki, T., Hirano, T. and Uzawa, K. (2024) Statistical tensile and flexural fatigue lives of unidirectional CF/PP laminates. *Journal of Reinforced Plastics and Composites*, **43**, 195–204.

4 Miyano, Y. and Nakada, M. (2024) Integrated accelerated testing methodology for CFRP durability. *Composites Part B*, **281**, 111527.

19

Application 5 of Integrated Accelerated Testing Methodology: Prediction of Creep Failure Life for Unidirectional Carbon-fiber-reinforced Polymer with Heat-resistant Epoxy Resin as Matrix Exposed to High Temperature Under Tension Load

19.1 Introduction

The authors focus on carbon fiber-reinforced polymer (CFRP), which has come to be used in structures requiring heat resistance. It is believed that CFRP will undergo heat degradation when exposed to high temperatures for extended periods. The long-term strength decline of CFRP due to chemical and heat degradation is an important factor in assessing durability. In order to predict the long-term creep strength of CFRP, it is important to evaluate how the viscoelastic behavior of the matrix resin changes as chemical degradation progresses, and it is even more important to scientifically elucidate how these changes affect the durability of CFRP [1].

The purpose of this chapter is to predict the long-term creep rupture time of unidirectional CFRPs with a heat-resistant epoxy resin matrix using the integrated accelerated testing methodology (Integrated ATM) [2]. First, carbon fiber strands impregnated with a heat-resistant epoxy resin (CFRP strands) are molded using a filament winding system to prepare virgin unidirectional CFRP specimens. Second, thermally aged specimens are fabricated by exposing the virgin specimens to accelerated conditions determined based on the modified time–temperature superposition principle (modified TTSP) for heat degradation, assuming a service temperature of 110 °C and a 10-year period. Third, the creep strength of virgin and thermally aged CFRP strands is statistically predicted based on Integrated ATM described in Chapter 14, and the effect of heat degradation on the long-term life of the CFRP strands is evaluated. A key feature of this chapter is that the heat degradation of the matrix resin is considered as a parameter change in the strength design of CFRP. The authors examine the effect of heat degradation on the long-term creep rupture time of CFRP and discuss the applicability of Integrated ATM.

Properties and Testing of Fiber-Reinforced Polymers, First Edition. Yasushi Miyano and Masayuki Nakada. © 2026 Wiley-VCH GmbH. All rights reserved, including rights for text and data mining and training of artificial intelligence technologies or similar technologies.
Published 2026 by Wiley-VCH GmbH.

19.2 Experiments

19.2.1 Specimens

The carbon fiber used in this chapter is a high-strength PAN-based carbon fiber (T700-6000, Toray Industries, Inc.). Table 19.1 shows its mechanical properties, as listed in the catalog [3]. CFRP strands were molded by the filament winding method using this carbon fiber and a heat-resistant epoxy resin (XNR6813, Nagase ChemteX Corporation). The epoxy resin composition of the CFRP strands is shown in Table 19.2. As shown in Table 19.2, two curing schedules were selected for preparing virgin and thermally aged strands. For the thermally aged strands, an additional heat treatment at 140 °C for 3000 hours was performed on the virgin strands, for the reasons explained at the end of Section 19.2.3.

19.2.2 Testing Method

As shown in Figure 19.1, the gauge length of the CFRP strand was approximately 180 mm. The glass transition temperature of virgin epoxy resin, $T_g = 202\,°C$, was determined from the shoulder drop temperature at which the storage modulus suddenly drops in the temperature characteristics of the storage modulus measured at 1 Hz using a dynamic mechanical analysis (DMA) testing machine. The fiber volume fraction of the CFRP strand, $V_f = 55\%$, was determined from the weight of the CFRP strand. The original static testing machine equipped with a temperature chamber achieved a constant elongation rate at various constant temperatures from 25 to

Table 19.1 Carbon fiber and the mechanical properties.

Carbon fiber product name	Company	Density (g/cm^3)	Tex (g/km)	Elastic modulus (GPa)	Tensile strength (MPa)
Torayca T700-6000	Toray Industries Inc.	1.80	400	230	4900

Table 19.2 Composition and curing schedule of CFRP strands.

Strand	Carbon fiber	Resin composition (weight ratio)	Resin product (company)	Cure schedule
Virgin	T700-6000	Epoxy (220)	XNR6813 (Nagase ChemteX Corp.)	120 °C × 6 h + 180 °C × 4 h + 200 °C × 2 h
Heat-degraded		Hardener (232)	XNH6813 (Nagase ChemteX Corp.)	120 °C × 6 h + 180 °C × 4 h + 200 °C × 2 h + 140 °C × 3000 h

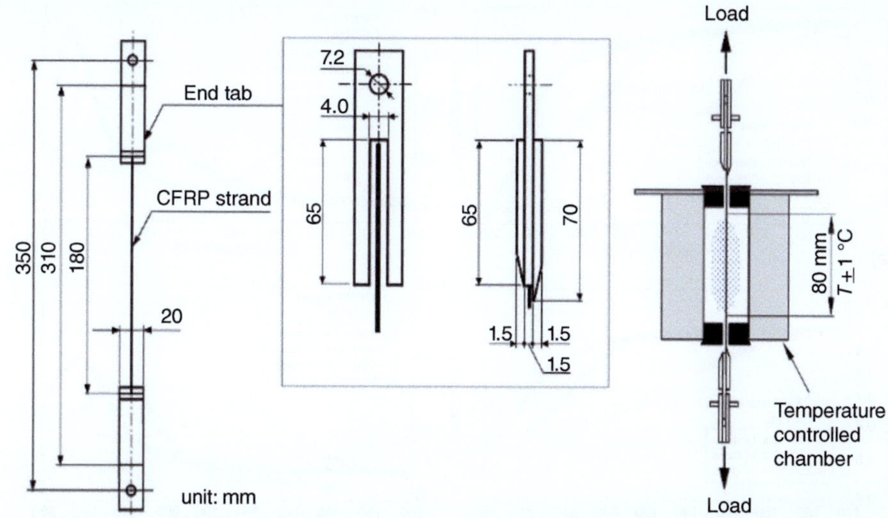

Figure 19.1 Specimen configuration and temperature-controlled chamber.

200 °C based on the fact that the specimen was not visually confirmed to be pulled out of the end tabs during loading. It was also confirmed that the load-elongation curve was nearly linear in all tests. The original creep testing machine 20 sets equipped with a temperature chamber achieved a constant load of 1 kN and various constant temperatures from 25 to 200 °C.

The CFRP strand tensile strength σ_s is calculated using the following equation.

$$\sigma_s = \frac{W_{max}}{t_e}\rho \tag{19.1}$$

In this equation, W_{max} represents the maximum load (N), and ρ and t_e represent the density of carbon fiber (kg/m^3) and the tex of carbon fiber strand (g/km), respectively.

19.2.3 Heat Degradation Treatments

A virgin matrix resin beam measuring 1.6 mm thick, 15 mm wide, and 50 mm long, and a virgin CFRP strand measuring 0.30 mm thick, 2.0 mm wide, and 50 mm long were subjected to an additional heat degradation treatment. The temperature dependence of the storage modulus and loss tangent of the matrix resin and CFRP strand after the additional heat degradation treatment was measured using DMA testing machine. Figure 19.2 shows the temperature dependence of the storage modulus E' and loss tangent δ at 1 Hz after three cases of heat degradation treatment for the matrix resin and CFRP strand. From Figure 19.2, the shoulder drop temperature of E' and the peak temperature of tan δ were investigated for various treatments, and the results are listed in Table 19.3. The shoulder drop temperature is then obtained as the temperature at which E' begins to decrease rapidly with increasing temperature.

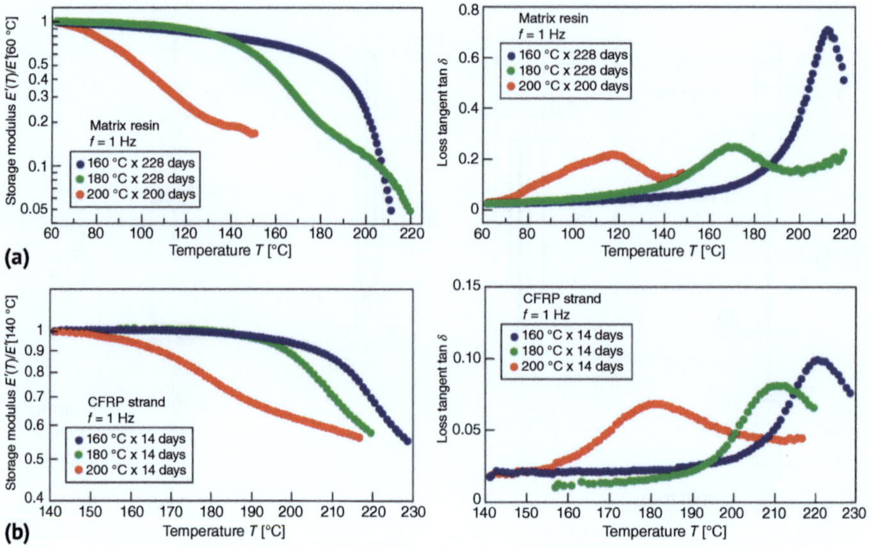

Figure 19.2 Temperature dependences of storage modulus and loss tangent for matrix resin and CFRP strand after various heat degradation treatments: (a) matrix resin and (b) CFRP strand.

Table 19.3 Shoulder drop temperature of storage modulus and peak temperature of loss tangent at various treatments: (a) matrix resin and (b) CFRP strand.

(a)

	Cure time *t* (days)							
	14	63	200	228	14	63	200	228
Cure temperature (°C)	Shoulder drop temperature of E'(°C)				Peak temperature of tan δ (°C)			
160	204	205	–	199	220	219	–	212
180	198	208	–	141	211	221	–	168
200	158	145	73	–	182	155	117	–

(b)

	Cure time *t* (days)							
	0.88	3.6	14	64	0.88	3.6	14	64
Cure temperature (°C)	Shoulder drop temperature of E'(°C)				Peak temperature of tan δ (°C)			
160	199	198	210	205	212	210	221	219
180	197	197	198	–	211	212	211	–
200	207	187	160	–	219	201	181	–

The left side of each of the four graphs in Figure 19.3 shows the relationship between shoulder drop temperature and peak temperature versus cure time at various cure temperatures, taken from Table 19.3. The right side of each of the four graphs in Figure 19.3 shows master curves for the reduced cure time at the reference cure temperature. The time–temperature shift factors for these master curves are shown in Figure 19.4. The results clearly demonstrate that the time–temperature superposition principle applies to the thermal aging behavior of the matrix resin and CFRP strands, as the curves at various temperatures are smoothly superimposed, as shown in Figure 19.3. A single order of magnitude time compression of

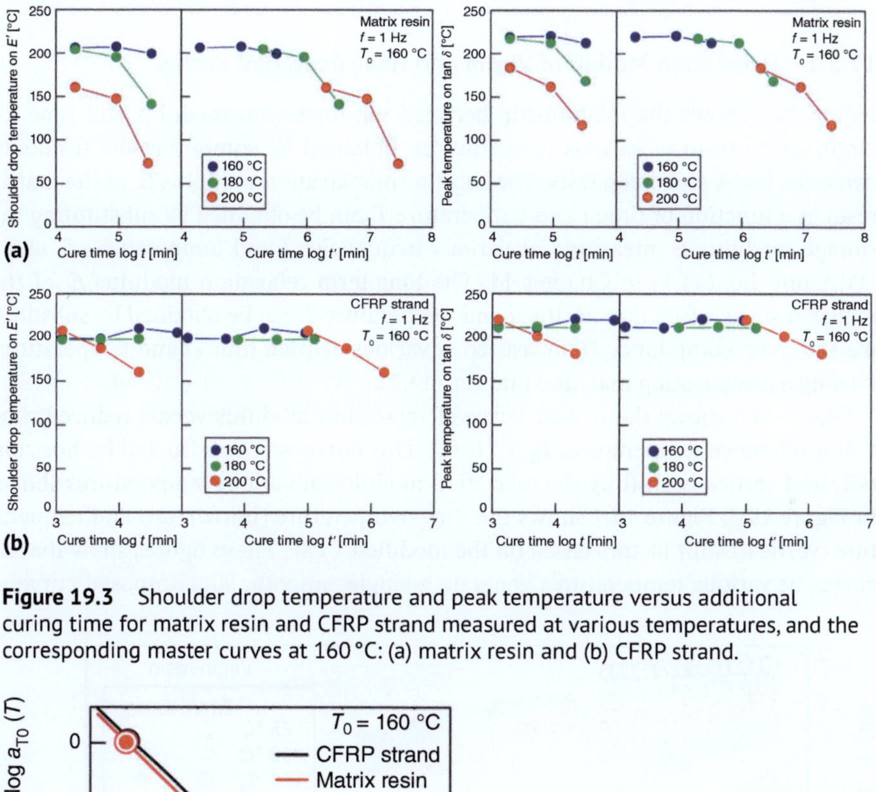

Figure 19.3 Shoulder drop temperature and peak temperature versus additional curing time for matrix resin and CFRP strand measured at various temperatures, and the corresponding master curves at 160 °C: (a) matrix resin and (b) CFRP strand.

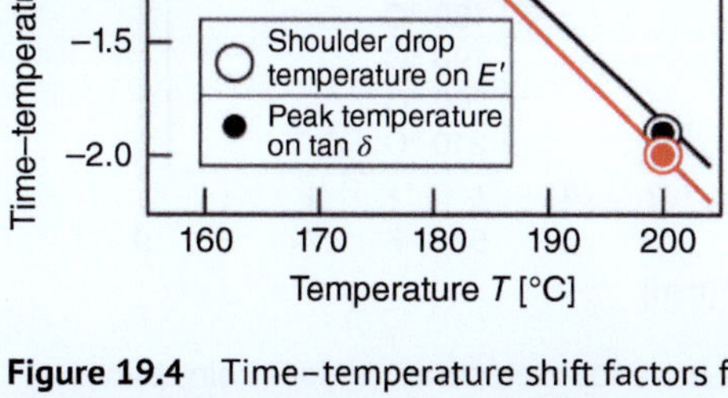

Figure 19.4 Time–temperature shift factors for shoulder drop temperature and peak temperature for matrix resin and CFRP strand.

20 °C applies to this behavior, as shown in Figure 19.4. The 3000-hour additional cure treatment at 140 °C for the thermally aged CFRP strands shown in Table 19.2 corresponds to approximately 10 years of exposure at 110 °C, the expected operating temperature for a practical composite structure.

19.3 Results and Discussion

All the experimental results obtained were processed according to the integrated ATM presented in Chapter 14.

19.3.1 Relaxation Moduli of Virgin and Heat-degraded Resins

Figure 19.5 shows the relationship between the relaxation modulus and time for virgin epoxy resin at various temperatures, obtained by combining the results of two tests, DMA and creep tests. The short-term relaxation modulus E_r of the matrix resin as a function of time t and temperature T can be obtained by substituting the storage modulus E' measured at various frequencies f and temperatures T using DMA into Eq. (14.1) in Chapter 14. The long-term relaxation modulus E_r of the matrix resin as a function of time t and temperature T can be obtained by substituting the creep compliance D_c measured at various elapsed times t and temperatures T using a creep testing machine into Eq. (14.2).

Figure 19.6 shows the master curve of relaxation modulus versus reduced time t' at a reference temperature, $T_0 = 110$ °C. This curve was constructed by horizontally and vertically shifting the relaxation modulus at various temperatures shown in Figure 19.5. Figure 19.7 shows the time–temperature (horizontal) and temperature (vertical) shift factors based on the modified TTSP. These figures show that all curves at various temperatures generate a single, smooth, superimposed curve at

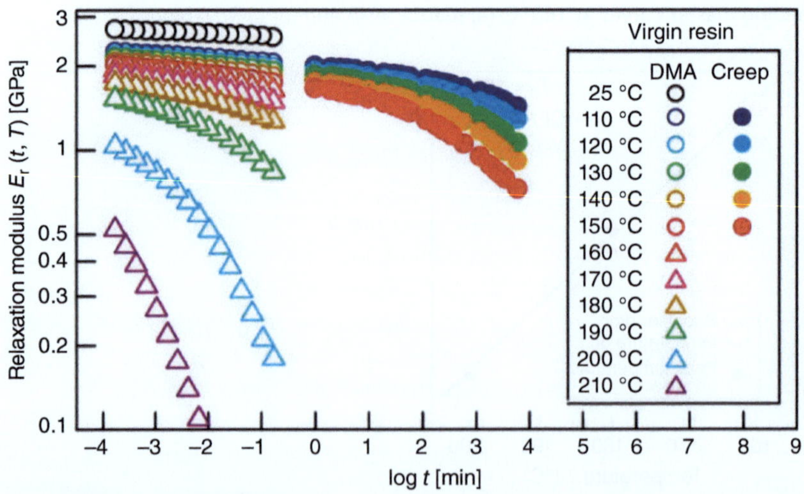

Figure 19.5 Relaxation moduli versus time at various temperatures for virgin resin measured using DMA and creep tests.

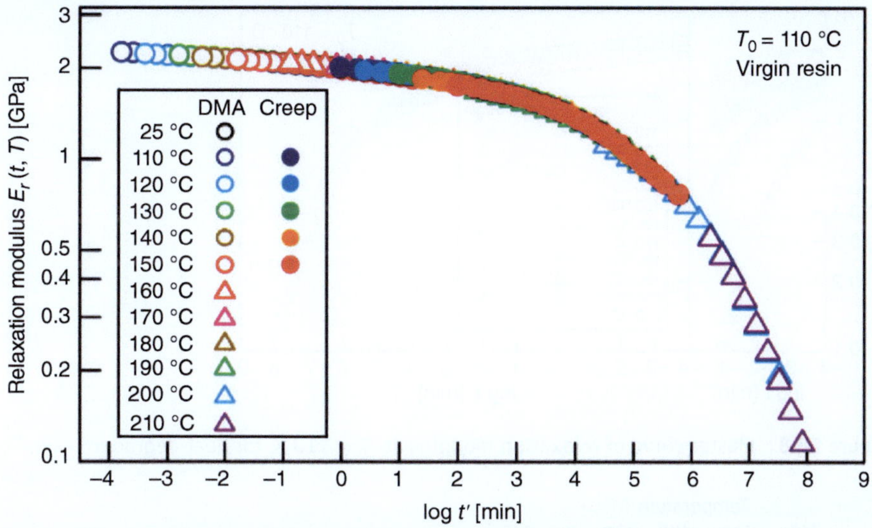

Figure 19.6 Master curve of the relaxation modulus at $T_0 = 110\,°C$ for virgin resin.

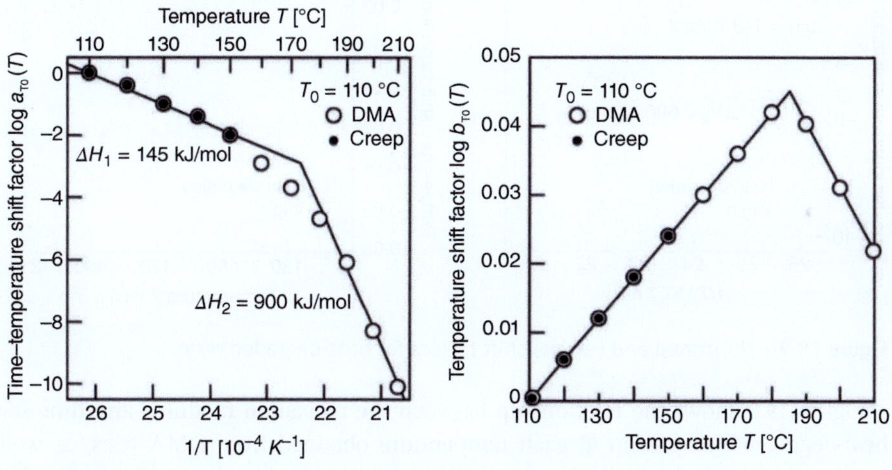

Figure 19.7 Horizontal and vertical shift factors for virgin resin.

the reference temperature. The figures also demonstrate that the modified TTSP is applicable to the viscoelastic properties of this resin. Therefore, as shown in Figure 19.6, the master curve can be used to predict the relaxation modulus over very long periods. For example, for the resin shown in Figure 19.6, the relaxation modulus can be predicted over periods ranging from several years to several decades at 110 °C. Figure 19.7 also shows that both the time–temperature shift factors and the temperature shift factors are expressed as two bilinear curves with similar inflection temperatures. This can be viewed as a gradual transition from a glassy state at low temperatures to a rubbery state at high temperatures. Furthermore, the maximum slope in Figure 19.6 is $m_R = 0.58$, which is used to predict the creep strength of CFRP strands.

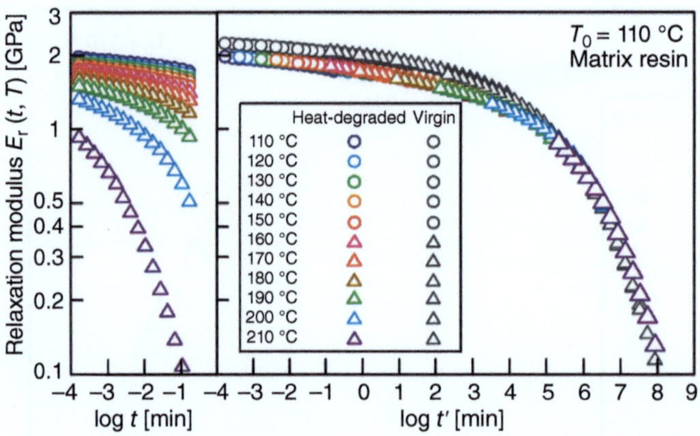

Figure 19.8 Master curve of relaxation modulus at $T_0 = 110\,°C$ for heat-degraded resin.

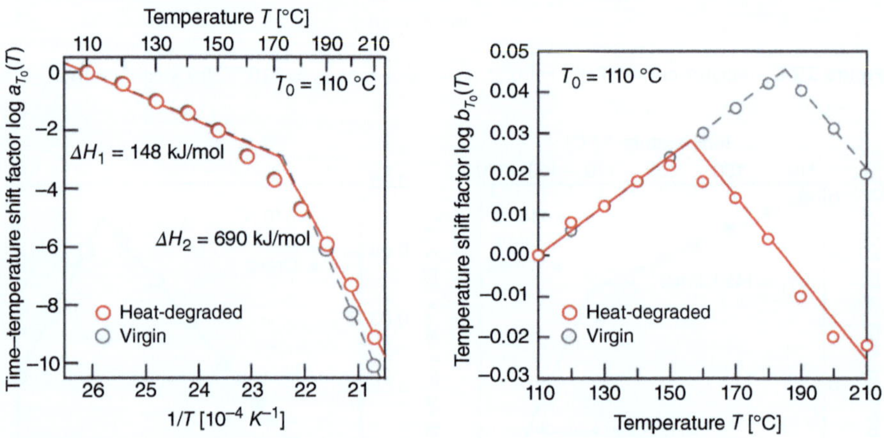

Figure 19.9 Horizontal and vertical shift factors for heat-degraded resin.

Figure 19.8 shows the relationship between the relaxation modulus and time for heat-degraded epoxy resin at each temperature obtained from DMA tests, as well as the relationship between the master curve and the reduced time t' at the reference temperature $T_0 = 110\,°C$. The horizontal and vertical shift factors are shown in Figure 19.9. Figures 19.8 and 19.9 show that the relaxation modulus and shift factors of the matrix resin change little with 10 hours of heat degradation at 140 °C. The decrease in the inflection point temperature of the vertical shift factors with heat degradation shown in Figure 19.9 is thought to be due to chemical changes in the resin.

19.3.2 Static Strengths of Virgin and Heat-degraded CFRP Strands at Various Temperatures

Figure 19.10 shows the relationship between static strength and temperature for virgin and heat-degraded CFRP strands. For both CFRP strands, the static strength decreases significantly as the temperature increases. After heat degradation at 140 °C for 3000 hours, the static strength is clearly reduced across the entire temperature range compared to that of virgin CFRP strands.

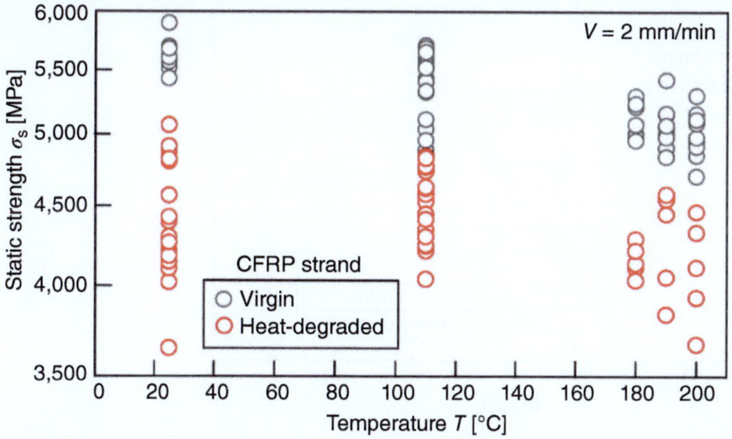

Figure 19.10 Static strengths versus temperature for virgin and heat-degraded CFRP strands.

Table 19.4 Weibull parameters of static tensile strength of CFRP strands at various temperatures.

	Virgin			Heat-degraded		
Temperature T (°C)	**Number of specimens**	**Shape parameter α_s**	**Scale parameter β_s (MPa)**	**Number of specimens**	**Shape parameter α_s**	**Scale parameter β_s (MPa)**
Room temperature	9	39	5686	20	13	4554
110	19	22	5490	20	21	4612
180	9	34	5190	5	–	4204
190	10	30	5150	5	–	4465
200	10	29	5099	5	–	4256

As shown in Table 19.4, the shape parameter α_s and scale parameter β_s of Weibull distribution were determined for the static strength of virgin and heat-degraded CFRP strands at each temperature. The static strength σ_s measured at each temperature was non-dimensionalized by dividing it by the scale parameter β_s. Figure 19.11 shows the Weibull distribution for the sum of σ_s/β_s at all temperatures for virgin and heat-degraded CFRP strands. From this figure, it is clear that the shape parameter α_s of non-dimensional static strength decreases due to heat-degradation treatment. The values of α_s shown in the table in Figure 19.11 and β_s at the reference temperature $T_0 = 110\,°C$ shown in Table 19.4 were chosen as α and σ_0, respectively, for virgin and heat-degraded CFRP strands in Eqs. (14.7) and (14.8) in Chapter 14.

Figure 19.12 shows the fractographs of virgin and heat-degraded CFRP strands after static testing at room temperature. From these photographs, it can be inferred that the interfaces between carbon fiber and matrix resin have deteriorated due to the heat-degradation treatment.

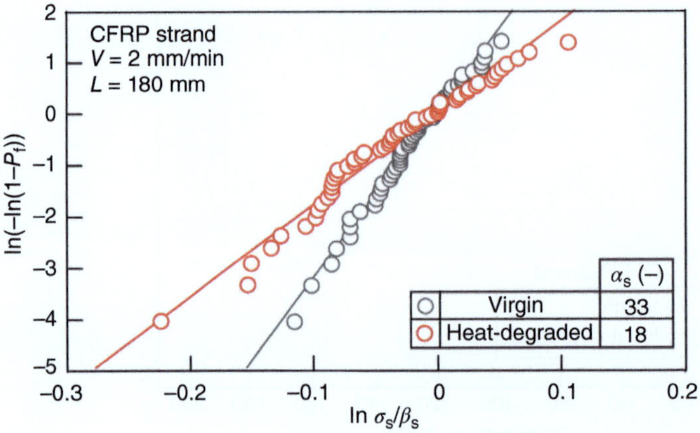

Figure 19.11 Weibull distributions of non-dimensional static tensile strength for virgin and heat-degraded CFRP strands.

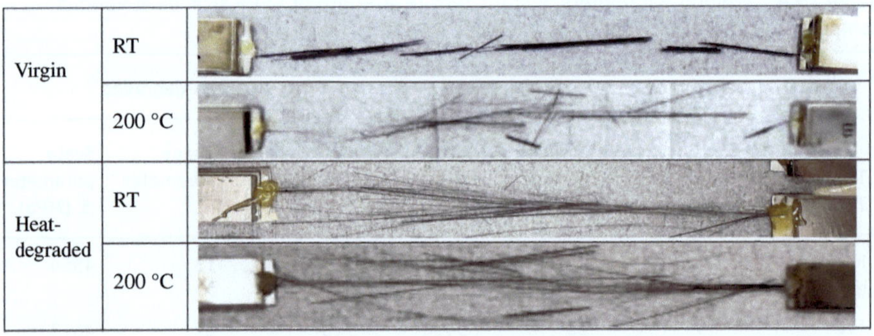

Figure 19.12 Fractographs after static tests for CFRP strands.

Figure 19.13 shows the relationship between the static strength σ_s of a CFRP strand and the ratio of viscoelastic modulus E_s^*/E_{r0} of the matrix resin under the same conditions of time and temperature. The relationship between σ_s and E_s^*/E_{r0} is shown by a single solid line with a slope of n_R, the viscoelastic parameter from Eqs. (14.7) and (14.8). All parameters in Eq. (14.8) for predicting the creep strength master curve were determined for a CFRP strand as shown in Table 19.5. These are the shape parameters α and scale parameter σ_0, the viscoelastic parameter n_R, and the slope of the relaxation modulus m_R of matrix resin at the reference temperature $T_0 = 110\,°C$ and the reference time to failure $t_0 = 1$ minute.

19.3.3 Statistical Creep Failure Times of Virgin and Heat-degraded CFRP Strands

Creep failure tests were conducted on virgin and heat-degraded CFRP strands at two constant stress levels and one constant temperature $T = 180\,°C$. The results of the creep failure tests are shown in Figure 19.14. The predicted curves were calculated

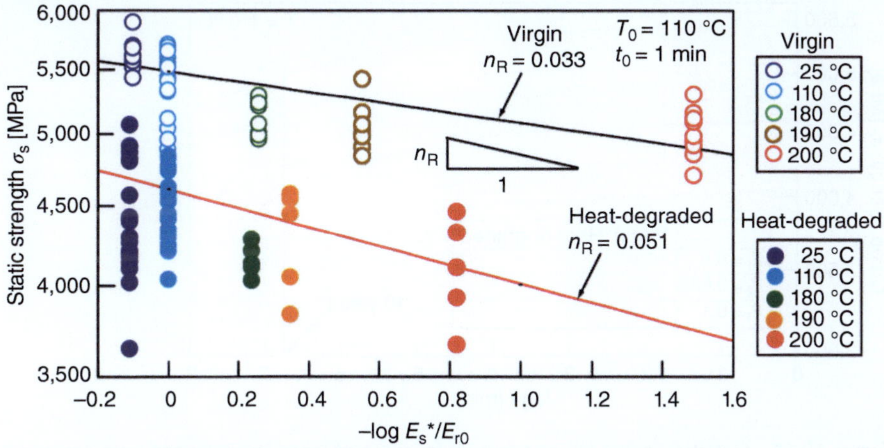

Figure 19.13 Static tensile strength of CFRP strands versus relaxation modulus of matrix resin.

Table 19.5 Parameters for statistical tensile creep strength of CFRP strand.

	σ_0 (MPa)	α	n_R	m_R
Virgin	5490	33	0.033	0.58
Heat degraded	4612	18	0.051	0.59

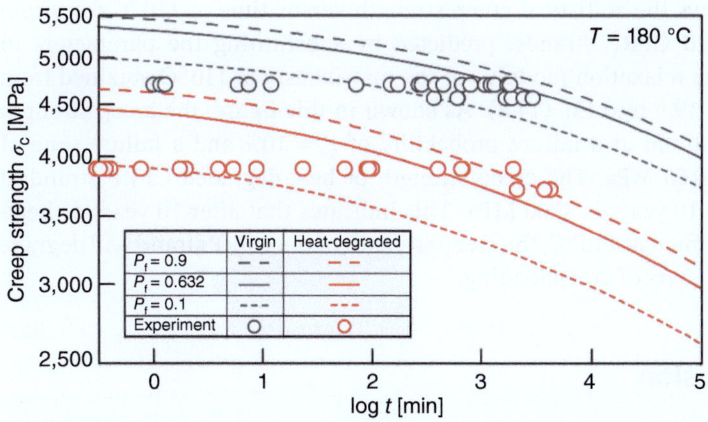

Figure 19.14 Statistical creep failure times under constant loads for virgin and heat-degraded CFRP strands at 180 °C.

by substituting the parameters in Table 19.5 into Eq. (14.8). The predicted statistical creep failure times were in good agreement with the experimental data for virgin and heat-degraded CFRP strands, respectively. These results demonstrate that the statistical creep failure times for virgin and heat-degraded CFRP strands can be predicted using the master curves of the relaxation modulus of the matrix resin shown

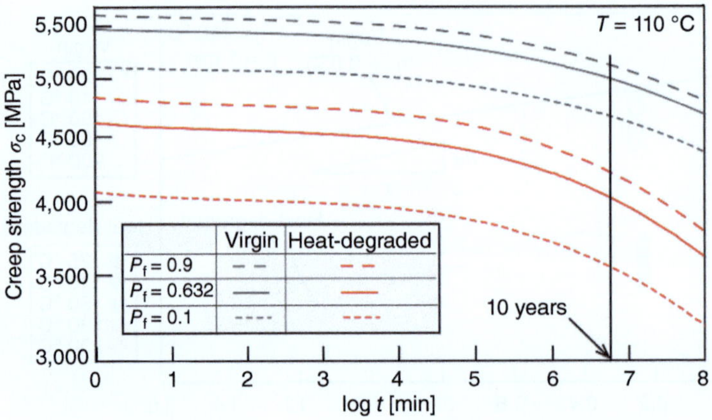

Figure 19.15 Statistical creep failure times under constant loads for virgin and heat-degraded CFRP strands at 110 °C.

in Figures 19.6 and 19.8 and the parameters listed in Table 19.5. It was revealed that the scale parameter σ_0 and shape parameter α of the tensile creep strength of the CFRP strands changed significantly after heat degradation treatment, but the relaxation modulus of the matrix resin remained unchanged.

19.3.4 Long-term Prediction of Statistical Creep Strength for Heat-degraded CFRP Strands

Figure 19.15 shows the statistical creep strength versus time at 110 °C for virgin and heat-degraded CFRP strands, predicted by substituting the parameters in Table 19.5 and the relaxation modulus of the matrix resin at 110 °C obtained from Figures 19.8 and 19.9 into Eq. (14.8). As shown in this figure, the creep strength of virgin CFRP strand at a failure probability of $P_f = 10\%$ and a failure time of $t = 1$ minute is 5150 MPa. The creep strength of heat-degraded CFRP strand at $P_f = 10\%$ and $t = 10$ years is 3800 MPa. This indicates that after 10 years of heat-degradation treatment at 110 °C, the creep strength of the CFRP strand will degrade by 74% due to 10 years of creep loading.

19.4 Conclusion

The integrated ATM for statistical failure life prediction of CFRP was applied to that under tension loading of unidirectional CFRP with a heat-resistant epoxy resin matrix. The results are shown below.

1. Two types of unidirectional CFRP were used: virgin CFRP strands and heat-degraded CFRP strands. The latter were produced by heat-treating virgin CFRP strands at 140 °C for 3000 hours. Heat treatment at 140 °C for 3000 hours corresponds to thermal aging conditions at 110 °C for 10 years.

2. The relationship between statistical creep failure strength and time to failure of virgin and heat-degraded CFRP strands at an operating temperature of 110 °C was predicted using the integrated ATM.
3. It was clarified that the creep strength of CFRP strand at the failure probability $P_f = 10\%$ and failure time $t = 1$ min at 110 °C is decreased to 74% by creep loading of 10 years at 110 °C after heat degradation treatment of 10 years at 110 °C.

References

1 Miyano, Y., Nakada, M., Morisawa, Y., Matsuno, J. and Kageta, S. (2024) Prediction of creep failure life for unidirectional CFRP with heat-resistant epoxy resin as matrix exposed to high temperature under tension load. *Journal of Composite Materials*, **58**, 1051–1062.

2 Miyano, Y. and Nakada, M. Integrated accelerated testing methodology for CFRP durability. *Composites Part B*, **281**, 111527.

3 Toray Industries Inc. Web. https://www.cf-composites.toray/resources/data_sheets/#anc1, Torayca T700S Standard Modulus (Accessed 09 February 2024).

20

Application 6 of Integrated Accelerated Testing Methodology: Effects of Annealing on Statistical Creep Life for Carbon-fiber-reinforced Polymer Strands with Thermoplastic Epoxy Resin as Matrix

20.1 Introduction

Annealing of resins and resin-based composite materials can be defined as a process in which they are held at temperatures higher than room temperature for a relatively long period of time as a means of improving their mechanical properties [1].

This chapter presents specific examination of thermoplastic epoxy resin (TPEP) impregnated carbon fiber strands (CF/TPEP strands) and evaluates the annealing effects on the long-term creep life of CF/TPEP strands under tension load. First, the creep life characteristics of CF/TPEP strands are evaluated using integrated accelerated testing methodology (Integrated ATM) [2] with the viscoelasticity of matrix TPEP when a constant tensile load is applied to the strands at a constant temperature. Second, the viscoelastic behaviors of TPEP are measured when exposed to three annealing conditions. Furthermore, the progress of annealing caused by exposing TPEP to an arbitrary temperature for a long period of time is evaluated as a change in viscoelastic behavior. Third, Integrated ATM, which incorporates consideration of the progress of annealing during loading, was applied to predict the strand creep life. The predicted values were then compared with the experimentally obtained values.

20.2 Formulations

Integrated ATM for evaluating the statistical time- and temperature-dependent creep and fatigue strengths of CFRP and structures was introduced in Chapter 14. In this section, the formulation is focused on the longitudinal creep life of unidirectional CFRP, including annealing process based on Integrated ATM.

20.2.1 Matrix Resin Viscoelasticity

The master curve of relaxation modulus of matrix resin can be found by horizontal and vertical shifting of the relaxation moduli measured at various temperatures

Properties and Testing of Fiber-Reinforced Polymers, First Edition. Yasushi Miyano and Masayuki Nakada. © 2026 Wiley-VCH GmbH. All rights reserved, including rights for text and data mining and training of artificial intelligence technologies or similar technologies.
Published 2026 by Wiley-VCH GmbH.

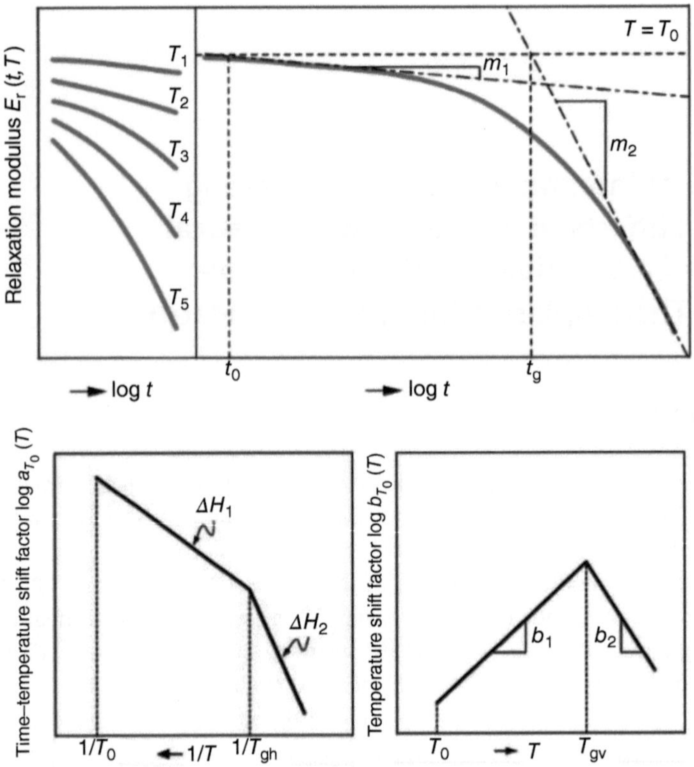

Figure 20.1 Master curve of the relaxation moduli and horizontal and vertical shift factors for matrix resin.

based on the modified time–temperature superposition principle (modified TTSP), as presented in Figure 20.1.

The master curve of relaxation modulus E_r of matrix resin can be formulated as Eq. (20.1) and time–temperature (horizontal) shift factor a_{T_0} and temperature (vertical) shift factor b_{T_0} can also be formulated as Eqs. (20.2) and (20.3).

$$\log E_r(t,T_0) = \log E_r(t_0,T_0) - \log\left[\left(\frac{t}{t_0}\right)^{m_1} + \left(\frac{t}{t_g}\right)^{m_2}\right] \tag{20.1}$$

$$\begin{aligned}\log a_{T_0}(T) &= \frac{\Delta H_1}{2.303G}\left(\frac{1}{T} - \frac{1}{T_0}\right)H(T_{gh} - T)\\ &+\left[\frac{\Delta H_1}{2.303G}\left(\frac{1}{T_{gh}} - \frac{1}{T_0}\right) + \frac{\Delta H_2}{2.303G}\left(\frac{1}{T} - \frac{1}{T_{gh}}\right)\right](1 - H(T_{gh} - T))\end{aligned} \tag{20.2}$$

$$\begin{aligned}\log b_{T_0}(T) &= b_1(T - T_0)H(T_{gv} - T)\\ &+[b_1(T_{gv} - T_0) + b_2(T - T_{gv})](1 - H(T_{gv} - T))\end{aligned} \tag{20.3}$$

Therein, t_0 and t_g represent the reference and glass transition times, respectively. In addition, m_1 and m_2 denote the slopes in the short and long time regions, respectively. Also, ΔH_1 and ΔH_2 stand for the activation energies at low and high temperature

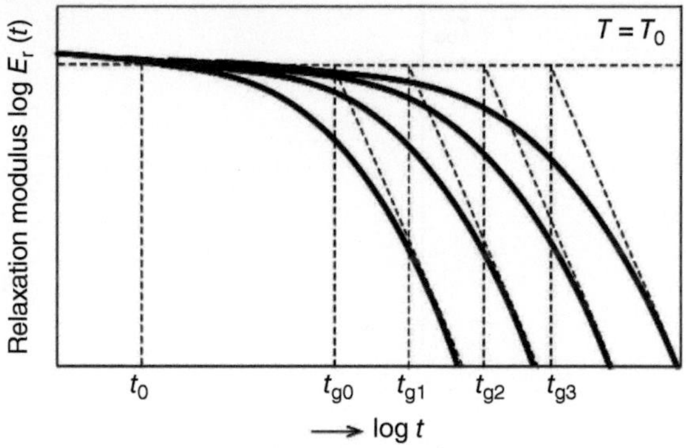

Figure 20.2 Master curves of relaxation moduli for various annealing conditions.

regions, respectively. H is the Heaviside step function. G is the gas constant. T_0 is a reference temperature. T_{gh} and T_{gv} respectively express the glass transition temperatures for a_{T_0} and b_{T_0}. Also, b_1 and b_2 are the parameters for vertical shift factor.

20.2.2 Effect of Annealing on Matrix Resin Viscoelasticity

The glass transition time t_g, which is an indicator of the annealing of a resin, is definable as the time at which the relaxation modulus at the reference temperature T_0 drops sharply with increasing time, as shown in the upper diagram of Figure 20.1. The t_g is determined by the curing time and temperature during molding. It increases along with the operating time and temperature.

Figure 20.2 shows how the master curve of the relaxation modulus at a reference temperature changes with the progress of annealing. It can be inferred from the figure that only the glass transition time of the master curve increases considerably as annealing progresses. The t_{g0} value in the figure is the glass transition time of the master curve immediately after curing. Also, t_{g1}, t_{g2}, and t_{g3} stand for the glass transition times that increase as the operating process progresses.

Figure 20.3 presents the increase in glass transition time with operating time t_{op} during the operating process at a constant temperature T_{op}. The horizontal axis of the figure shows the curing time t_{cu} in the curing process, plus the operating time t_{op} in the subsequent operating process. The curing time at the curing temperature in the curing process is converted to the equivalent curing time t'_{cu} at temperature T_{op} of the operating process. This amount is added to the operating time t_{op} of the operating process. The acceleration rate used here is the 10 °C doubling rule commonly used in industry to measure the degradation of mechanical properties of resins and resin-based composites. This rule states that the rate of degradation doubles for every 10 °C increase in temperature.

20.2.3 Statistical Static and Creep Strengths of CFRP

The statistical static strengths under an arbitrary constant strain rate and temperature are given by the following equation from Integrated ATM.

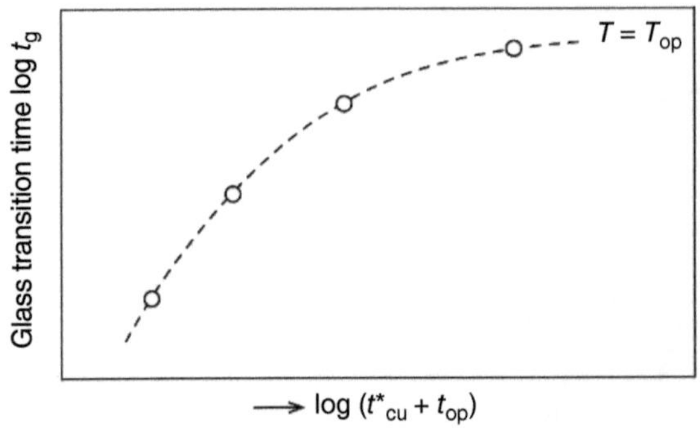

Figure 20.3 Glass transition time versus annealing time at operating temperatures.

$$\log \sigma_s(t,T) = \log \sigma_0(t_0,T_0) + \frac{1}{\alpha}\log[-\ln(1-P_f)] + n_R \log\left[\frac{E_s(t,T)}{E_r(t_0,T_0)}\right]$$
$$E_s(t,T) = E_r(t/2,T) \tag{20.4}$$

Therein, t and t_0 represent the failure time and reference failure time, respectively, T and T_0 denote the temperature and reference temperature, respectively, and P_f stands for the failure probability. Furthermore, σ_0 and α denote the scale and shape parameters of static strength at t_0 and T_0, respectively. Also, n_R is a viscoelastic parameter determined by the failure mode. E_r is the relaxation modulus of the matrix resin.

Parameters σ_0, α, and n_R in Eq. (20.4) show time- and temperature-dependent statistical static strengths obtained from static tests of CFRP conducted at various temperatures under a constant strain rate.

Statistical creep strength σ_c can be ascertained by shifting the static strength with $\log A$ based on Christensen's viscoelastic crack kinetic theory. Therefore, creep strength σ_c can be denoted by subscript "s" replaced with "c" in Eq. (20.4), as presented in Eq. (20.5). In this equation, E_c represents the viscoelastic modulus for a constant stress load. It has been clarified that the master curve of creep strength is obtainable by shifting the master curve of static strength by $\log A$, as portrayed in Figure 20.4.

$$\log \sigma_c(t,T) = \log \sigma_0(t_0,T_0) + \frac{1}{\alpha}\log[-\ln(1-P_f)] + n_R \log\left[\frac{E_c(t,T)}{E_c(t_0,T_0)}\right]$$
$$E_c(t,T) = E_s(At,T) = E_r(At/2,T) \tag{20.5}$$

The master curve of viscoelastic modulus E_c at reference temperature T_0 can be shown by the following equation using Eqs. (20.1) and (20.5).

$$\log E_c(t,T_0) = \log E_r\left(\frac{A_1 t_0}{2}, T_0\right) - \log\left[\left(\frac{A_1 t}{2t_0}\right)^{m_1} + \left(\frac{A_2 t}{2t_g}\right)^{m_2}\right] \tag{20.6}$$

For the case of tension loading applied to unidirectional CFRP, the shifting amount $\log A$, as ascertained from slope k_R of the logarithmic static strength against the logarithmic failure time curve, is calculated using the following equation.

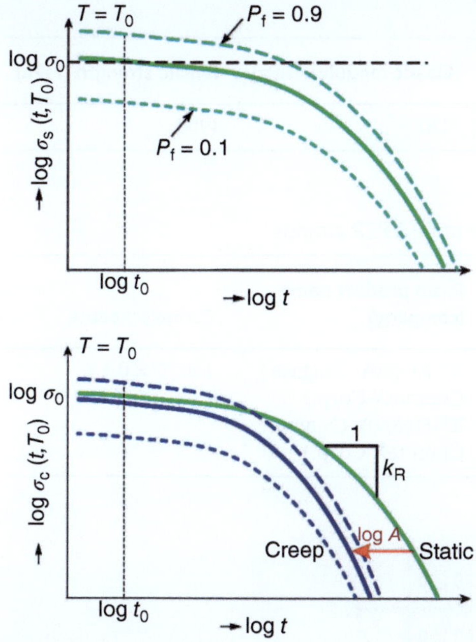

Figure 20.4 Master curves of statistical tensile static and creep strengths versus failure time for unidirectional CFRP.

$$\log A_i = \log\left(1 + \frac{1}{k_{Ri}}\right), \quad k_{Ri} = m_i n_R, \quad i = 1, 2 \tag{20.7}$$

Therein, m_i represents the slope of the logarithmic relaxation modulus of the matrix resin against the logarithmic time curve presented in Figure 20.1.

20.3 Experimental Procedures

A CF/TPEP strand consisting of high-strength type PAN-based carbon fiber and thermoplastic epoxy was molded using the pultrusion method. Table 20.1 presents the characteristics of carbon fiber and the mechanical properties used for this study. Table 20.2 shows the matrix epoxy composition and the curing conditions for the CF/TPEP strand.

In this study, the CF/TPEP strands were manufactured by using a thermoplastic epoxy resin monomer mixture diluted with methyl ethyl ketone. Figure 20.5 presents a conceptual diagram of the pultrusion process by dipping into a resin bath filled with thermoplastic epoxy resin diluted with an organic solvent. For the case in which the dry fiber fed to the resin bath is a thick and/or twisted bundle, the undiluted resin is commonly difficult to impregnate in the narrow gaps of fibers that are closely tightened. In that case, dilution with a solvent effectively helps the resin soak into the fiber bundle. After squeezing out the excess resin with double rolls, the bundle is put in an oven to evaporate the solvent and to polymerize the thermoplastic epoxy resin impregnated in the fiber bundle. While almost all solvent was

Table 20.1 Carbon fiber specifications.

Carbon fiber	Density (g/cm^3)	Tex (g/km)	Elastic modulus (GPa)	Tensile strength (MPa)
T700-6000	1.80	400	230	4900

Table 20.2 Composition and cure schedule of CF/TPEP strands.

Carbon fiber	Resin composition (weight ratio)	Resin product name (company)	Curing schedule
T700-6000	Thermoplastic epoxy (100) Cure accelerator (6.5)	XNR6850V (Nagase ChemteX Corp.) XNH6850V (Nagase ChemteX Corp.)	150 °C × 0.5 h

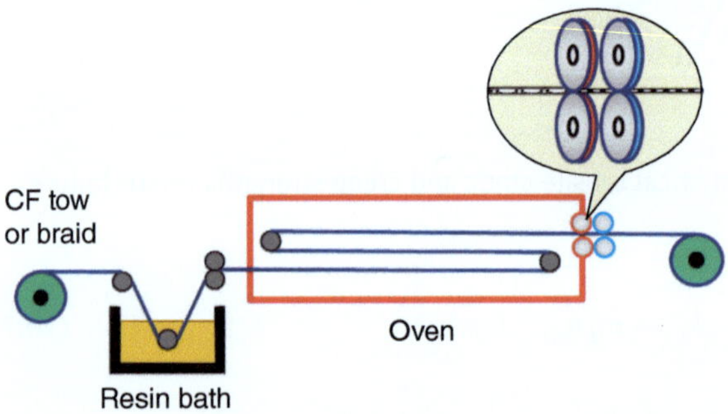

Figure 20.5 Pultrusion process to produce CF/TPEP strands using the dipping method into a resin bath filled with thermoplastic epoxy resin diluted with organic solvent.

volatilized during the strand-manufacturing process, just a slight amount of solvent can remain in the CF/TPEP strands. During the subsequent operation process, the evaporation of the residual solvent and the extension of the molecular chains through polymerization will occur, which will result in a change in the viscoelastic properties of the TPEP resin.

The CF/TPEP strand specimen configuration is presented in Figure 20.6. The diameter and gage length of CF/TPEP strands are approximately 1 and 200 mm, respectively. Static and creep tension tests for CF/TPEP strands were conducted.

20.4 Results and Discussion

20.4.1 Master Curve of the Relaxation Modulus of TPEP

Figure 20.7 portrays the relaxation moduli at various temperatures for thermoplastic epoxy resin (TPEP) immediately after curing processing, along with the master curve at the reference temperature $T_0 = 25\,°C$. The time–temperature (horizontal) shift factor $a_{T_0}(T)$ and the temperature (vertical) shift factor are portrayed in Figure 20.8.

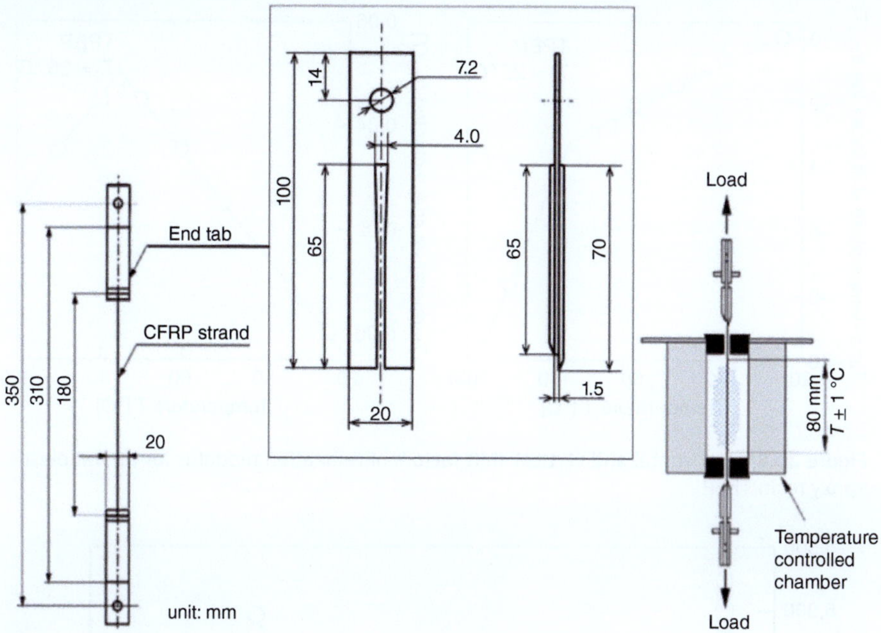

Figure 20.6 Tensile static and creep strength tests for CF/TPEP strands.

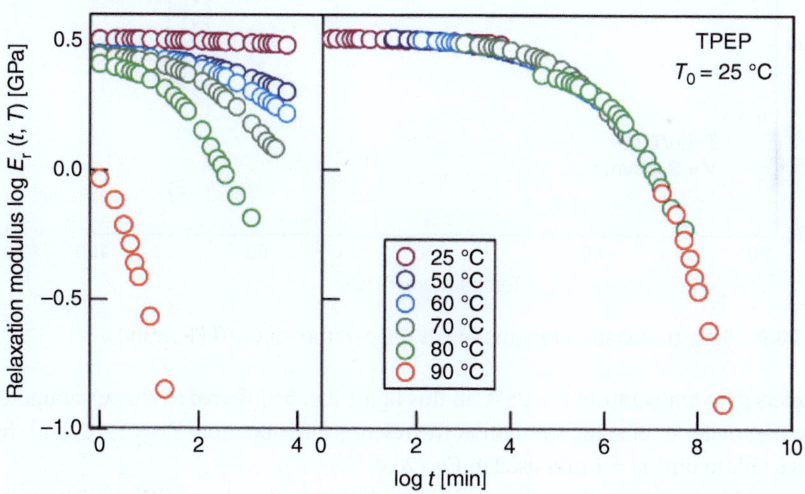

Figure 20.7 Master curve of relaxation modulus for thermoplastic epoxy resin TPEP.

20.4.2 Statistical Static Strength of CF/TPEP Strands

Figure 20.9 depicts the static strength versus temperature, which decreases concomitantly with increasing temperature.

The Weibull distributions for static strengths at various temperatures for CF/TPEP strands are depicted in Figure 20.10. In this figure, α_s is the shape parameter; β_s is the scale parameter for the static strength of CF/TPEP strand. The shape parameter maintains an almost constant value for the static strength. Shape parameters α_s and scale

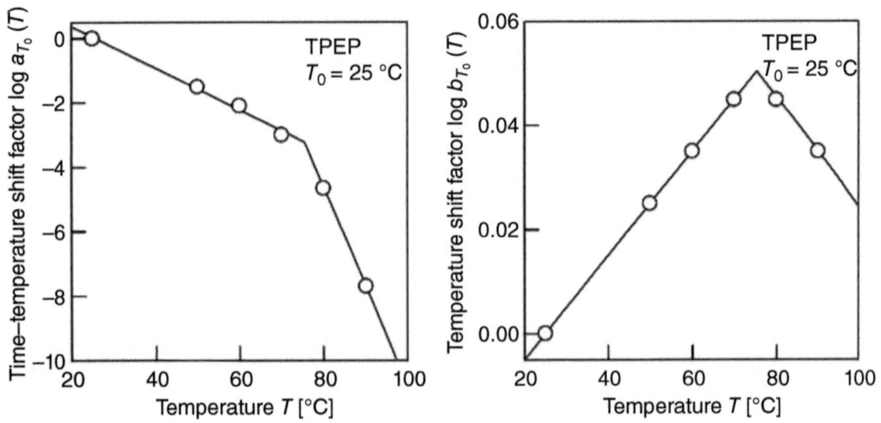

Figure 20.8 Horizontal and vertical shift factors of relaxation modulus for thermoplastic epoxy resin TPEP.

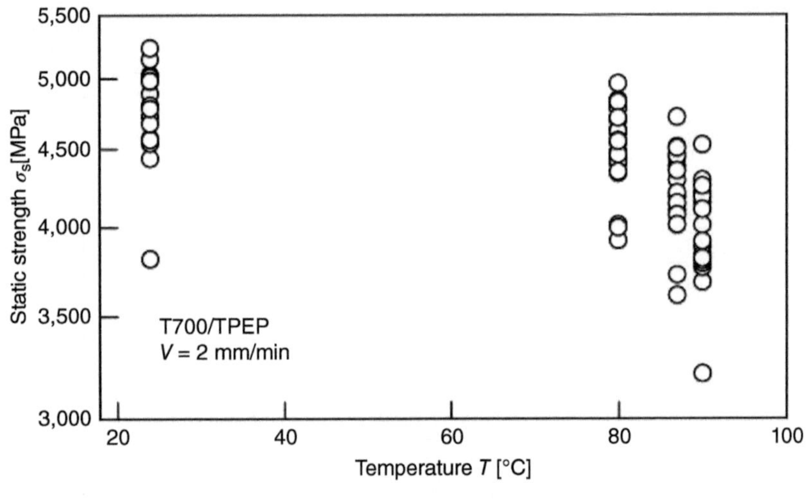

Figure 20.9 Statistical static strengths versus temperature of CF/TPEP strands.

parameters β_s at temperature $T = 25°C$ in this figure can be inferred as shape parameter α, scale parameter σ_0 of static strength at the reference temperature $T_0 = 25°C$, and the reference failure time $t_0 = 1$ min used in Eq. (20.4).

Figure 20.11 presents the non-dimensional static strength σ_s/σ_0 of a CF/TPEP strand against the non-dimensional viscoelastic modulus of matrix resin E_s^*/E_{r0} for the same conditions of time and temperature. The relation of σ_s/σ_0 against E_s^*/E_{r0} can be represented as one solid straight line with slope of n_R, which is the viscoelastic parameter in Eq. (20.4). The viscoelastic parameters n_R obtained from the experiments agree well with those determined theoretically from substituting the strength shape parameter of carbon fiber $\alpha_c = 4.59$ into the equation shown by Rosen's model [3].

Additionally, the shifting amounts A_i in Eq. (20.7) are obtained from the viscoelastic parameter n_R and the slopes of relaxation modulus of matrix resin m_i.

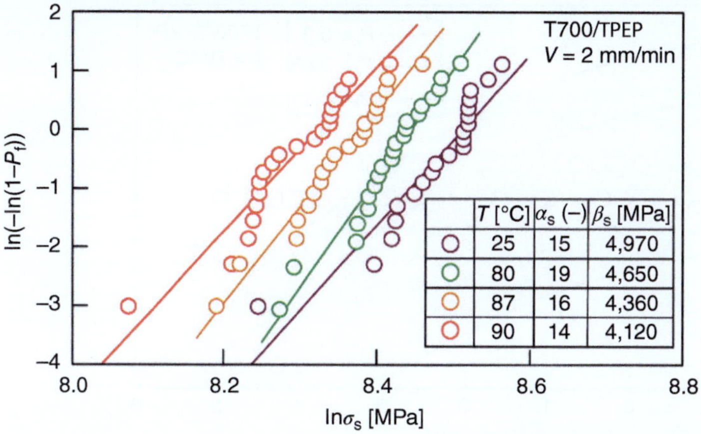

Figure 20.10 Weibull distributions of statistical static strength of CF/TPEP strands.

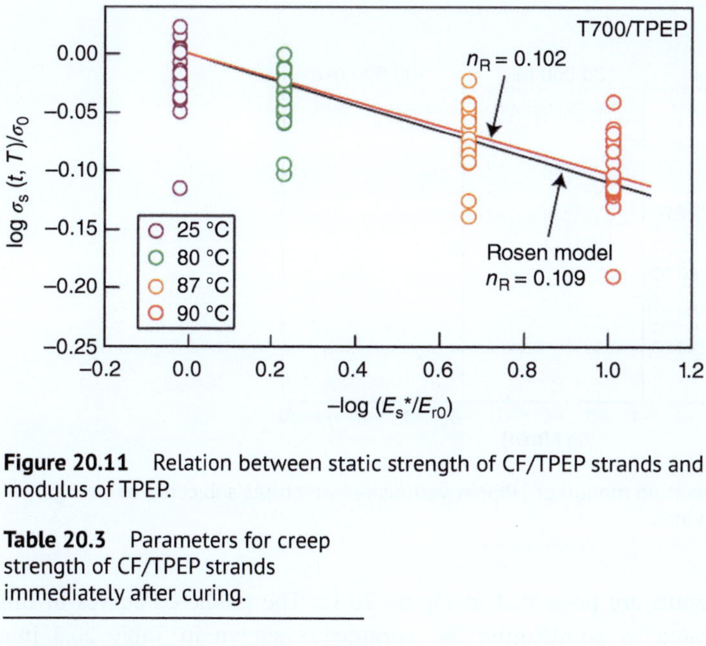

Figure 20.11 Relation between static strength of CF/TPEP strands and viscoelastic modulus of TPEP.

Table 20.3 Parameters for creep strength of CF/TPEP strands immediately after curing.

σ_0 (MPa)	α	n_R	A_1	A_2
4,970	15	0.102	491	30

All parameters in Eq. (20.5) for the master curve of creep strength are found for CF/TPEP strands, as presented in Table 20.3.

20.4.3 Statistical Creep Strength of CF/TPEP Strands

Creep failure tests of CF/TPEP strands after curing at $T = 150\,°C$ for 30 minutes were conducted at three arbitrary constant stress levels and at one constant temperature

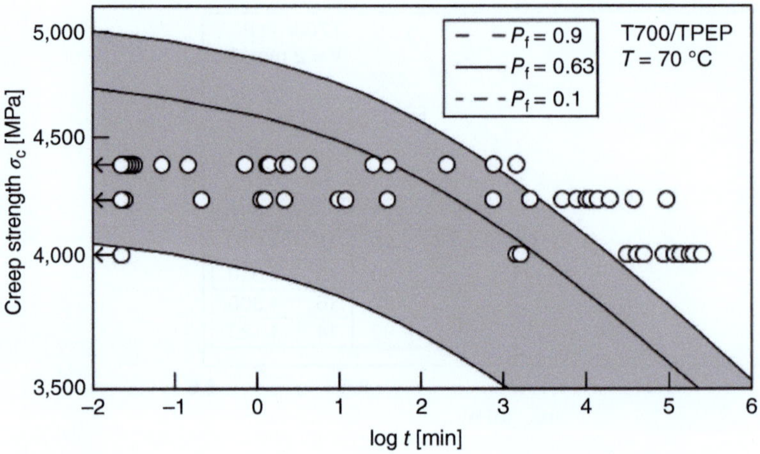

Figure 20.12 Predicted and measured tensile creep strength of CF/TPEP strand at $T = 70$ °C.

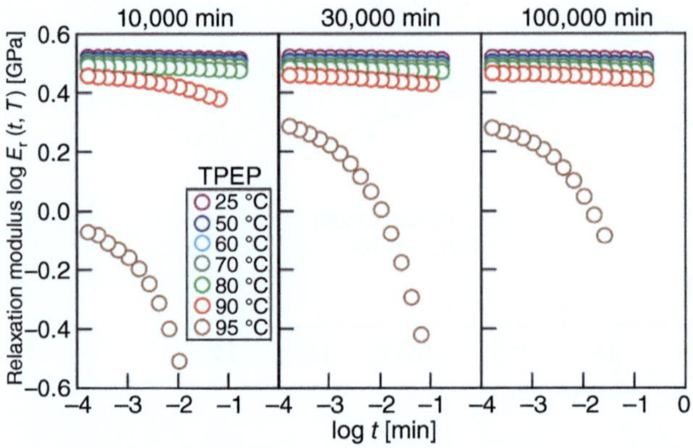

Figure 20.13 Relaxation moduli of TPEP at various temperatures subjected to curing processes of three kinds.

$T = 70$ °C. The results are presented in Figure 20.12. The predicted curves in this figure are calculated by substituting the parameters shown in Table 20.3 into Eq. (20.4) after calculating the relaxation moduli at $T = 70$ °C. Results clarified that the predicted results do not agree with the experimentally obtained findings.

20.4.4 Progress of Annealing of TPEP During the Operating Process

It is assumed that the crosslinking reaction of the TPEP resin progresses during the operating process following the curing process, leading to a change in the resin viscoelasticity. Figure 20.13 shows the relaxation modulus calculated from the dynamic mechanical analysis (DMA) test results at various temperatures for TPEP resin samples that had been cured at 150 °C for 30 minutes, and were then subjected to curing processes at 70 °C for 10,000, 30,000, and 100,000 minutes.

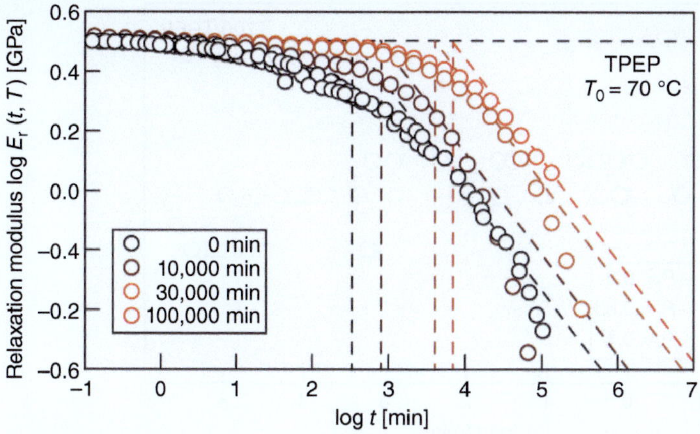

Figure 20.14 Master curves of relaxation modulus of TPEP after various operating durations.

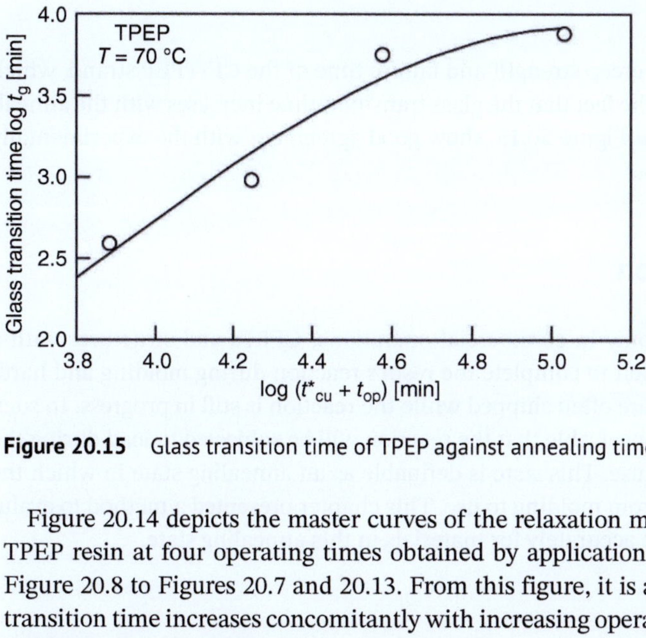

Figure 20.15 Glass transition time of TPEP against annealing time at $T = 70\,°C$.

Figure 20.14 depicts the master curves of the relaxation modulus at $T = 70\,°C$ of TPEP resin at four operating times obtained by application of the shift factors of Figure 20.8 to Figures 20.7 and 20.13. From this figure, it is apparent that the glass transition time increases concomitantly with increasing operating time. Resin cross-linking occurs during both the curing process and the operating process. This process is definable as resin annealing. The relation between this annealing time and the resin's glass transition time is presented in Figure 20.15. The t_{op} of the horizontal axis is the operation time. Also, t^*_{cu} is the curing time at the curing temperature converted to time at the operating temperature based on the 10 °C doubling rule.

20.4.5 Statistical Creep Strength of CF/TPEP Strands Attributable to Annealing Progress During the Operating Process

Figure 20.16, similar to Figure 20.12, shows measured data and predictions of the creep strength against failure time for CF/TPEP strands. The predicted results of

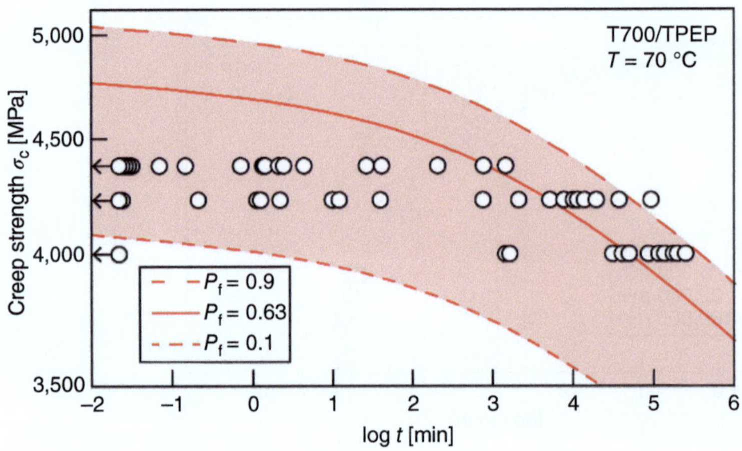

Figure 20.16 Predicted and measured tensile creep strength attributable to annealing progress during the operating process.

the relation between creep strength and failure time of the CF/TPEP strand, which was predicted using the fact that the glass transition time increases with the annealing time, presented in Figure 20.15, show good agreement with the experimentally obtained results.

20.5 Conclusion

For reasons of economy in commercial operations, CFRPs and structures with a resin matrix are not left to complete the resin's reaction during molding and hardening. The products are often shipped while the reaction is still in progress. In such cases, it is entirely conceivable that the reaction will be subjected to load during the operating process of use. This state is definable as an annealing state in which the reaction progresses from molding to use. This chapter presented a method to evaluate the creep load life accurately for materials in this annealing state.

References

1 Nakada, M., Miyano, Y., Morisawa, Y., Motohashi, T., Nishikawa, K., Nishida, H. and Uzawa, K. (2025) Effects of annealing on statistical creep life for unidirectional CFRTP. *Materials System*, **42**, 27–33.

2 Miyano, Y. and Nakada, M. (2024) Integrated accelerated testing methodology for CFRP durability. *Composites Part B*, **281**, 111527.

3 Rosen, B.W. (1964) Tensile failure of fibrous composites. *AIAA Journal*, **2**, 1985–1991.

Appendix A

Effect of Physical Aging on the Creep Deformation of an Epoxy Resin

A.1 Introduction

The purpose of this study is to shed light on the role of physical aging and temperature on creep deformation and its mechanism by measuring the creep compliance of an epoxy resin that is heat-treated at various combinations of time and temperature [1].

A.2 Creep Deformation for Aged Polymers

The specific volume of polymer resins consists of the occupied volume and the free volume [2]. The former is the volume occupied by molecular chains, and the latter is the space that permits movement of molecular chains. Figure A.1 shows the dependence of the occupied volume and free volume on temperature. In rapid cooling from temperature above T_g, the specific volume of polymer resin follows the route of A, G, and B, where G stands for the specific volume at T_g while A and B represent the specific volume at a temperature above and below T_g, respectively. As the cooling rate decreases, the glass-transition temperature drops, say, from T_g to T_g'' and the route changes from AGB to AG″B″. Therefore, the state at B is not in thermodynamic equilibrium. When the material is kept at a constant temperature below T_g, the free volume decreases moderately and the thermodynamic equilibrium state is approached with the passage of time, as shown by the arrowhead line. This behavior is called physical aging.

We now consider the effect of aging and temperature T on the creep compliance. The aging of a polymer depends on the time exposed to a temperature history and may be represented by a single parameter β, which will be defined shortly. Here, the specimens are aged by exposing them to a temperature T_a for a time t_a. We refer T_a and t_a as the aging temperature and aging time, respectively. We denote the reference values of T, t_a, and T_a by T_0, t_{a0}, and T_{a0}. For polymers obeying time–temperature superposition principle, the mechanical response under a certain temperature

Properties and Testing of Fiber-Reinforced Polymers, First Edition. Yasushi Miyano and Masayuki Nakada. © 2026 Wiley-VCH GmbH. All rights reserved, including rights for text and data mining and training of artificial intelligence technologies or similar technologies.
Published 2026 by Wiley-VCH GmbH.

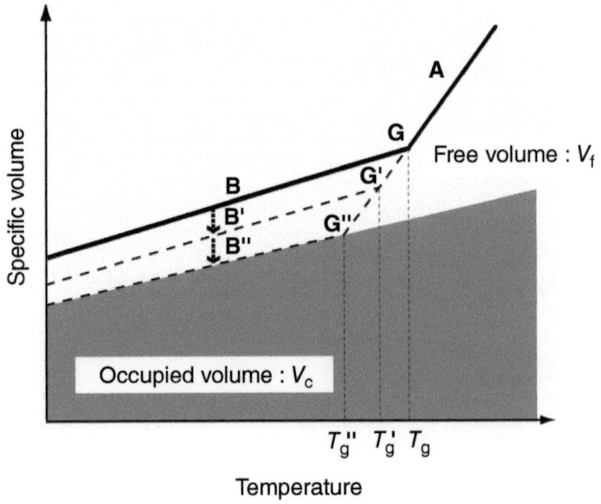

Figure A.1 Explanation of physical aging.

history can be mapped onto the response at the reference temperature T_0, which is called the master curve, and the time scale at T_0 is called the reduced time t'.

For a pair of t_a and T_a, that is, $\beta(t_a,T_a)$, we can construct the master compliance curve $D_c(t';\beta,T_0)$ from $D_c(t;\beta,T)$ tested at various temperatures, which are plotted on log–log scales. As schematically shown on the top of Figure A.2, we translate $D_c(t;\beta,T)$ at temperatures nearest to T_0 horizontally and vertically so that they overlap on $D_c(t;\beta,T_0)$; the overlapped curve is a portion of the master curve $D_c(t';\beta,T_0)$. In this manner, we determine the time–temperature shift factor $a_{T_0}(T)$ and the temperature shift factor $b_{T_0}(T)$ experimentally, where

$$\log t - \log t' = \log a_{T_0}(T) \tag{A.1}$$

$$\log D_c(t;\beta,T) - \log D_c(t';\beta,T_0) = \log b_{T_0}(T) \tag{A.2}$$

We construct the master curves for various aging conditions, pair of t_a and T_a. At this point, we speculate that the time–temperature superposition principle also holds for t_a and T_a. As shown at the middle of Figure A.2, we translate the master curves $D_c(t';\beta,T_0)$ horizontally so that they overlap on the master curve $D_c(t';\beta_0, T_0)$ at the reference aging parameter β_0. We now define the aging parameter $\beta(t_a,T_a)$ as the amount of horizontal shift toward $D_c(t';\beta_0,T_0)$, implying $\beta_0 = \beta(t_{a0},T_{a0}) = 0$. If the perfect match is achieved, the superposition principle for t_a and T_a holds. We call the curve thus obtained the master-master compliance curve $D_c(t'';\beta_0,T_0)$ and the reduced-reduced time t'' where

$$\beta(t_a,T_a) = \log t' - \log t'' \tag{A.3}$$

The horizontal shift (A.3) is well known for the creep compliance at a constant temperature. If the superposition principle holds for the aging time t_a and aging temperature T_a, we define the reduced aging time t_a' the aging time at reference aging temperature T_{a0} and the aging time–temperature shift factor $a_{T_{a0}}(T_a)$ by

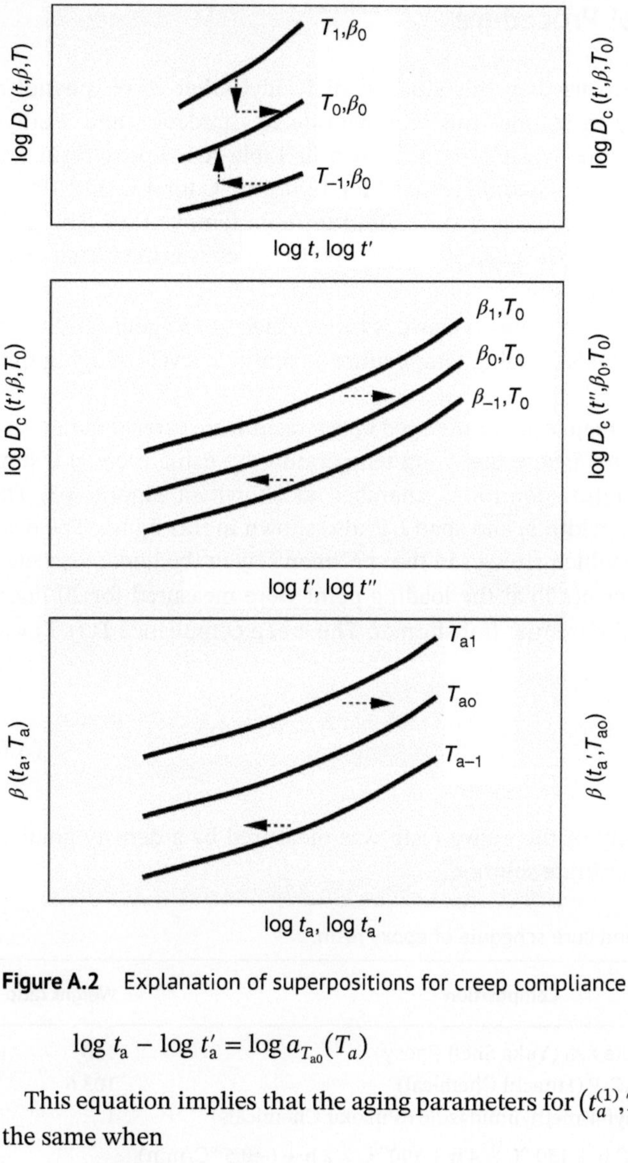

Figure A.2 Explanation of superpositions for creep compliance D_c and aging parameter β.

$$\log t_a - \log t'_a = \log a_{T_{a0}}(T_a) \tag{A.4}$$

This equation implies that the aging parameters for $(t_a^{(1)}, T_a^{(1)})$ and for $(t_a^{(2)}, T_a^{(2)})$ are the same when

$$\frac{t_a^{(1)}}{a_{T_{a0}}(T_a^{(1)})} = \frac{t_a^{(2)}}{a_{T_{a0}}(T_a^{(2)})} \tag{A.5}$$

We plot $\beta(t_a, T_a)$ versus t_a for various T_a as shown at the bottom of Figure A.2. We speculate that the curves for $\beta(t_a, T_a)$ collapse on $\beta(t'_a, T_{a0})$ by a suitable horizontal shift, and we introduce the master curve for the aging parameter $\beta(t'_a, T_{a0})$ and reduced aging time. This speculation is confirmed experimentally in the next section.

A.3 Experimental Procedure

The composition of resin used in this study is diglycidyl ether of bis-phenol A (Epikote 828) for epoxy resin, metylnadic anhydride for hardener, and 2-ethyl-4-methyl-imidazole for cure accelerator, as shown in Table A.1. Epoxy resin was cured by casting with a cure schedule of 70 °C for 12 h, 150 °C for 4 h, and 190 °C for 2 h, followed by a slow cooling at 0.5 °C/min to room temperature. The glass transition temperature T_g for the resin system is 149 °C, which was measured using differential scanning calorimetry.

The epoxy resin cured as mentioned above is heat-treated at 15 aging conditions, which combined three levels of aging temperature T_a and five levels of aging time t_a as shown in Table A.2.

Three-point bending creep tests for the aged epoxy resin were carried out at various constant temperatures below the aging temperature by using a creep testing machine with a temperature-controlled chamber, as shown in Figure A.3. The specimen of thickness h, width b, and span L is also shown in this figure. The dead load W_0 is 58.8 N below which stresses in the specimen stay in the linear viscoelastic range. The deflections $\delta(t,T)$ at the loading point were measured for 20 hours using a linear variable differential transformer. The creep compliance $D_c(t,T)$ was calculated by

$$D_c = \frac{4bh^3\delta(t,T)}{L^3W_0} \tag{A.6}$$

In addition, the density of the epoxy resin was measured by a density gradient column using a calcium nitrate solution.

Table A.1 Composition and cure schedule of epoxy resin.

Composition		Weight ratio
Epoxy	Epikote 828 (Yuka Shell Epoxy)	100
Hardener	MHAC-P (Hitachi Chemical)	103.6
Cure accelerator	2-Ethyl-4-methylimidazole (Shikoku Chemicals)	1
Cure schedule: 70 °C × 12 h + 150 °C × 4 h + 190 °C × 2 h + (−0.5 °C/min)		
Glass transition temperature: 149 °C		

Table A.2 Aging conditions of epoxy resin.

Aging temperature T_a (°C): 80, 100, 120
Aging time t_a (min): 1×10^3, 3×10^3, 1×10^4, 3×10^4, 1×10^5

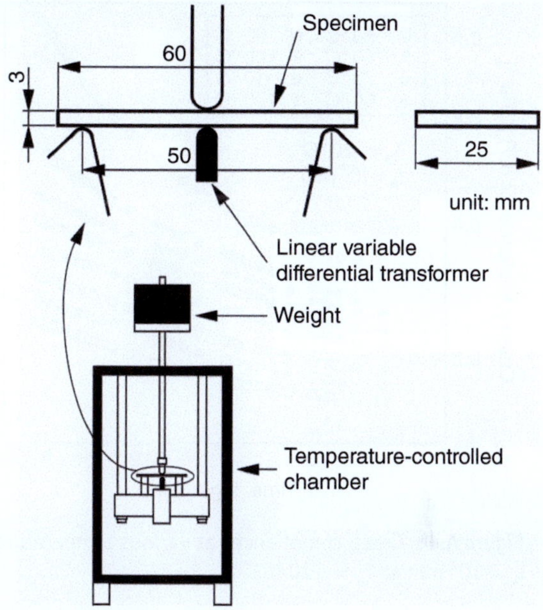

Figure A.3 Specimen configuration and test apparatus.

A.4 Results and Discussion

A.4.1 Creep Compliance

The creep compliances for the epoxy resin aged for $t_a = 10^3$ min at $T_a = 120\,°C$ were measured at various temperatures and are shown in Figure A.4. The master curve at a reference temperature T_0 in Figure A.5 can be obtained by horizontally and vertically shifting the creep compliances at various temperatures based on the modified time–temperature superposition principle. Figure A.6 shows the time–temperature shift factor $a_{T_0}(T)$ and temperature shift factor $b_{T_0}(T)$, where the solid curves are the least-squares average of the data for the 15 combinations of t_a and T_a on Table A.2. Here, Arrhenius equation with $\Delta H = 125$ kJ/mol represents the horizontal shift factor, while a straight line represents the vertical shift factor:

$$\log a_{T_0}(T) = \frac{\Delta H}{2.303R}\left(\frac{1}{T} - \frac{1}{T_0}\right) \tag{A.7}$$

$$\log b_{T_0}(T) = K(T - T_0) \tag{A.8}$$

where ΔH, R, and K are activation energy, gas constant, and material constant, respectively. The circles are data corresponding to $t_a = 10^3$ min at $T_a = 120\,°C$. We observed only an approximate fit for the horizontal shift and wide scattering for the vertical shift; the data spread shown in the figure is one of the worst.

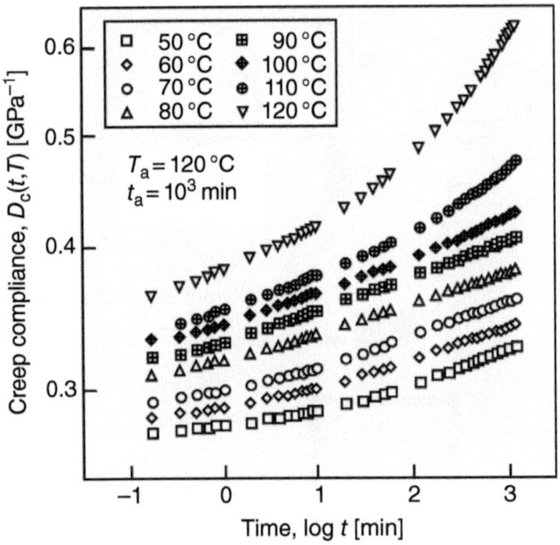

Figure A.4 Creep compliances at various temperatures for epoxy resin aged for $t_a = 10^3$ min at $T_a = 120\,°C$.

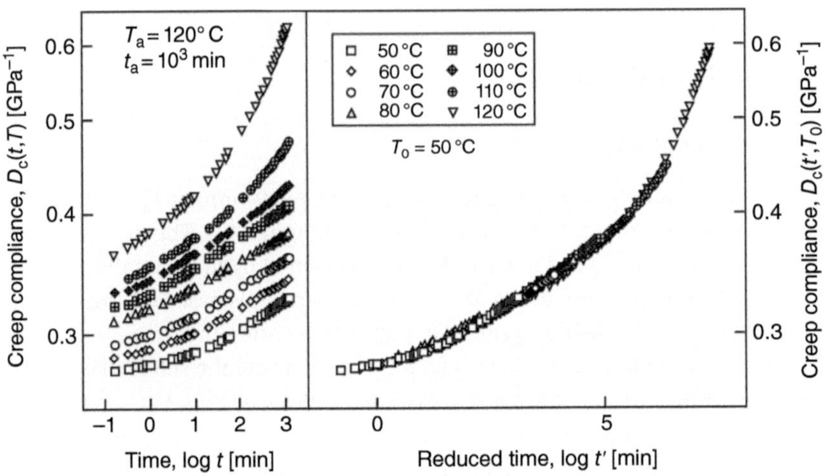

Figure A.5 Master curve of creep compliance for epoxy resin aged for $t_a = 10^3$ min at $T_a = 120\,°C$.

Furthermore, after the creep test, it was also confirmed by a Fourier transform infrared spectrometer (using KBr disk) that no chemical aging of these epoxy resins had occurred.

A.4.2 Effect of Physical Aging on Creep Compliance

The master curves for epoxy resins aged under 15 conditions were constructed, of which nine cases are shown on the left side of Figure A.7. The creep compliance is

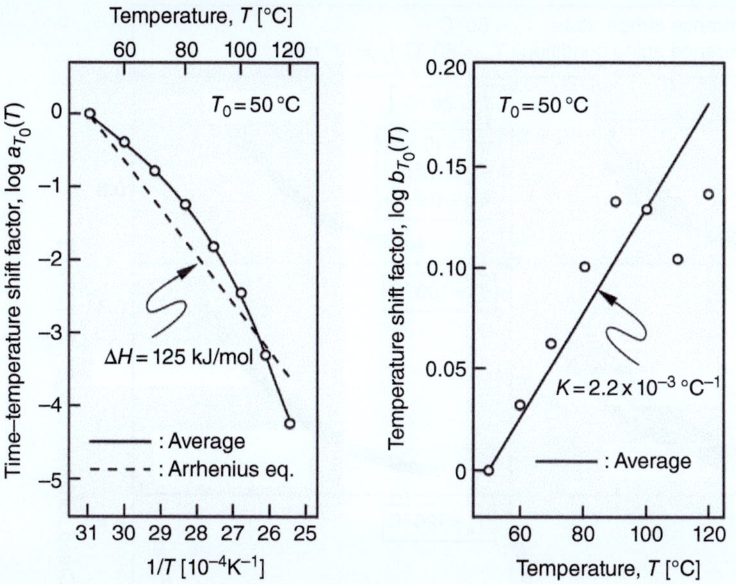

Figure A.6 Shift factors $a_{T_0}(T)$ and $b_{T_0}(T)$ for epoxy resin aged for $t_a = 10^3$ min at $T_a = 120\,°C$.

suppressed as the aging time and temperature increase; the creep compliance shifts to the right with increasing aging time and temperature. Each curve on the right side of Figure A.7 shows the master–master curve $D_c(t'',T_0)$ at the reference aging conditions $t_{a0} = 10^3$ min and $T_{a0} = 80\,°C$ obtained from $D_c(t',T_0)$ by shifting horizontally for various aging times at a constant aging temperature.

The aging parameter β versus the aging time t_a at various aging temperatures T_a is shown on the left side of Figure A.8. As stated in Section A.2, the aging parameter $\beta(t_a,T_a)$ obeys the superposition principle for t_a and T_a, and thus the master curve of β versus the reduced aging time t'_a can be constructed by shifting $\beta(t_a,T_a)$ horizontally at the reference aging temperature T_{a0}. The smooth master curve constructed on the right-hand side of Figure A.8 confirms the validity of the superposition principle for the aging parameter.

The densities ρ for epoxy resins aged under 15 aging conditions, are shown on the left side of Figure A.9. The superposition principle also applies to the density. The smooth master curve of density at a reference aging temperature $T_{a0} = 80\,°C$ shown on the right side of this figure was obtained by horizontally shifting the densities at various aging temperatures.

Figure A.10 shows the time–temperature shift factors for the aging parameter β and density ρ where two points overlap at $T_{a0} = 80\,°C$. These shift factors for both β and ρ agree with each other and the Arrhenius equation with $\Delta H = 70$ kJ/mol.

Figure A.11 shows the relationship between β and ρ for 15 aging conditions. This figure shows that the aging parameter and the density satisfy the one-to-one correspondence regardless of the aging time and aging temperature.

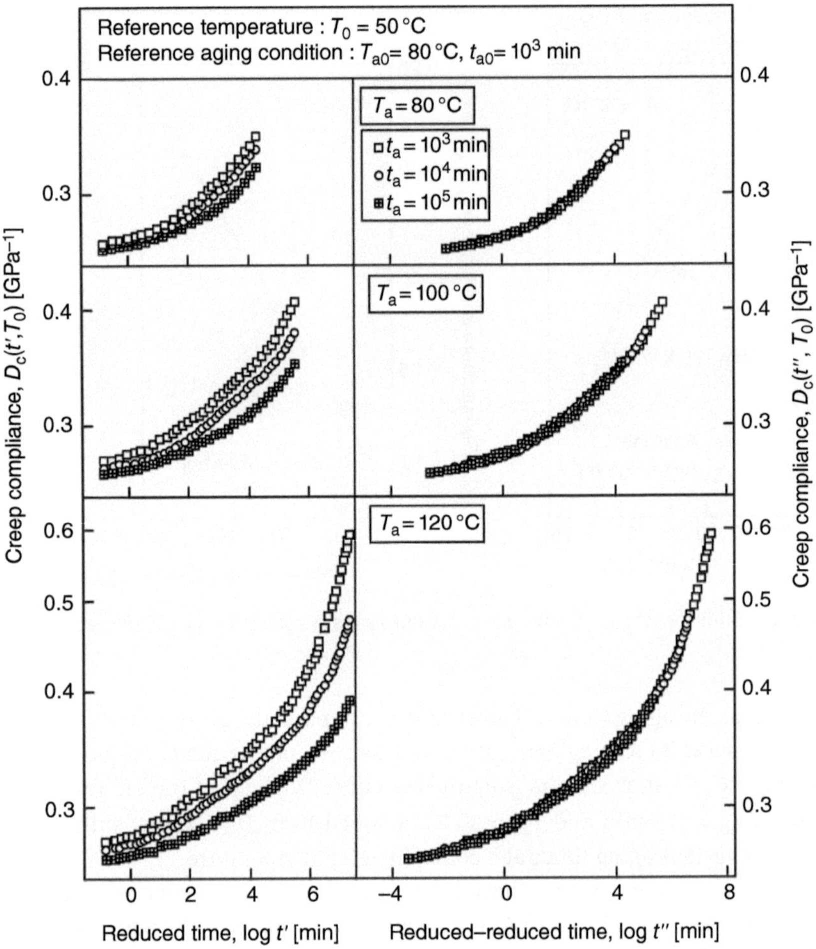

Figure A.7 Master–master curves for creep compliance of epoxy resin for various aging times and aging temperatures.

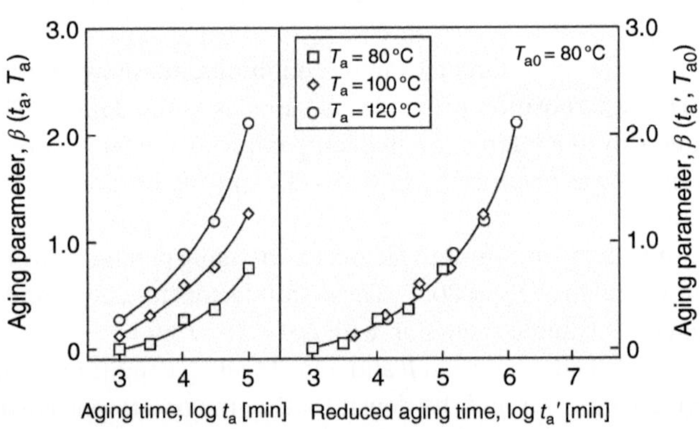

Figure A.8 Aging parameter β versus aging time at various aging temperatures and the master curve at reference aging temperature $T_{a0} = 80\,°C$.

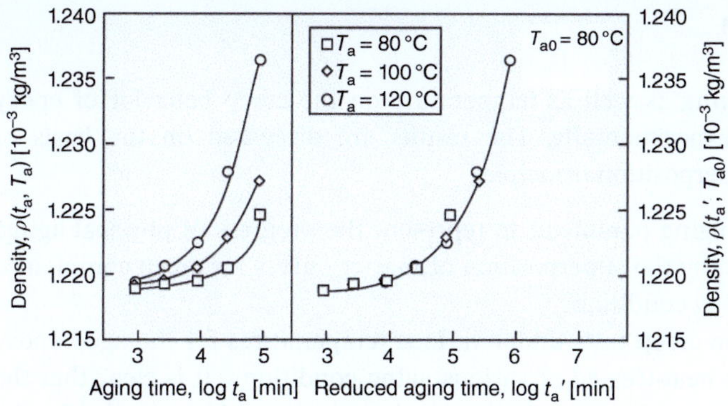

Figure A.9 Density ρ versus aging time at various aging temperatures and the master curve at reference aging temperature $T_{a0} = 80\,°C$.

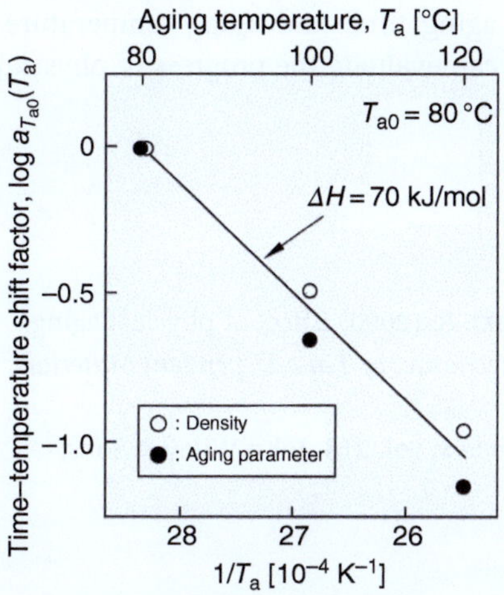

Figure A.10 Aging time–temperature shift factor $a_{T_{a_0}}(T_a)$ for aging parameter and density.

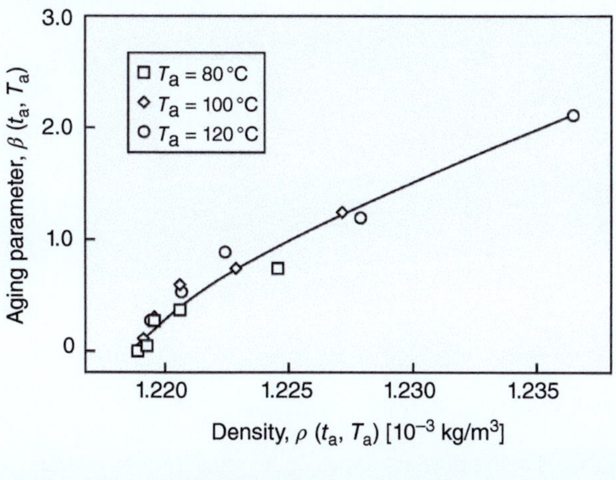

Figure A.11 The relationship between aging parameter β and density ρ.

A.5 Conclusion

Effects of physical aging as well as temperature on the creep behavior of epoxy resin were studied experimentally. The results are discussed on the basis of time–temperature superposition principle.

1. We proposed an aging parameter to represent the progress of physical aging, which is obtained by the superposition of master curves for creep compliance under various aging conditions.
2. We conducted the creep tests under various temperatures for the aged epoxy resin, which was heat-treated at various aging conditions. It is clear that the increase in creep compliance with time and temperature is controlled by the progress of physical aging.
3. We have shown that there is one-to-one correspondence between the aging parameter obtained by the above experiments and the density, both of which obey the same Arrhenius equation for aging time and aging temperature. Therefore, our proposed aging parameter can evaluate the progress of physical aging.

References

1 Miyano, Y., Nakada, M., Kasamori, M. and Muki, R. (2000) Effect of physical aging on the creep deformation of an epoxy resin, *Mechanics of Time-Dependent Materials*, **4**, 9–20.

2 Ferry, J.D. (1961) *Viscoelastic Properties of Polymers*, vol. 218, John Wiley & Sons, Inc., New York.

Appendix B

Reliable Test Method for Tensile Strength in Longitudinal Direction of Unidirectional Carbon-fiber-reinforced Polymers

B.1 Introduction

We have investigated techniques for the precise evaluation of the tensile properties of carbon fiber-reinforced polymers (CFRPs) under a constant temperature environment [1], particularly addressing a resin-impregnated carbon fiber strand (CFRP strand) [2–4]. In this method, the CFRP strand is molded by a filament winding technique, and then the end tabs are bonded to the CFRP strand (post-bonded CFRP strand). Recently, high-strength carbon fiber strands with numerous filaments have become commercially available. Such strands with a great rupture load have the important shortcoming that the CFRP strand slips at the end tabs before failure. As a result, the measured tensile strengths are likely to be lower and scattered.

In this appendix, we report the development of co-cured CFRP strand specimens with end tabs to improve the bonding strength of the end grips with the CFRP strand. The performance of co-cured CFRP strand specimens was demonstrated for two types of carbon fiber strands.

B.2 Evaluation of Tensile Strength Using Post-bonded CFRP Strand Specimen

Two types of CFRP strands, which consist of high-strength carbon fiber and a general-purpose epoxy resin, were molded using a filament winding machine, as shown in Figure B.1. Two types of carbon fiber strands, T300-3000 and T800-12000 (Toray Industries Inc.), were used. The composition of the epoxy resin (Mitsubishi Chemical Corp.) and the cure condition of the CFRP strands are shown in Table B.1. The glass transition temperature T_g of the epoxy resin is approximately 160 °C. The fiber volume fraction of the CFRP strand is 50%. The diameter and length of CFRP strands are 1 and 310 mm, respectively. The end tabs were bonded to the CFRP strand using an epoxy resin adhesive (Figure B.2).

Properties and Testing of Fiber-Reinforced Polymers, First Edition. Yasushi Miyano and Masayuki Nakada. © 2026 Wiley-VCH GmbH. All rights reserved, including rights for text and data mining and training of artificial intelligence technologies or similar technologies.
Published 2026 by Wiley-VCH GmbH.

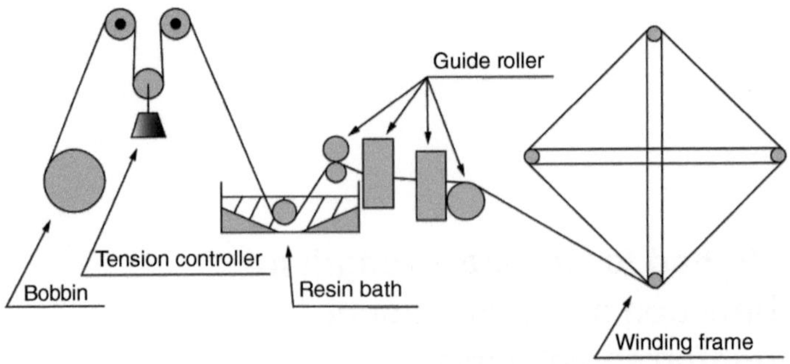

Figure B.1 Configuration of winding machine for resin-impregnated carbon fiber strand (CFRP strand).

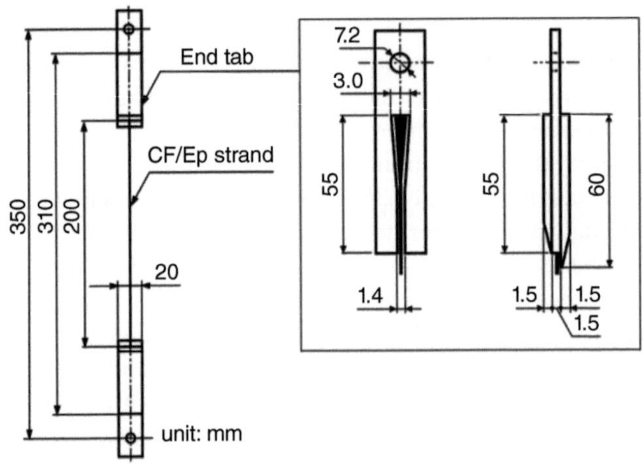

Figure B.2 Configuration of post-bonded CFRP strand specimen.

Table B.1 Carbon fiber strands and resin system.

Carbon fiber strand	Composition of resin (weight ratio)	Cure schedule
T300-3000 T800-12000	Epoxy: jER828 (100) Hardener: MHAC-P (103.6) Cure accelerator: 2E4MZ (1)	100 °C × 5 h + 150 °C × 2 h

The tensile tests for the post-bonded CFRP strand were conducted at room temperature using a specially designed universal testing machine, shown in Figure B.3. The cross-head speed was 2 mm/min.

The tensile strength of the CFRP strand, σ_s, is defined as

$$\sigma_s = \frac{P_{max}}{t_e}\rho \tag{B.1}$$

where P_{max} is the maximum load (N). ρ and t_e are the density of carbon fiber (kg/m^3) and tex of carbon fiber strand (kg/m), respectively.

Figure B.4 shows the load versus elongation diagrams of two types of post-bonded CFRP strands. In this figure, the load versus elongation diagrams for each specimen

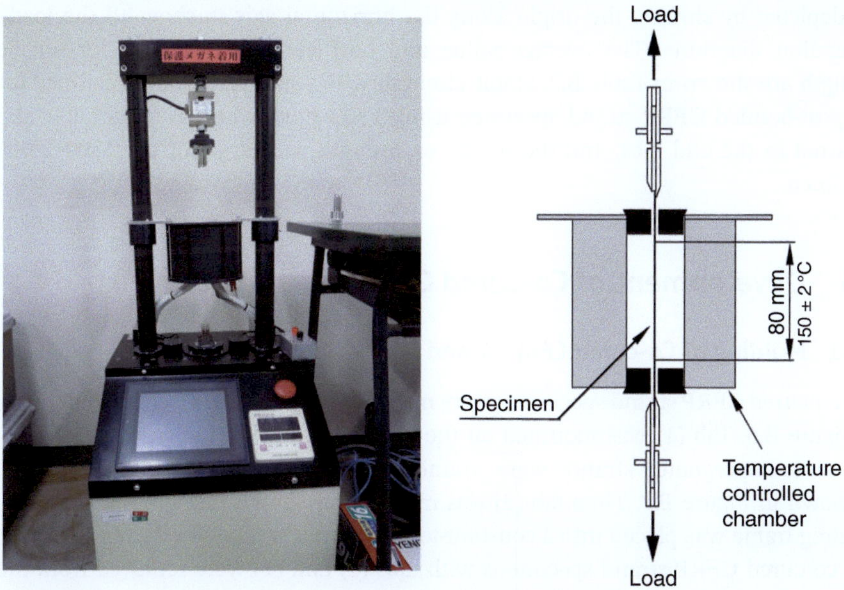

Figure B.3 Universal tensile testing machine for CFRP strand specimen.

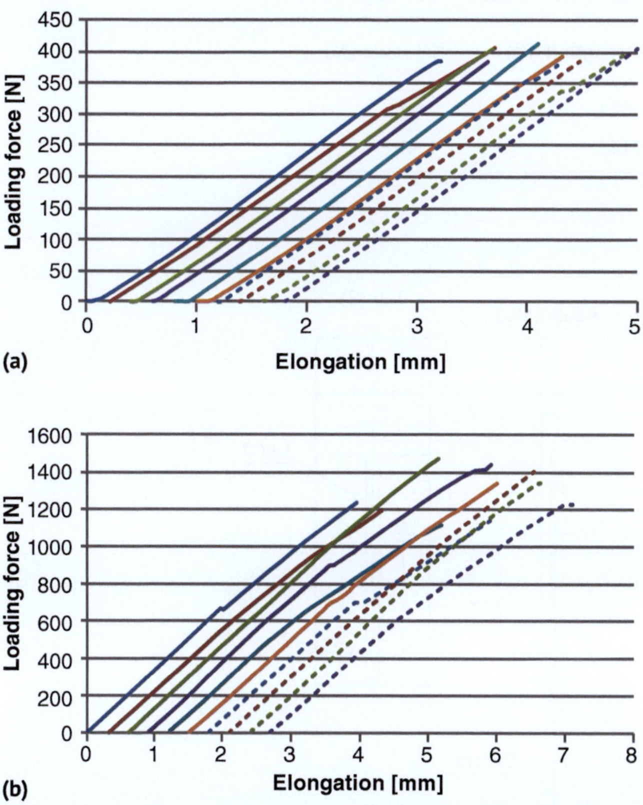

Figure B.4 Load–elongation diagrams for the post-bonded CFRP strand specimen: (a) T300-3000 and (b) T800-12000.

are depicted by shifting the origin along the horizontal axis to show all the load–elongation diagrams. The average value and coefficient of variation for tensile strength are shown in Table B.2. Stable strength with small variation is obtained for the post-bonded CFRP strand specimen using T300-3000, whereas remarkable slip occurred in the end tabs, and the scatter of strength was large for the T800-12000 specimen.

B.3 Development of Co-cured CFRP Strand Specimen

B.3.1 Molding of Co-cured CFRP Strand Specimen

The co-cured CFRP strand specimens were molded by using tabs (1) and (2), as shown in Figure B.5. Tab (1) was mounted on the winding frame as shown in Figure B.6, and resin impregnated strands were wound around the grooved portion of tab (1) as shown in Figure B.7. Then tab (2) was mounted as shown in Figure B.8, and the winding frame was placed into a constant-temperature chamber for curing the resin. The co-cured CFRP strand specimens with tabs (1) and (2) were removed from the

Table B.2 Static test results for the post-bonded specimen.

Carbon fiber strand	Strength (MPa)	CV value (%)
T300-3000	3514	2.99
T800-12000	5143	9.58

CV, coefficient of variation.

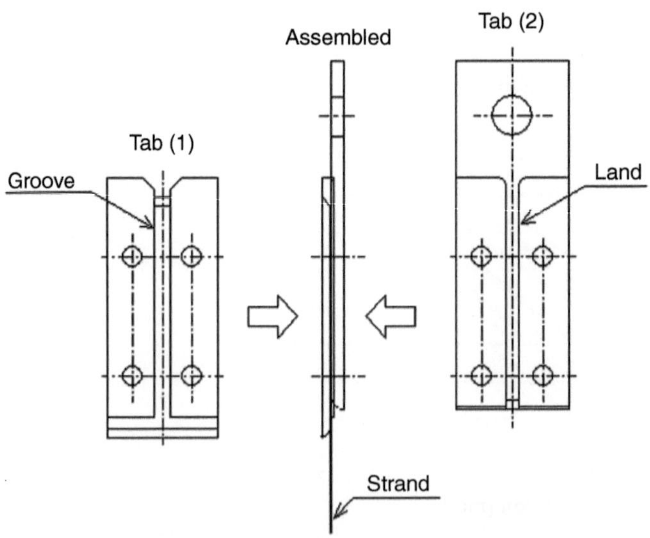

Figure B.5 Configuration of co-cured CFRP strand specimen.

Figure B.6 Winding frame with tab (1).

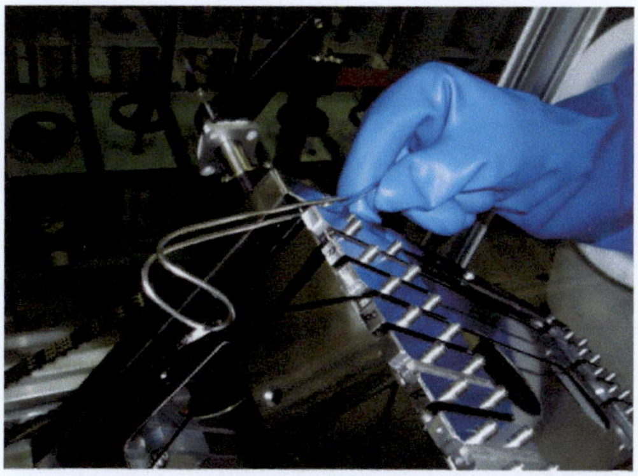

Figure B.7 Hand-winding of the strand to winding frame.

Figure B.8 Assembling tab (2) to the winding frame before curing.

winding frame after curing. Tabs (1) and (2) were then fastened with bolts to enhance the bonding strength between the CFRP strand and the tabs, as shown in Figure B.9.

Figure B.10 shows the load–elongation diagrams of two types of co-cured CFRP strand specimens. The average value and coefficient of variation for tensile strength are shown in Table B.3. It was verified that this configuration would prevent slip in

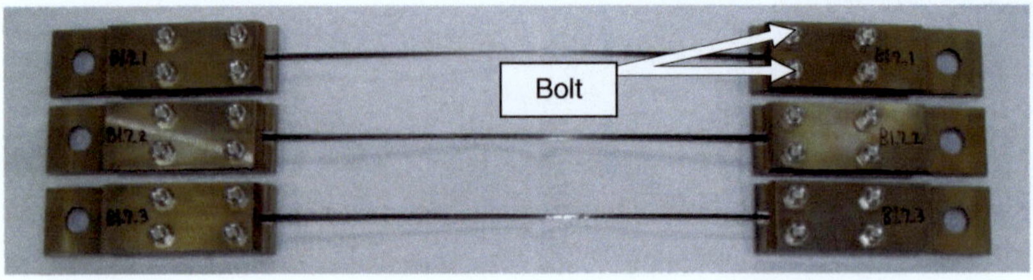

Figure B.9 Bolt fastening between tabs (1) and (2) after curing.

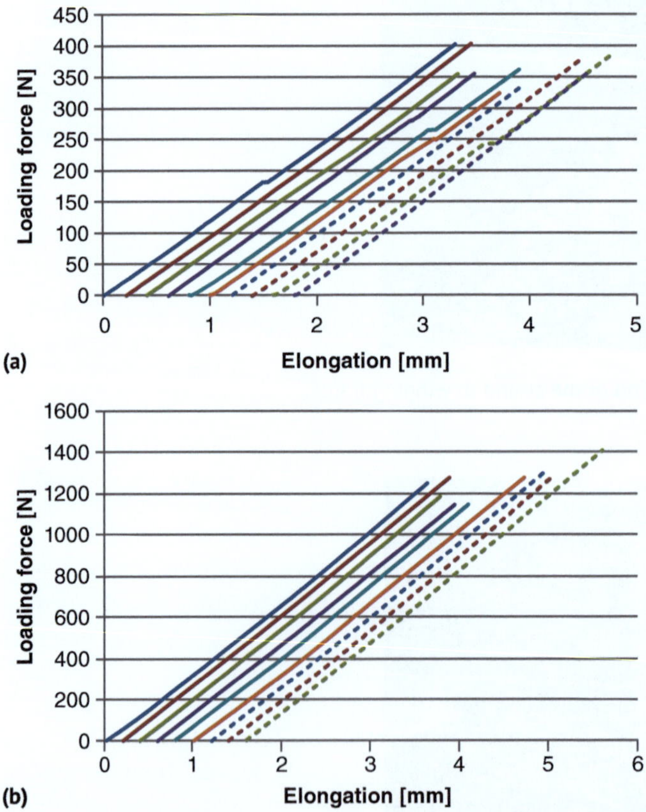

Figure B.10 Load–elongation diagrams for co-cured CFRP strand specimen: (a) T300-3000 and (b) T800-12000.

Table B.3 Static test results for the co-cured specimen.

Carbon fiber strand	Strength (MPa)	CV value (%)
T300-3000	3216	7.93
T800-12000	5211	6.75

CV, coefficient of variation.

the end tabs of T800-12000. However, the scatter of strength is still large. For the co-cured CFRP strand specimen using T300-3000, the tensile strength is lower and the scatter of tensile strength is large compared to that for post-bonded CFRP strand specimen.

B.3.2 Improvement of Co-cured CFRP Strand Specimen

The causes of strength reduction in the co-cured CFRP strand specimen were investigated, and suitable measures were taken to remedy these. The assumed dominant causes of strength reduction in a co-cured CFRP strand specimen include the damage to carbon fiber during hand-winding shown in Figure B.7, stress concentration in the CFRP strand near the end tabs, damage of carbon fiber by the thermal expansion of the winding frame during the curing process, and the damage of CFRP strand by handling after curing.

We improved the co-cured CFRP strand specimen based on the assumptions mentioned above. The causes of damage and the measures for preventing the damage are listed in Table B.4. To prevent damage during winding the strand, we modified the winding machine by employing a traverse setup, as shown in Figure B.11a. By using the traverse setup, the strand can be wound on the winding frame without manual winding. To prevent the stress concentration in the CFRP strand near the end tabs, we modified the end tabs with tapers and added epoxy resin adhesion at the end of tabs as well as in the tapered portions of the tabs, as shown in Figure B.11b. To prevent damage to the CFRP strand due to the thermal expansion of the winding frame, we cut the strand between specimens before curing, as shown in Figure B.11c. To prevent the damage of the CFRP strand by handling, we used holding plates, as shown in Figure B.11d. By using the holding plates, the CFRP strand specimens were not bent while removing them from the winding frame.

Figure B.12 shows the load–elongation diagrams of the improved co-cured CFRP strand specimen. The average value and coefficient of variation for the tensile strength are shown in Table B.5. The catalog data for the tensile strength of carbon fiber are also shown in this table. A higher tensile strength compared to the catalog data and the small variation of tensile strength can be obtained by using the improved co-cured CFRP strand specimens.

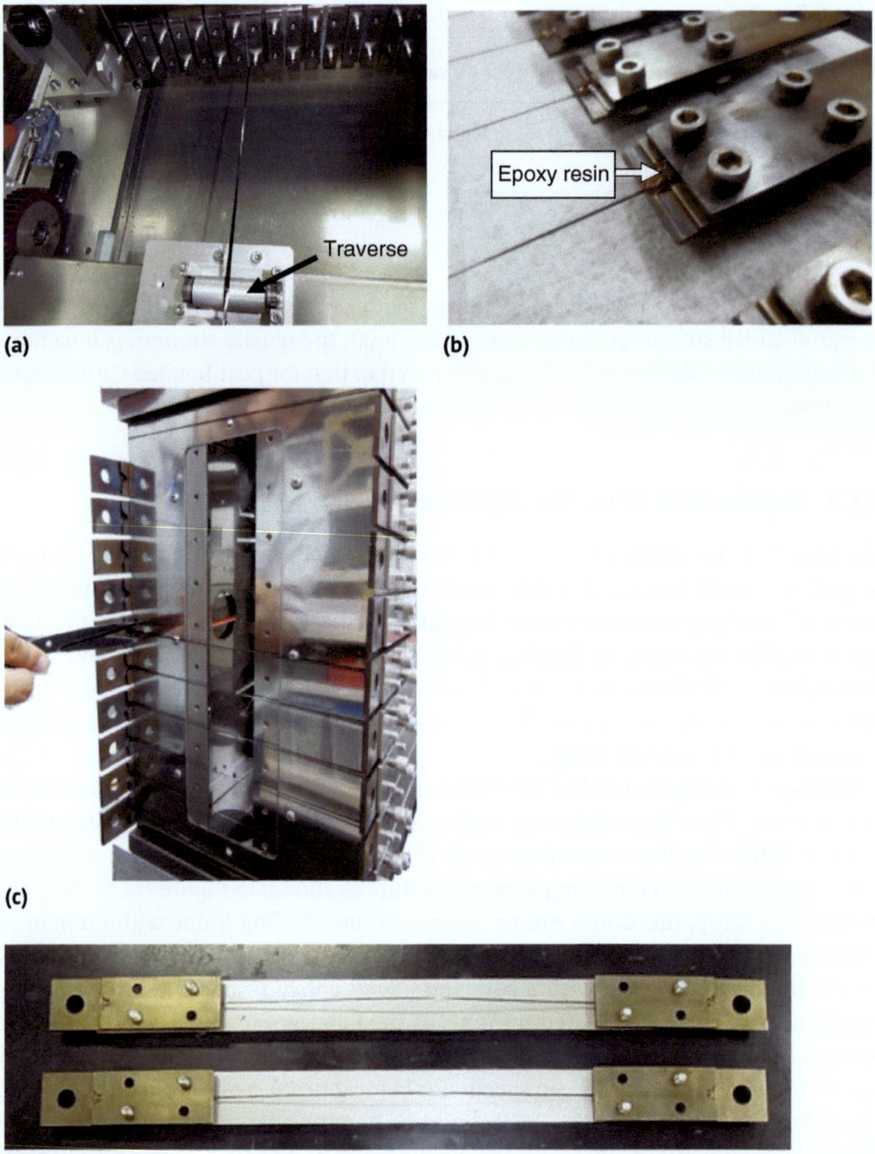

Figure B.11 Improvements on co-cured CFRP strand specimen: (a) without hand-winding of strand by using traverse, (b) added epoxy resin adhesion at the end of tabs, (c) cutting strands between specimens before curing, and (d) holding plates for co-cured CFRP strand specimen.

Table B.4 Improvement work on the co-cured specimen.

Cause of damage	Measure for improvement
(1) Damage during winding	Employing a traverse to prevent damage from hand-winding
(2) Stress concentration at end tabs	Having tapers of 2° × 15 mm on end tabs and adding epoxy resin adhesion around the tapers to avoid the influence of stress concentration
(3) Damage by thermal expansion of the winding frame	Cutting strands between specimens before curing to avoid the large tension in strand due to thermal expansion of the winding frame
(4) Damage to the cured CFRP strand by handling	Using holding plates to avoid the bending of CFRP strands while removing the CFRP strand from the winding frame.

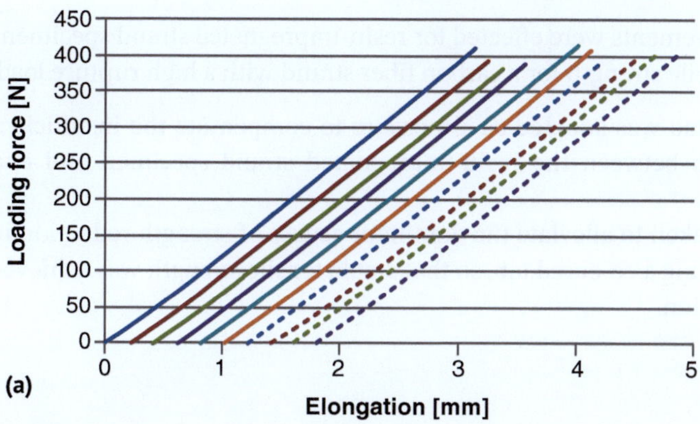

(a)

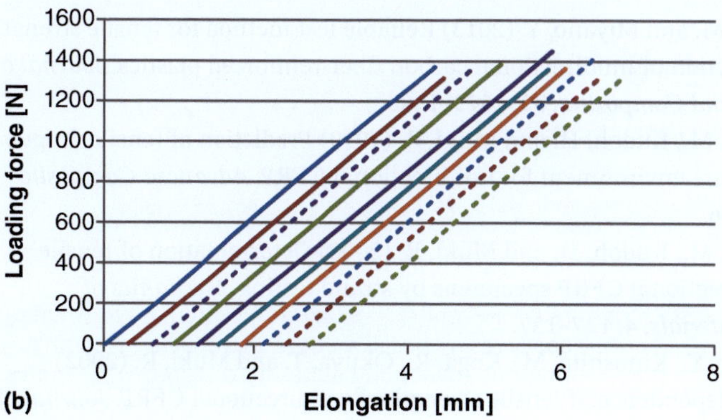

(b)

Figure B.12 Load–elongation diagrams for improved co-cured CFRP strand specimen: (a) T300-3000 and (b) T800-12000.

Table B.5 Static test results for the improved co-cured specimen.

Carbon fiber strand	Specimen	Strength (MPa)	CV value (%)	Toray catalog data [10] (MPa)
T300-3000	Post-bonded	3514	2.99	3530
	Co-cured	3216	7.93	
	Co-cured (improved)	3536	2.87	
T800-12000	Post-bonded	5143	9.58	5490
	Co-cured	5211	6.75	
	Co-cured (improved)	5561	3.09	

CV, coefficient of variation.

B.4 Conclusion

The following improvements were effected for resin-impregnated strand specimens to obtain reliable tensile strength for a carbon fiber strand with a high rupture load.

1. A co-cured end tab was adopted as a measure to compensate the insufficient bonding strength between the resin-impregnated strand specimen and end tabs.
2. Measures were taken to alleviate the dominant causes of strength reduction in a specimen by using a co-cured tab, so that stable tensile strength was achieved with small variation.

References

1 Okuya, T., Nakada, M. and Miyano, Y. (2013) Reliable test method for tensile strength in longitudinal direction of unidirectional carbon fiber-reinforced plastics. *Journal of Reinforced Plastics and Composites*, **32**, 1579–1585.

2 Miyano, Y., Nakada, M., Kudoh, H. and Muki, R. (1999) Prediction of tensile fatigue life under temperature environment for unidirectional CFRP. *Advanced Composite Materials*, **8**, 235–246.

3 Miyano, Y., Nakada, M., Kudoh, H. and Muki, R. (2000) Determination of tensile fatigue life of unidirectional CFRP specimens by strand testing. *Mechanics of Time-Dependent Materials*, **4**, 127–137.

4 Nakada, M., Miyano, Y., Kinoshita, M., Koga, R., Okuya, T. and Muki, R. (2002) Time-temperature dependence of tensile strength of unidirectional CFRP. *Journal of Composite Materials*, **36**, 2567–2581.

Appendix C

Size Dependence on Tensile Strength for Resin-impregnated Carbon Fiber-reinforced Polymer Strands

C.1 Introduction

Christensen, Miyano, and Nakada established a formulation for the effects of length and radius on the Weibull distribution of tensile strength along the longitudinal direction of unidirectional carbon fiber-reinforced polymer (CFRP) [1], and experimentally confirmed the validity of the formulation for the effect of length [2].

In this appendix, we experimentally verify the validity of the formulation for the size dependence of strength, theoretically established by Christensen, Miyano, and Nakada for isotropic materials and oriented fiber composites by examining the length dependence of single fibers and CFRP strands, as well as the radius dependence of the tensile strength of CFRP strands. We statistically measured the tensile strengths of single fibers with different fiber lengths and CFRP strands with different lengths and filament counts. The size dependence of these tensile strengths was compared with the results predicted by substituting the shape parameters of the Weibull distribution for these tensile strengths in Christensen's formulation.

C.2 Formulations

Christensen, Miyano, and Nakada introduced a theoretical formulation for the size dependence of strength for isotropic materials and oriented fiber composites [2]. The effective size-dependent scale parameter σ_s in the Weibull distribution of single fiber tensile strength is given by the following equation:

$$\sigma_s = \left(\frac{L_0}{L}\right)^{\frac{1}{m}} \sigma_{s0} \tag{C.1}$$

where L and L_0 are the fiber length and reference fiber length, respectively, m is the shape parameter of the tensile strength of a single fiber, and σ_{s0} is the reference scale parameter of tensile strength of a single fiber at the reference fiber length L_0.

Properties and Testing of Fiber-Reinforced Polymers, First Edition. Yasushi Miyano and Masayuki Nakada. © 2026 Wiley-VCH GmbH. All rights reserved, including rights for text and data mining and training of artificial intelligence technologies or similar technologies. Published 2026 by Wiley-VCH GmbH.

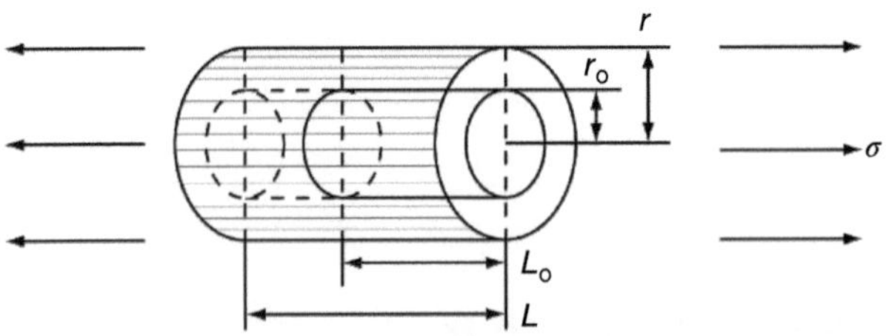

Figure C.1 Volume enlargement for aligned fiber composites.

Table C.1 Mechanical properties of carbon fiber strands employed in this study.

Carbon fiber strand	Density (g/cm^3)	Tex (g/km)	Elastic modulus (GPa)	Tensile strength (MPa)
T300B-50B-1000	1.76	66	230	3530
−3000		198		
−6000		396		
−12000		800		

Figure C.1 shows the volume expansion of an oriented fiber composite subjected to longitudinal tensile stress. The length coordinate L extends to the fiber length, but the radial coordinate r does not extend. The material is transversely isotropic, meaning it is isotropic in the plane perpendicular to the axis of symmetry. The effective size-dependent scale parameter σ_s of the tensile strength of an oriented fiber composite is given by the following equation.

$$\sigma_s = \left(\frac{L_0}{L}\right)^{\frac{1}{m}} \left(\frac{r_0}{r}\right)^{\frac{1}{2m}} \sigma_{s0} \tag{C.2}$$

where L and L_0 are the length and reference length of the composite material, respectively, and r and r_0 are the radius and reference radius. m is the shape parameter of the tensile strength of the composite material, and σ_{s0} is the reference scale parameter of the tensile strength of the composite material at the reference length L_0 and radius r_0. The ratio of the radius r to reference radius r_0 is defined as the ratio of the number of filaments n in the carbon fiber strand to the reference number of filaments n_0, as follows:

$$\frac{r}{r_0} = \left[\frac{n}{n_0}\right]^{\frac{1}{2}} \tag{C.3}$$

C.3 Experimental Procedures

C.3.1 Carbon Fibers and Resin-impregnated Carbon Fiber Strands

In this study, four types of carbon fiber strand bobbins (Toray Industries, Inc., high-strength carbon fiber T300B-50B-1000, 3000, 6000, and 12000) were used, as shown in Table C.1. The strand product names "1000," "3000," "6000," and "12000" indicate

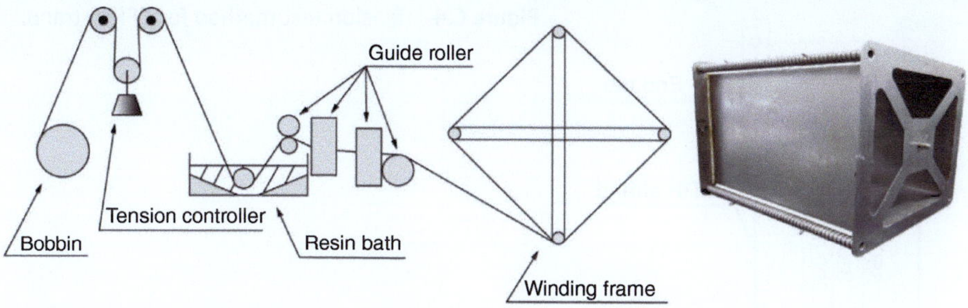

Figure C.2 Filament winding system for resin-impregnated CFRP strands.

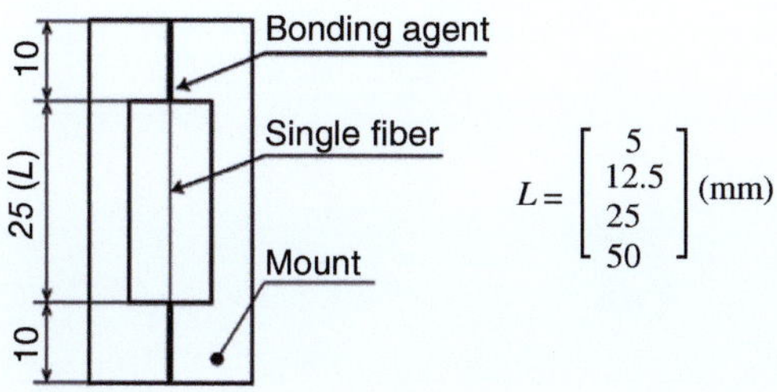

Figure C.3 Tension test method for a single carbon fiber.

Table C.2 Composition and cure schedule of resin-impregnated CFRP strands.

CFRP strand	Molding method	Composition of resin (Weight ratio)	Cure schedule
T300/EP	Filament winding	Epoxy: jER828 (100)	70 °C × 12 h
		Hardener: MHAC-P (103.6)	+ 150 °C × 4 h
		Cure accelerator: 2E4MZ (1)	+ 190 °C × 2 h

the number of filaments in each strand. This table shows the mechanical properties listed in Toray's catalog.

As shown in Figure C.2, four types of resin-impregnated carbon fiber strands (CFRP strands) consisting of four types of carbon fibers and general-purpose epoxy resin jER828 (manufactured by Mitsubishi Chemical Corporation) were molded using a filament winding system developed by the authors. In practice, 100–200 test specimens were molded at one time for each of the four types of CFRP strands using this system. The epoxy resin compositions and the curing schedules for the CFRP strands are shown in Table C.2.

C.3.2 Tension Tests for Single Carbon Fiber and CFRP Strand

As shown in Figure C.3, tensile tests were conducted at room temperature on carbon fiber monofilaments obtained from four types of carbon fiber strand bobbins to determine the Weibull shape and scale parameters of the single fiber tensile strength. Also, as shown in the table in Figure C.3, the tensile strength of T300B-50B-3000 carbon fiber monofilaments was tested at four different filament lengths. The tensile test speed was 1 mm/min. The tensile strength σ_s of carbon fiber was calculated using the following equation, assuming that the cross-sectional area of a single fiber is uniform.

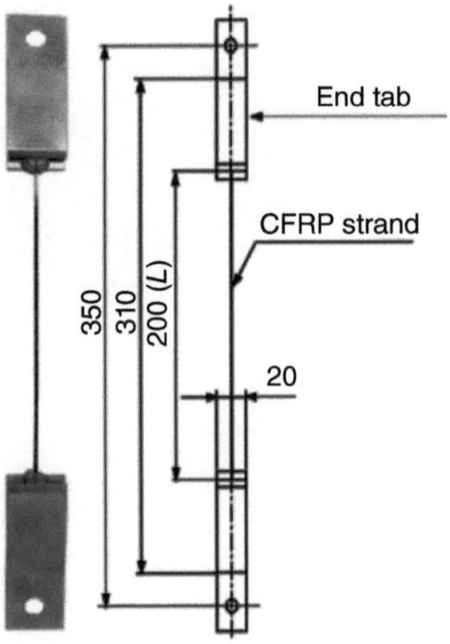

Filament number	L (mm)
3000	20
	60
	200
1000	200
3000	
6000	
12000	

Figure C.4 Tension test method for CFRP strand.

$$\sigma_s = \frac{P_{max}}{t_e} \rho N \tag{C.4}$$

where P_{max} is the maximum load (N), ρ is the density of the carbon fiber (kg/m^3), N is the number of filaments in the strand, and t_e is the mass of 1000 m of carbon fiber strand (g/1000 m), defined as the tex of the strand as given in Table C.1.

As shown in the table in Figure C.4, the tension tests were conducted at room temperature to determine the shape and scale parameters in the Weibull distribution of tensile strength of CFRP strands with three different strand lengths and four different strand filament numbers. The tensile test speed was 1 mm/min. The tensile strength σ_s of the CFRP strands was calculated using the following equation.

$$\sigma_s = \frac{P_{max}}{t_e} \rho \tag{C.5}$$

C.4 Experimental Results

C.4.1 Statistical Tensile Strength of Carbon Monofilament

Figure C.5 shows the Weibull distribution of single fiber tensile strength for the four types of strand lots. The shape parameter m of these fibers is in good agreement with

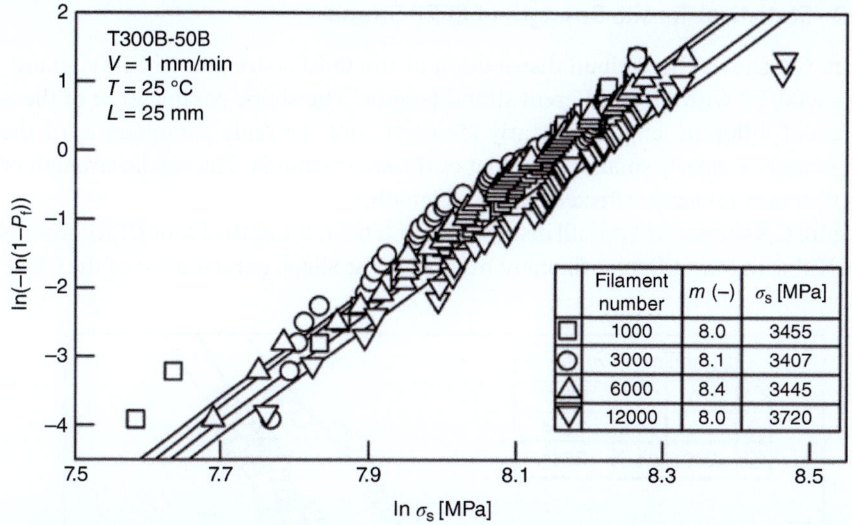

Figure C.5 Weibull distributions for the tensile strength of single carbon fiber for various strand lots of T300B-50B.

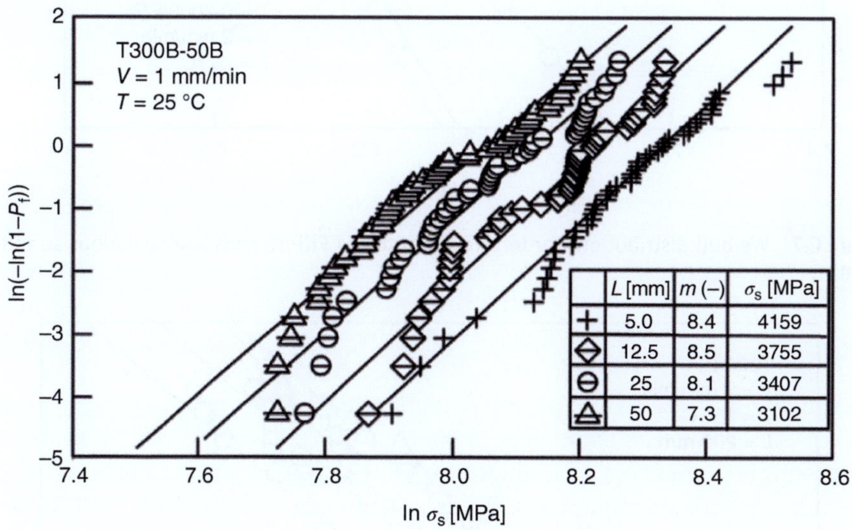

Figure C.6 Weibull distributions for the tensile strength of single carbon fiber having various lengths.

each other, and the difference between the scale parameters σ_s is also small. These facts indicate that the four types of strand lots were manufactured stably and reliably.

Figure C.6 shows the Weibull distribution of single fiber tensile strength for four fiber lengths of T300B-50B-3000. The shape parameter m for these different fiber lengths is in good agreement with each other, and the scale parameter σ_s for long fibers is clearly smaller than that for short fibers. It is clearly shown that single fiber tensile strength depends on fiber length.

C.4.2 Statistical Tensile Strength of CFRP Strand

Figure C.7 shows the Weibull distribution of the tensile strength of CFRP strands T300-3000/EP with three different strand lengths. The shape parameter m of these strands of different lengths is nearly identical, and the scale parameter σ_s of the long strands is clearly smaller than that of the short strands. The tensile strength of CFRP strands is clearly affected by strand length.

Figure C.8 shows the Weibull distribution of tensile strength of four CFRP strands T300/EP with four different filament numbers. The shape parameter m of the CFRP

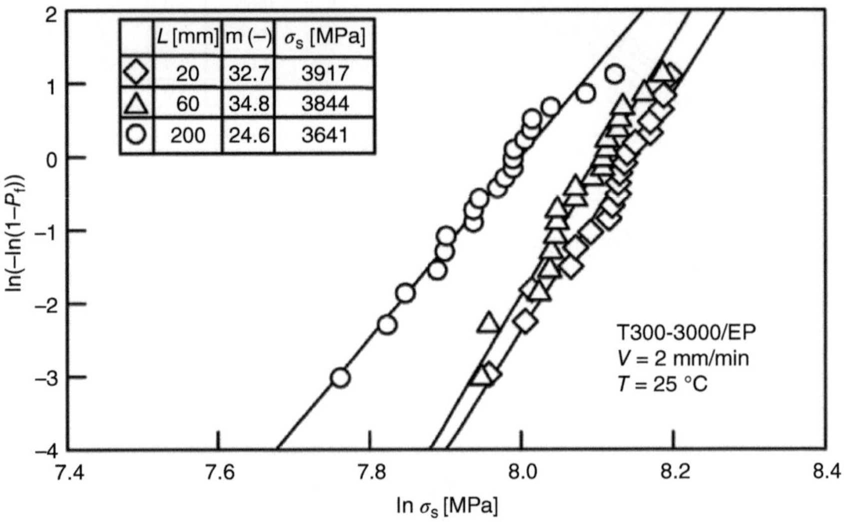

Figure C.7 Weibull distributions for tensile strength of CFRP strands having various strand lengths.

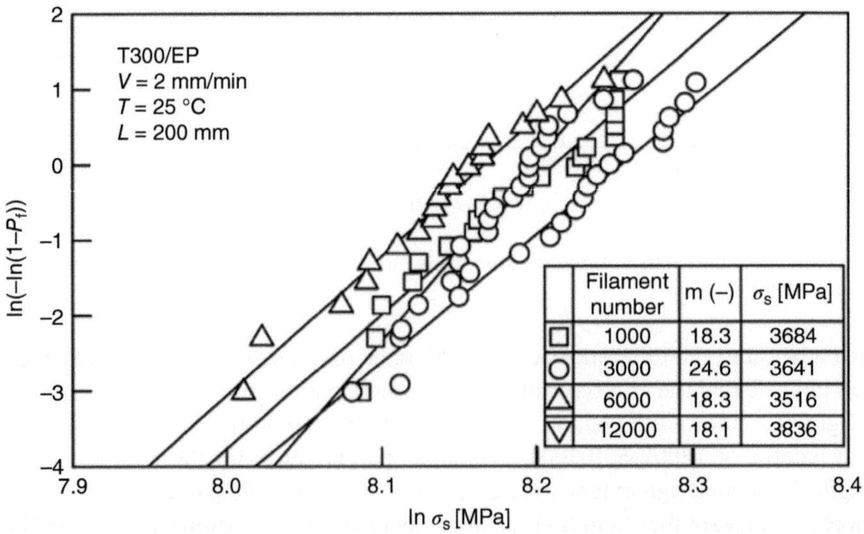

Figure C.8 Weibull distributions for tensile strength of CFRP strands having filament number.

strands with four different filament numbers is nearly identical, but the scale parameter σ_s does not show a clear trend depending on the filament number. The dependence of the filament number on the tensile strength of CFRP strands will be discussed in the next section.

C.5 Evaluation of Size Effect

Table C.3 shows the effect of length on the tensile strength of a single fiber and a CFRP strand, where the reference fiber length and strand length are 25 and 200 mm, respectively. The table shows the measured tensile strength σ_s, the measured and predicted strength ratio σ_s/σ_{s0} for the length ratio L/L_0 for each single carbon fiber and CFRP strand. The predicted strength ratio σ_s/σ_{s0} is calculated by substituting the length ratio L/L_0 and the shape parameter m at the reference fiber length and strand length into Eq. (C.1) and Eq. (C.2), respectively.

Table C.4 shows the effect of the number of filaments in the tensile strength of the CFRP strand. Here, the reference filament number is 3000. The radius ratio for each filament number is calculated using Eq. (C.3). This table shows the measured

Table C.3 Length dependence on tensile strength for carbon fiber and CFRP strand.

	Length L (mm)	Length ratio L/L_0	Tensile strength σ_s (MPa)	Measured σ_s/σ_{s0}	Predicted σ_s/σ_{s0}
Single carbon fiber (L_0 = 25 mm)	5.0	0.2	4159	1.22	1.22
	12.5	0.5	3755	1.10	1.09
	25	1	3407	1	1
	50	2	3102	0.927	0.918
CFRP strand (L_0 = 200 mm)	20	0.1	3917	1.098	1.076
	60	0.3	3844	1.055	1.056
	200	1	3641	1	1

Table C.4 Filament number dependence on tensile strength for CFRP strands.

Filament number	Radius ratio r/r_0	Measured strength σ_s (MPa)	Corrected strength σ_s (MPa)	Corrected σ_s/σ_{s0}	Predicted σ_s/σ_{s0}
1000	0.57	3684	3632	0.998	1.03
3000	1	3641	3641	1	1
6000	1.414	3516	3477	0.954	0.98
12000	2	3836	3513	0.965	0.96

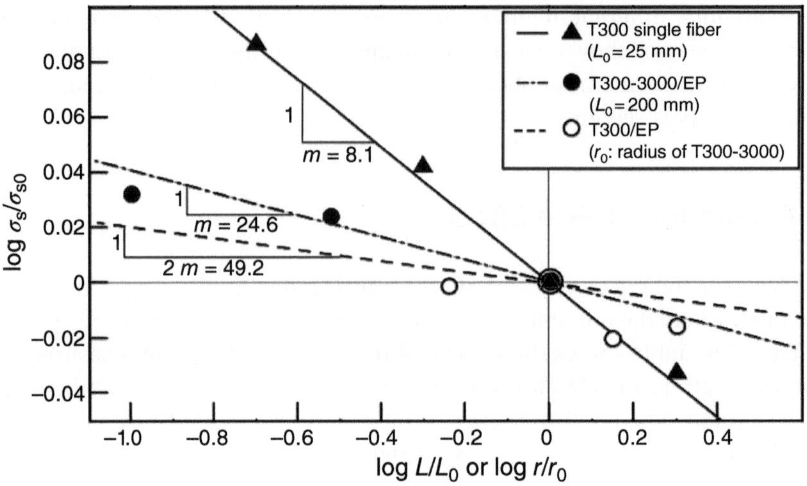

Figure C.9 Size dependence of tensile strengths for a single carbon fiber and CFRP strand.

and corrected strengths σ_s, the corrected and predicted strength ratios σ_s/σ_{s0} for each filament number and radius ratio r/r_0 of the CFRP strand. As shown in Figure C.5, the corrected strength for each filament number is obtained by normalizing the measured strength for each carbon fiber strength to the strength for the reference filament number of 3000. The predicted strength ratio σ_s/σ_{s0} is calculated by substituting the radius ratio r/r_0 and the shape parameters of strength m at the reference filament number into Eq. (C.2).

Figure C.9 shows the measured and predicted strength ratios as a function of length or radius ratio. The measured strength ratios, indicated by open circles, represent the corrected strength ratios. These prediction lines are obtained by substituting the shape parameter m of the carbon fiber and CFRP strands into Eqs. (C.1) and (C.2), respectively. The predicted size dependence of the three strengths agrees well with the measured results and is completely controlled by the shape parameter of the Weibull distribution of the tensile strength in the fiber direction.

C.6 Conclusion

In this appendix, we experimentally verified the theoretical formulation for the size dependence of strength established by Christensen et al. for isotropic materials and oriented-fiber composites by examining the tensile strength of monofilament carbon fiber and resin-impregnated CFRP strands. We statistically measured the tensile strength of monofilament carbon fiber with different fiber lengths and resin-impregnated CFRP strands with different lengths and filament numbers. The size dependence of these tensile strengths agreed well with the results predicted by Christensen's formulation, substituting the shape parameters of the Weibull distribution for these tensile strengths. We found that this size dependence is completely controlled by the shape parameters of the Weibull distribution of the tensile strength in the fiber direction.

References

1 Miyano, Y. Nakada, M. and Kageta, S. (2020) Size Dependence on tensile strength for resin impregnated CFRP strand. *Materials System*, **37**, 59–63.

2 Christensen, R.M., Miyano, Y. and Nakada, M. (2015) The size dependence of tensile strength for brittle isotropic materials and carbon fiber composite materials. *Composites Science and Technology*, **106**, 9–14.

Index

Note: Page numbers in *italics* and **bold** refers to figures and tables respectively.

Properties and Testing of Fiber-Reinforced Polymers, First Edition. Yasushi Miyano and Masayuki Nakada. © 2026 Wiley-VCH GmbH. All rights reserved, including rights for text and data mining and training of artificial intelligence technologies or similar technologies.
Published 2026 by Wiley-VCH GmbH.

g

h